Mass Spectrometry

(in two parts) **Part B**

PRACTICAL SPECTROSCOPY

A SERIES

Edited by Edward G. Brame, Jr.
Elastomer Chemicals Department
Experimental Station
E. I. du Pont de Nemours and Co., Inc.
Wilmington, Delaware

Volume 1: Infrared and Raman Spectroscopy (in three parts)
edited by Edward G. Brame, Jr. and Jeanette G. Grasselli

Volume 2: X-Ray Spectrometry
edited by H. K. Herglotz and L. S. Birks

Volume 3: Mass Spectrometry (in two parts)
edited by Charles Merritt, Jr. and Charles N. McEwen

Volume 4: Infrared and Raman Spectroscopy of Polymers
H. W. Siesler and K. Holland-Moritz

ADDITIONAL VOLUMES IN PREPARATION

Mass Spectrometry

(in two parts) **Part B**

edited by Charles Merritt, Jr.

Food Sciences Laboratory
U.S. Army Natick Research and Development Command
U.S. Department of Army
Natick, Massachusetts

Charles N. McEwen

Central Research and Development
Department
Experimental Station
E. I. du Pont de Nemours and Co., Inc.
Wilmington, Delaware

MARCEL DEKKER, INC. New York and Basel

Library of Congress Cataloging in Publication Data

Main entry under title:

Mass spectrometry.

(Practical spectroscopy; v. 3)
Includes bibliographies references and indexes.
Includes indexes.
1. Mass spectrometry. I. Merritt, Charles, [Date] II. McEwen, Charles N., [Date] III. Series.
QD96.M65M37 543'.0873 78-24085
ISBN 0-8247-6947-3

MARCEL DEKKER, INC.
270 Madison Avenue, New York, New York 10016

Current printing (last digit):
10 9 8 7 6 5 4 3 2 1

PRINTED IN THE UNITED STATES OF AMERICA

PREFACE

This book is Part B of the Mass Spectrometry volume of Practical Spectroscopy and, like its predecessor, Part A, and companion volumes, aims to bring together a number of topics of practical value. In this part, the application of mass spectrometry to studies of drug metabolism by Reinhold and Costello and the use of chemical derivatization by P. Vouros are treated in detail and with comprehension by authors whose experience and involvement with their subjects well qualify them to write. As in Part A, a chapter on collisional activation by Bente and McLafferty has been included which deals with newly developing techniques in which practical applications may soon be realized. Finally, potentially powerful methods of negative ion chemical ionization are described in a chapter organized by J. R. Hass. This last chapter reflects a culmination of current practice, since it has been organized from a collection of papers presented at a workshop on the subject held at NIEHS in Research Triangle Park, N.C. The authors and editors wish, in particular, to acknowledge the contributions of those participants whose work is reported herein.

The editors and publisher wish especially in this preface to acknowledge the contribution of the cover design by Dr. Berndt Soltmann, and apologize for this omission in the preface to Part A.

Charles Merritt, Jr.
Charles N. McEwen

CONTRIBUTORS TO PART B

Vernon N. Reinhold, Department of Biological Chemistry, Harvard Medical School, Boston, Mass. 02115

Catherine E. Costello, Department of Chemistry, Massachusetts Institute of Technology, Cambridge, Mass. 02139

Paul Vouros, Department of Chemistry and Institute of Chemical Analysis, Applications and Forensic Science, Northeastern University, Boston, Mass. 02115

Paul F. Bente, III, Research and Development Department, Hewlett Packard Company, Avondale, Pa. 19311

Fred W. McLafferty, Department of Chemistry, Cornell University, Ithaca, New York 14850

J. Ronald Hass, Laboratory of Environmental Chemistry, National Institute of Environmental Health Sciences, P.O. Box 12233, Research Triangle Park, N.C. 27709

CONTENTS

CONTENTS OF PART A

Chapter 1

MASS SPECTRAL APPROACHES TO THE STUDY OF DRUG METABOLISM

VERNON N. REINHOLD

Department of Biological Chemistry
Harvard Medical School
Boston, Massachusetts

CATHERINE E. COSTELLO

Department of Chemistry
Massachusetts Institute of Technology
Cambridge, Massachusetts

I. INTRODUCTION

The qualitative and quantitative analysis of drug metabolites is a task for which mass spectrometry is particularly well suited. The sensitivity and specificity of the technique make possible the determination of metabolites even in the presence of similar entities--the original drug and other metabolites--and of the many other components of biological extracts. The success of the mass spectrometric approach requires an understanding of the routes of drug disposition, metabolism, and excretion, and of the chemical manipulations that are most effective for recovery of the metabolic products. Equally important is an understanding of the range of analytical procedures made possible by the utilization of mass spectrometry and its adjunct techniques.

With the consideration that the analysis of drug metabolites is an interdisciplinary problem we have summarized in this review the pharmacology of administered drugs. We have done this to aid the analysts in selecting the body fluid or tissue to be examined for the presence of metabolites and to assist them in predicting the likely chemical modifications of drugs by the living system. Particular attention is focused on the contributions that mass spectrometric techniques have made in the elucidation of these pathways. All literature references cited are those where mass spectrometry has played a major role in the elucidation of structure. In addition, a survey of the chemical and instrumental techniques presented in the recent literature that are suitable for this type of analysis is included in order to help the mass spectrometrist to make a judicious choice among them or, perhaps, to improve upon them.

The application of mass spectrometry to drug metabolism studies is a rapidly expanding area of research. For this review we have drawn upon original sources and our own experimental results and have included references primarily from 1973 to the present. Earlier work is summarized in the review articles cited and these are grouped with the general methods. Publications that deal with the analysis of specific drugs and their metabolites have been classed according to the type of drug involved (with some poetic license). Those publications concerned with methods development are assigned according to the approach that they describe. We hope this organization will permit the reader quickly to survey area(s) of particular interest.

II. PRINCIPLES OF DRUG DISTRIBUTION

A. Absorption

To gain access to body compartments drugs must cross one or more barriers and this transfer is usually accomplished by either passive diffusion (a simple process leading to equilibrium), facilitated diffusion, active transport, or pinocytosis. It is frequently observed that passive diffusion depends on the degree of lipid solubility, and that compounds that are highly soluble in lipids diffuse rapidly whereas those that are relatively lipid-insoluble diffuse more slowly. Many drugs, especially hydrophilic molecules, show peculiarities in their migration through a membrane. For some drugs the rate of penetration is greater than would be expected given either limited lipid solubility, the presence of a charge, or, in some cases, the differences between optical isomers. This unexpectedly facile permeation has been defined as facilitated diffusion and is driven by a concentration gradient as with passive diffusion. In accounting for this absorption, a hypothetical receptor or carrier concept has been introduced in which binding to this carrier facilitates transmembrane movements. Active transport, on the other hand, is an energy-requiring process and is defined as such when drugs are observed to move against a concentration gradient. The rate of transport is dependent on the binding capacity of the solute to its hypothetical carrier and is limited by the availability of that carrier. Pinocytosis is the mechanism whereby drugs are absorbed even though they are not in a true solution. Droplets and solid particles are engulfed by processes comparable to phagocytosis. This occurs, for example, from the lumen of the intestinal tract across the intestinal epithelium into the lymphatic or venous capillaries. Other mechanisms of absorption have been recently delineated, such as convective (small molecules through pores) and ion-pair (transmembrane movement of electrochemically neutral complexes) absorption to account for passive diffusion of small water-soluble molecules and quaternary ammonium and sufonic acid absorption respectively. Various transport mechanisms are available in different organs and tissues and a drug may be absorbed by one or more of these mechanisms.

B. Disposition

The sequence of movement of a drug in the process of its distribution is the reverse of that involved in its absorption. Frequently, distribution occurs throughout the body; the drug is carried by the aqueous phase of the blood plasma and reaches tissue compartments at a rate determined by the blood flow through that organ. Arrival at the site of action thus assumes the drug, or its metabolite, has successfully passed through the surface epithelia, the capillary endothelia, and the plasma and intracellular membranes. These loci, the binding sites within and between these barriers, the size of the body's fluid compartments, and the overall physiochemical properties of the molecule itself are major factors accounting for the drug's ultimate distribution.

The extent of drug distribution is frequently expressed in terms of the water content of the body compartments, which are anatomically and functionally distinct. Thus, assuming a drug to be unbound to any tissue component, unmetabolized, not excreted, and found only in the extracellular fluid, its volume of distribution would be about 9-12 liters for adults. In the absence of active transport (assuming cell membrane permeability), transfer will proceed until equilibrium, thereby diluting the concentration of the drug equal to the total intracellular water space, approximately 28 liters. Consequently, a drug that can pass readily all biological barriers will have a volume of distribution equal to the total body water, about 40 liters (60% of body weight). Many lipid-soluble drugs have this volume of distribution. This oversimplified drug dilution ignores many known variables but does permit approximations of molecule concentration from the circulation. In many pharmacological studies the volume of distribution for certain drugs has been known to exceed the total body water, thus providing an indication that the drug either goes specifically to "deep" tissue in peripheral compartments or is stored or pooled in a peripheral compartment, such as fat.

C. Excretion

The processes of excretion, metabolism, and storage are the three mecahnisms whereby drugs are ultimately removed from their sites of action. Excretion at the kidneys, biliary system, intestine, and sometimes the lungs accounts for most drug elimination, and renal excretion is by far the most important of these routes.

The kidney is admirably suited to the task of drug elimination. About 130 ml of plasma water is filtered each minute through the glomerular membranes. Of this volume, only a small proportion is excreted as urine; the remainder is resorbed in the renal tubules. The actual site where elimination (and resorption) take place is the nephron. Each kidney has approximately 1 million nephrons, each consisting of a long, unbranched, tortuous tubule originating in the kidney cortex as a closed-ended structure known as Bowman's capsule. It has been estimated that the total length of the tubules of both kidneys is of the order of 75 miles. This huge surface area and the thinness of the barrier separating the tubular lumen from the internal environment make it obvious that the kidneys are ideally structured to provide maximum contact between the external and internal environments of the body. As blood flows through the Bowman's capsule (glomerulus), any drug that is free in the plasma will be filtered as will be other constituents. Only drugs bound to protein or drugs of excessively large molecular size will be retained in the bloodstream. The rate at which a drug will be eliminated in the urine is the net result of three renal processes: glomerular filtration, tubular secretion, and tubular readsorption. Glomerular filtration and tubular secretion are dependent on the concentration of drug in the plasma whereas resorption by the tubules is dependent on its concentration in the urine.

A second major route of drug excretion is the liver, which secretes 0.5 to 1.0 liter of bile daily into the duodenum through the common bile duct. However, the majority of agents reaching the small intestine in this way are not subsequently excreted in the feces. They are almost completely resorbed because their physiochemical properties are favorable for passive diffusion across the intestinal barrier. These agents then remain in the enterohepatic cycle (portobiliary circulation) until they are excreted in the urine.

Drugs may pass from the circulatory system into the gastrointestinal tract by different routes. The major pathway, as discussed above, is by biliary excretion, but drugs may also pass directly into the gastrointestinal tract or, more important, undergo salivary excretion. This transport mechanism is primarily passive diffusion of nonionized moieties.

D. Binding

When considering the distribution of many drugs in body compartments, attention should be focused on possibilities of protein binding.

Although the term "binding" lacks scientific description, it most frequently involves reversible drug-protein interactions in which several different types of bonding prevail. Since there exists a great diversity of drug and protein structural variations, the consequence of interaction can be expected to run the gamut of stability and attempts to classify these relationships physiochemically are meaningless. The most significant representation of this association must be the equilibrium constant. From kinetic studies and associated manipulations, much can be learned about the character of drug binding. Protein-bound drugs do not cross tissue boundaries (e.g., glomerulus, placenta, blood-brain barrier, renal tubular epithelium), and are not immediately available for metabolism. It should be kept in mind that binding is dynamic and rapidly reversible. Hence binding in no sense precludes further pharmacology of the drug.

E. Induction and Inhibition

Drugs and certain foreign chemicals have long been known to influence their own metabolism and that of therapeutic agents. The administration of phenobarbitol to various animal species markedly increases that animal's ability to metabolize other agents. These inducing agents likewise affect the metabolism of certain endogenous substances such as vitamins, estrogens, androgens, corticosteroids, and progestogens. Mechanisms of enzyme induction have been studied extensively and certain of these produce a proliferation of the endoplasmic reticulum, and thus increase synthesis of protein and cytochrome P-450 (see oxidations). Different types of enzyme inducers appear to act specifically but differentially on certain of the cytochrome constituents, increasing them in a selective manner. Others appear to be relatively nonspecific, affecting a wide range of enzymes that catalyze the metabolism of drugs.

Inhibitors of drug metabolism that result in potentiation of pharmacological effects have long been recognized. Most inhibitors are thought to act directly on the drug-metabolizing-enzyme system as a competitive substrate. The antischizophrenic action of chloropromazine (CPZ) may be due to the mono- and didesmethylated metabolites of

CPZ, which are excellent substrates for indolethylamine N-methyltransferases, thereby preventing the formation of psychotogenic substances [450].

Society's increasing use of tranquilizers combined with alcohol has caused concern for the additive or potentiating effect of these drugs. Ethanol is metabolized by mixed-function oxidases as is diazepam, a popular tranquilizer. It is therefore not surprising that investigators detected an inhibition of ^{14}C-diazepam metabolism when animals were pretreated with ethanol [152].

III. DRUG METABOLISM

Metabolism serves three principal purposes: (1) the supply of energy for body functions and maintenance; (2) catabolic and anabolic processes, the latter usually requiring energy; and (3) the conversion or biotransformation of foreign compounds to more polar, water-soluble, and ionized structures that can be eliminated more easily. Most drugs undergo enzyme catalyzed transformations but a few undergo spontaneous reaction when given the appropriate conditions such as pH or contact with a suitable physiological compound with which they can react. A few drugs are excreted unchanged.

The principal site of drug metabolism is the liver; less important are the kidney, muscle tissue, and gut wall. The liver consists primarily of parenchymal cells that form plates arranged in such a way that they provide walls around continuous spaces or lacunae. Within the lacunae are capillaries that differ from other capillaries in having larger pores, thus being more permeable to macromolecules. The walls of the capillaries are lined by Kupffer cells. Each cell has either one or two nuclei and a large number of mitochondria (about 2,000 per cell) and lysosomes. Metabolizing enzymes occur in the soluble, the mitochondrial, or the microsomal cell fractions. This extremely versatile group of enzymes is known collectively as the drug-metabolizing enzymes and requires as a major enzyme cofactor cytochrome P-450. Several other extrahepatic sites for drug metabolism have been identified, including the lung, intestine, and kidney, but their quantitative contribution to drug biotransformation is generally small.

As discussed earlier, in certain drug metabolic studies the activity of the drug-metabolizing enzymes has been shown to be either increased or decreased by the coadministration of any of a large number of other compounds. This increase in metabolic activity, enzyme induction, is an adaptation of the body to exogenous material. Decreased metabolic activity is thought to be due to substrate competition for the drug-metabolizing enzymes. Both processes of metabolism return to normal upon withdrawal of the coadministered drug.

Drug metabolic pathways are divided into two classifications based on nonsynthetic and synthetic modifications. The former, Phase I, are those that lead to oxidative, reductive, and hydrolytic products; Phase II reactions refer to the conjugative aspects of metabolism. The Phase I reactions usually introduce into the drug molecule functional groups such as OH, COOH, and NH_2, which enable the products to undergo the synthetic reactions of conjugation, Phase II. In both phases the polarity of the drug is usually increased.

A. Phase I Transformations

1. *Oxidation*

Oxidative enzymes are frequently classified by the mecahnism in which the enzyme produces an oxidized substrate. Since oxidation proceeds either through the addition of oxygen or the subtraction of hydrogen, all oxidative enzymes fall into two large classes, which are subdivided on the basis of number of oxygen atoms introduced and substrate acted upon:

1. Oxygenases
    a. Monooxygenases
    b. Dioxygenases
2. Dehydrogenases
    a. Oxidases

Oxygenases directly incorporate oxygen into the substrate molecule, whereas dehydrogenases catalyze the transfer of electrons and hydrogen atoms from the substrate to an electron acceptor. When the immediate electron acceptor in this reaction is oxygen, the enzymes are referred to as oxidases.

Liver microsomal monooxygenases are nonspecific with regard both to substrates and to position of the hydroxylated carbon atom. Activation of the O_2 molecule in the enzyme-substrate complex results in one atom of oxygen incorporated into the substrate, while the other oxygen atom is reduced by an electron donor to the oxidation state of water. Because of the resultant dual effect, monooxygenases have been termed mixed-function oxidases.

Most of the oxidative metabolic schemes discussed in this chapter are catalyzed by monooxygenases and membrane-bound heme protein(s) designated cytochrome P-450. The name P-450 derives from the observation that the reduced form (ferrous state) of the cytochrome heme protein combines with carbon monoxide to yield a complex that exhibits an absorption band at 450 nm (nanometers). The oxidation of substrates by aliphatic and aromatic hydroxylation, epoxidation, oxidation of nitrogen and sulfur, and oxidative dealkylation, as well as reduction of azo and nitro compounds, is frequently catalyzed by enzymes requiring the cytochrome P-450 cofactor.

a. *Aromatic oxidation.*

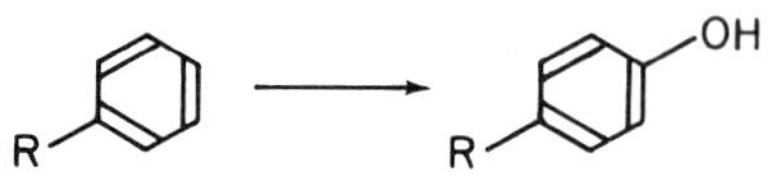

1 2

In the metabolism of propoxyphene, 3, using specifically labeled deuterium precursors [14], hydroxylation was shown to occur in only one of the two aromatic rings, 4.

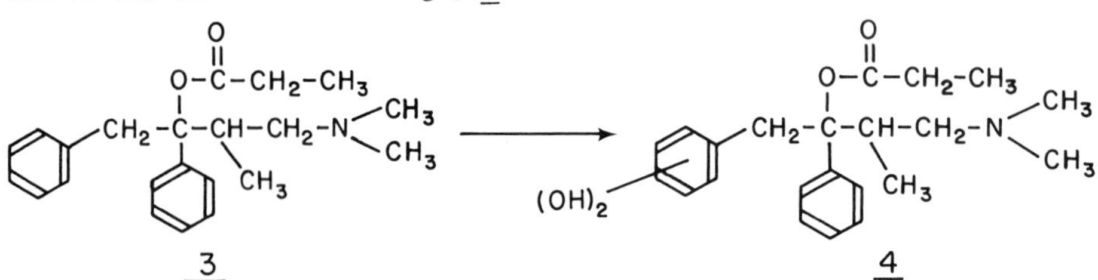

3 4

The anorexic agent phentermine, 5, has been shown to be hydroxylated to 6 in the para position [29].

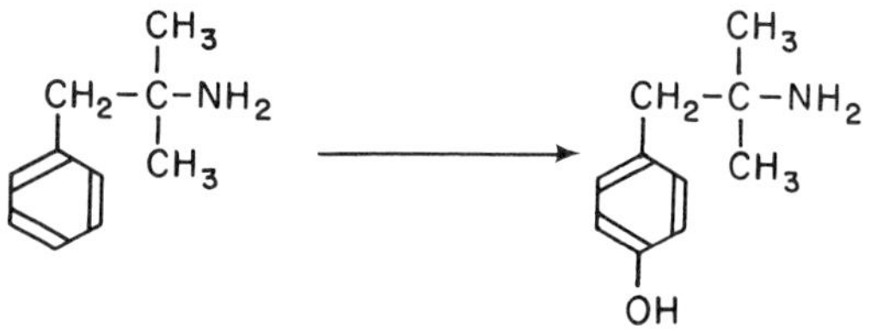

5 6

Studies of the mechanism of oxidative metabolism of the aromatic nucleus by monooxygenases have led to the discovery of intermediates that possess chemical properties with far-reaching implications for drug metabolism and chemical carcinogenicity. The consequence of one phase of this oxidative process was that the substituent initially present at the point of substitution by oxygen migrates during hydroxylation to an adjacent position on the aromatic ring. This phenomenon of intramolecular migration has been termed the NIH shift. Extensive studies on the migration and products of the monooxygenase system with a number of substrates have been rationalized by considering an arene oxide intermediate, 8, that can spontaneously isomerize to phenolic products 9, be enzymatically hydrated to transdihydrodiols 10, or react with a series of nucleophilic (nR) cellular constituents forming covalent adducts 11 [183].

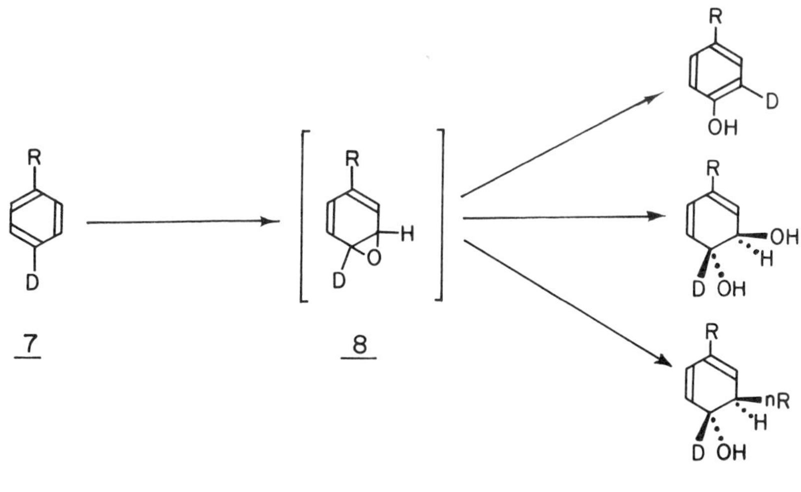

7 8 9 10 11

A large number of hydroxylated products can be considered as resulting from one or more steps of epoxidation, hydration, and conjugation. The ultimate product is determined by steric and electronic factors of the substrate as well as enzymes and nucleophilic agents present in the environment. This formation of phenols from aromatic hydrocarbons is widely accepted and has been shown to occur by the direct addition of an activated oxygen to the aromatic system, 8. Oxygen addition and subsequent rearrangement could explain the metabolism of 4-chloroamphetamine, 12, to the 3-chloro-4-hydroxy derivative, 13 [408].

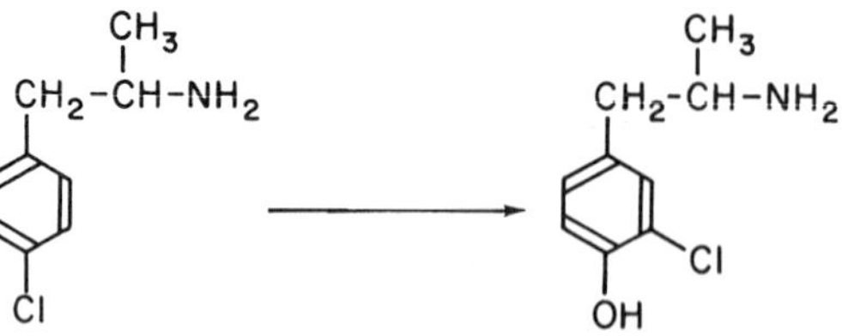

12 13

However, there are notable exceptions to this arene oxide pathway and the detection of conjugated products, 11, of aromatic substrates does not always implicate an epoxide intermediate. Hinson et al., in studying glutathionine conjugation of acetaminophen 14, interpretated their data as indicating N-hydroxylation, 15, followed by dehydration to an arylating species, acetimidoquinone 16, which was consistent with the absence of ^{18}O in the glutathione (GSH) conjugate 17 [16].

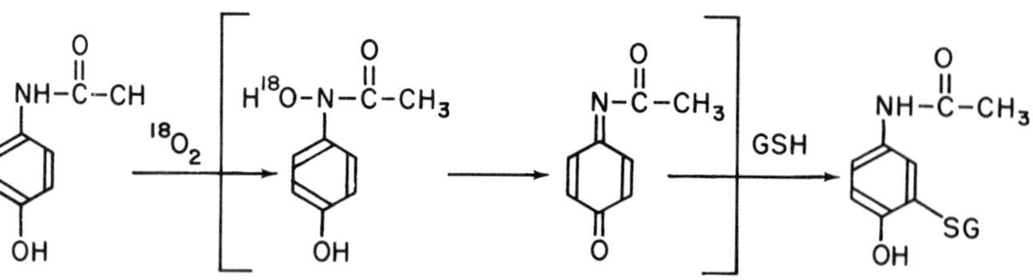

14 15 16 17

The oxidation of the antineoplastic agent ellipticine 18 shows hydroxylation, 19, in the absence of deuterium retention as expected for an NIH shift mechanism and pathway, 20 [183].

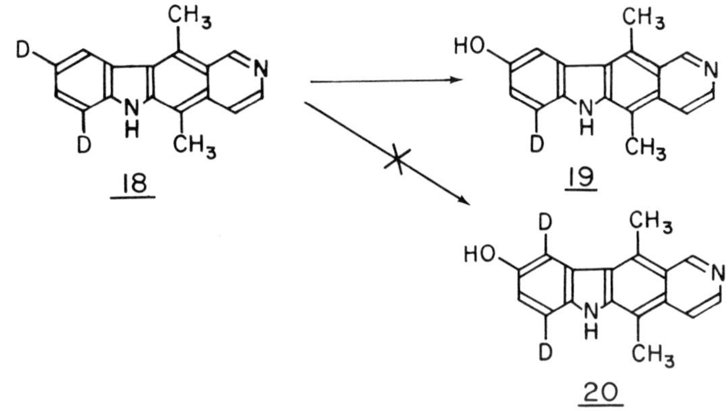

This product, 19, may show a primary isotope effect at the point of hydroxylation (comparable to enzymatic hydroxylation at saturated carbon) and this may suggest an oxygen insertion mechanism, 22 [185].

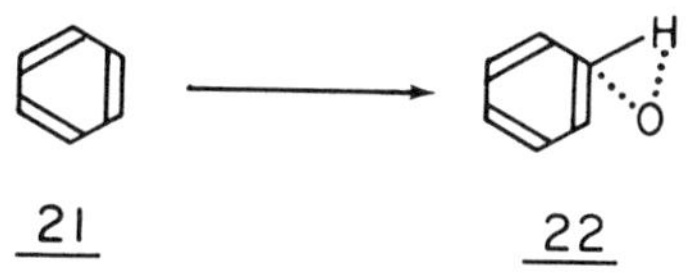

b. Aliphatic oxidation.

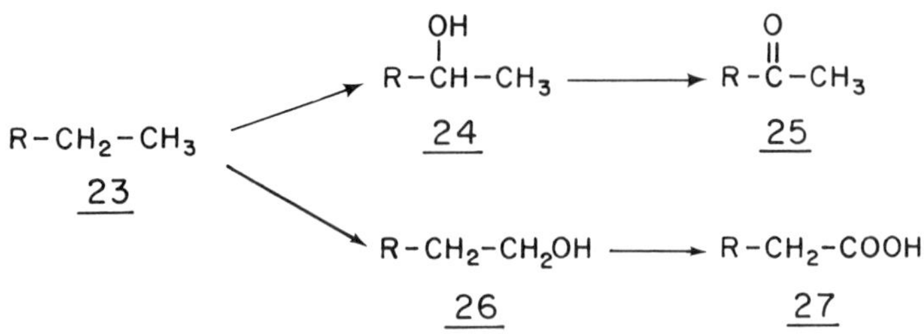

Enzymatic hydroxylation of a saturated carbon frequently shows a primary isotope effect and proceeds with a high degree of retention of configuration. Oxidation of 23 to the primary alcohol, 26, and the corresponding acid, 27, is referred to as ω-oxidation, with the latter step being catalyzed by soluble dehydrogenases.

The anticonvulsant drug sodium dipropylacetate, 28, has been shown to undergo ω-oxidation in rats yielding the corresponding alcohol, 2-n-propyl-5-hydroxypentanoic acid 29, and the diacid, 2-n-propylglutaric 30 [67].

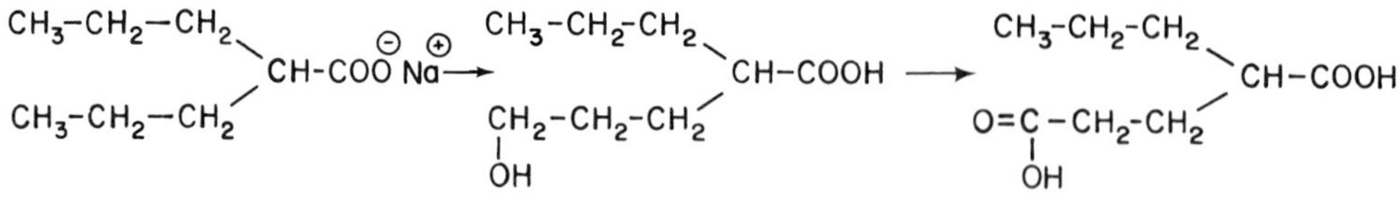

28 29 30

Many methylated aromatic or heterocyclic substrates are also metabolized by hydroxylation of the methyl substituent. A predominant oxidation of this type occurs in N-, O-, and S-dealkylation.

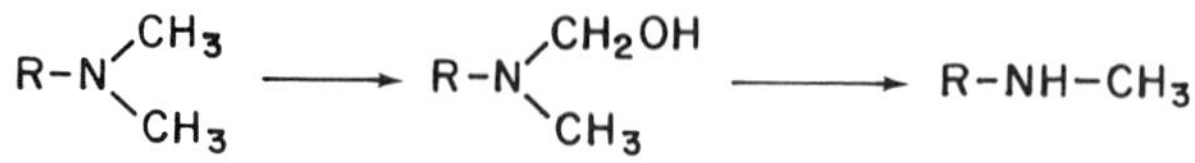

31 32 33

The infrequent detection of a hydroxylated intermediate such as 32 is most probably related to its chemical instability. However, products of this type are found and have been reported as conjugates with glucuronic acid. Hydroxylation alpha to the heteroatom in the nasal decongestant propylhexedrine 34 forms the primary alcohol 35 as well as a tertiary alcohol substituent, 36 [6].

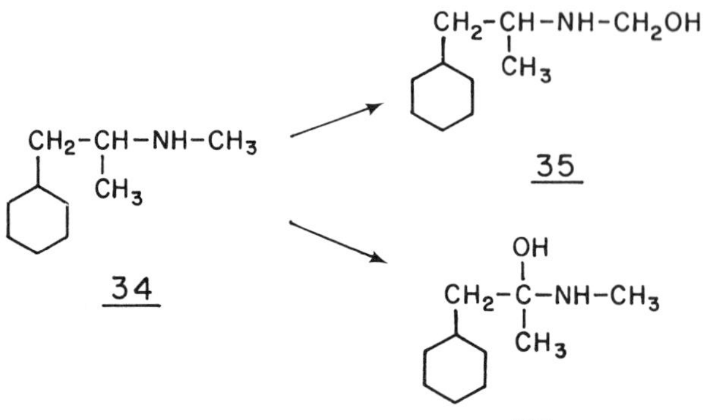

34 35 36

An enzymatic oxidation related to the alpha heteroatom and ω-oxidations are hydroxylations that occur at the penultimate methylene group, frequently called ω - 1 oxidations, 24. Liver monooxygenases exhibit severalfold higher specific activities toward aliphatic compounds than toward aromatic compounds. An increase in CH-bond strength has been shown to cause a decrease in oxidative activity. Thus, it appears that insertion of a heteroatom such as nitrogen, oxygen, or sulfur adjacent to the CH-bond induces facile attack by the active oxygen. The source of oxygen is molecular oxygen and not water.

The antiinflammatory drug phenylbutazone, 37, exhibits penultimate carbon oxidation (ω - 1), 38, in addition to aromatic oxidation, 39 [108].

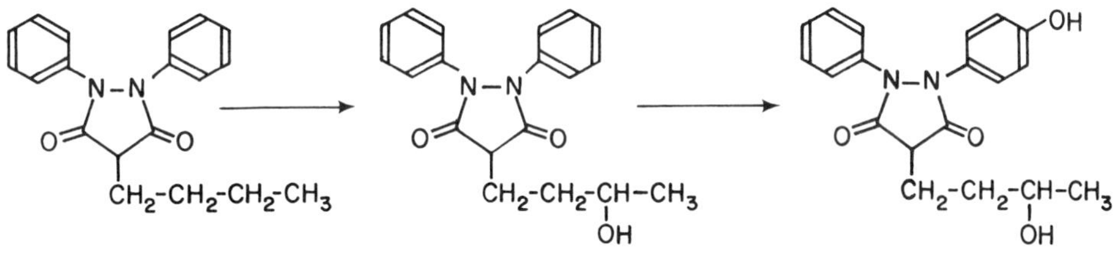

37 38 39

Secobutarbitone has shown similar ω - 1 oxidation in man [409].

c. N-Dealkylation.

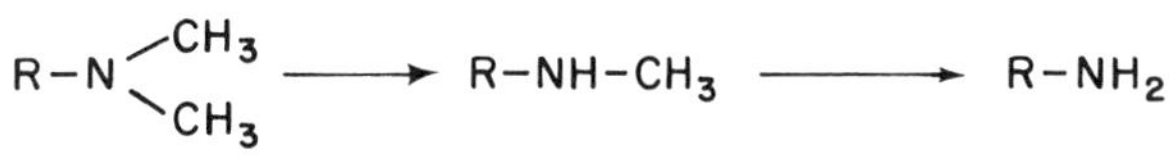

40 41 42

The liver microsomal system for N-dealkylation has broad substrate specificity, but, generally, lipophilic compounds serve as better substrates. Carbon-nitrogen bond cleavage occurs whether nitrogen has the form of amine (secondary or tertiary), amide, carbamate, or sulfonamide. Dealkylation appears to decrease as the alkyl side chain increases, i.e., methyl > ethyl > propyl > butyl. Studies indicate that with tertiary dimethyl amines the first methyl group is removed three to four times as rapidly as the second, and, as a result, the secondary amine accumulates in tissues. Carbon monoxide inhibits N-demthylation but this effect can be partially reversed by irradiation with monochromatic light at 450-nm wavelength. This would imply a reversal of the CO-inhibition of microsomal enzymes.

We noted earlier the susceptibility of carbon to oxidation when it is situated alpha to a heteroatom and the similarity of this enzymatic process to that of N-dealkylation should be kept in mind.

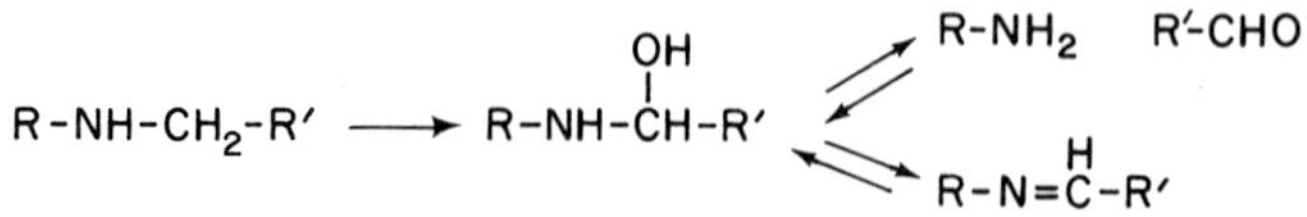

43 44 45

The carbinolamine 44 can be readily hydrated to the free amine and corresponding aldehyde representing the products of enzymatic N-dealkylation or dehydrated to the Schiff base, 45. Cyclophosphamide, 46, metabolites represent just such a case of carbinolamine hydration [128] in which R and R' are cyclized in the starting drug, 46.

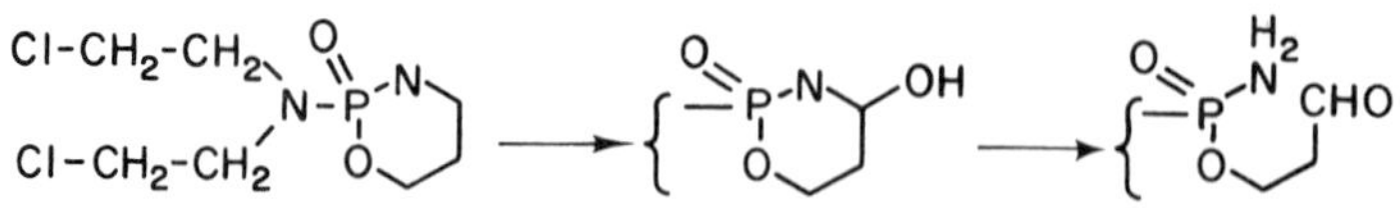

46 47 48

Although the enzymes involved in N-dealkylation show broad substrate specificity, the configuration of the substrate can influence the rate of dealkylation. These conclusions are difficult to assess because, overriding direct enzymatic kinetics on enantiomorphs, are the complications of lipid solubility (i.e., membrane permeability) and specific binding of substrates to extraneous protein and lipid. These complications are not lost in cell-free systems and have been known to be exacerbated in the 9,000xg or microsomal fractions. Whenever these problems have been taken into consideration or are known not to influence the kinetic constants, stereoselective dealkylation of enantiomers may be considered. The difference in the demethylation of nefopam, 49, isomers has been studied [83] and the excretion of nornefopam has been related to the enhanced specificity of the demethylase system for (+)-nefopam, 50.

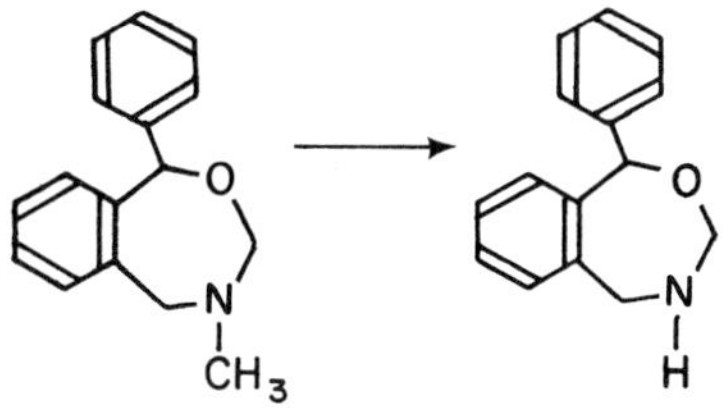

49 50

d. O-Dealkylation.

$$R\text{-}OCH_3 \longrightarrow R\text{-}OH$$

51 52

Rat microsomal studies have also shown a progressively decreasing rate of O-dealkylation as the alkyl substituent increased from methyl through n-butyl. Studies with $^{18}O_2$ and $H_2{}^{18}O$ indicated that O-demethylation was not accompanied by displacement of the intact methoxy group but rather by direct attack upon the methyl group itself. In man, dog, and rat, the coronary vasodilator cinepazide, 53, is metabolized by O-demethylation, 54, and subsequently conjugated with glucuronic acid, 55.

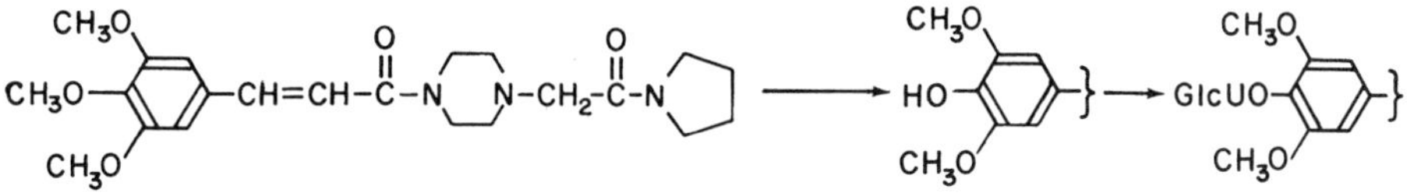

53 54 55

The microsomal enzyme fraction from liver possesses O-dealkylation activity, requires NADPH and oxygen, and has broad substrate specificity.

e. Deamination.

$$R-CH_2-NH_2 \longrightarrow R-CHO \longrightarrow R-COOH \;(58) \text{ or } R-CH_2OH \;(59)$$

56 57 58 59

Drug amine oxidases are frequently classified by their enzyme cofactor. Mammalian monoamine oxidases appear to be flavoproteins that show wide overlapping substrate specificity whereas the diamine oxidases are pyridoxal phosphate enzymes. They all catalyze oxidative deamination of biogenic amines and use molecular oxygen as an electron acceptor. Many of the aldehydes, 57, formed undergo spontaneous oxidation in the presence of air. Monoamine oxidase activity is present in many tissues including nervous tissue. This is in agreement with the assumption that substances such as tyramine, serotonin, and adrenaline are protected by the drug iproniazid from breakdown by the blocking of monoamine oxidase. Surprisingly little is known about the molecular properties of these enzymes. It is tempting to speculate that alkylated amines, like the vasodilator oxprenolol 60, are metabolized via ω - 1 oxidation followed by monoamine oxidase activity, 68 [460], to products such as 61.

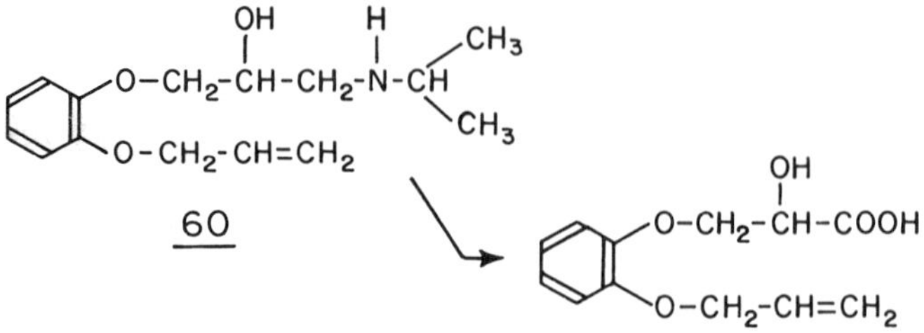

60 61

f. Sulfoxidation.

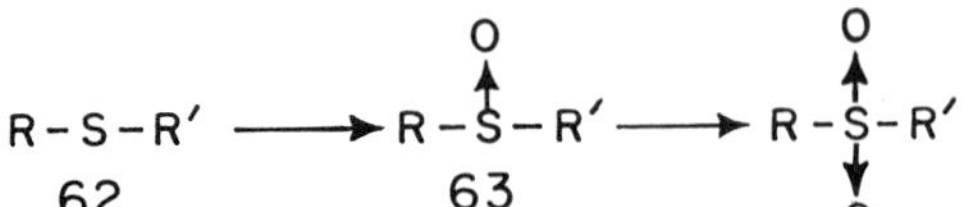

62 63 64

An interesting example of sulfoxidation has been reported to occur in rats with the sedative bromazepam 65 [150]. The nonsulfur-containing drug has been detected as the S-methyl conjugate, 65, along with two oxidized sulfur metabolites, the sulfoxide 67 and sulfone 68.

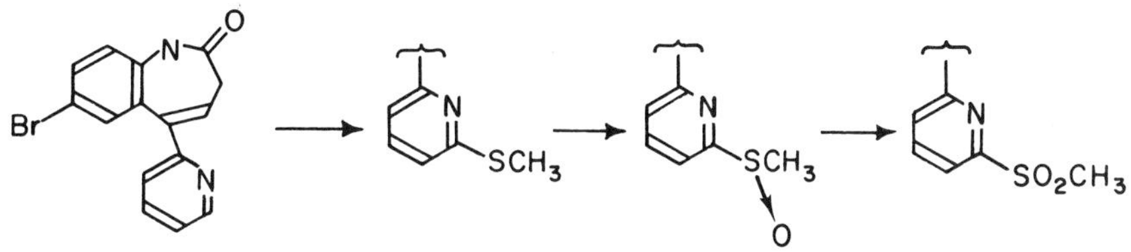

65 66 67 68

A more classic case of sulfoxidation in rats and dogs has been reported for the drug butaperazine 69, an agent for management of chronic schizophrenia.

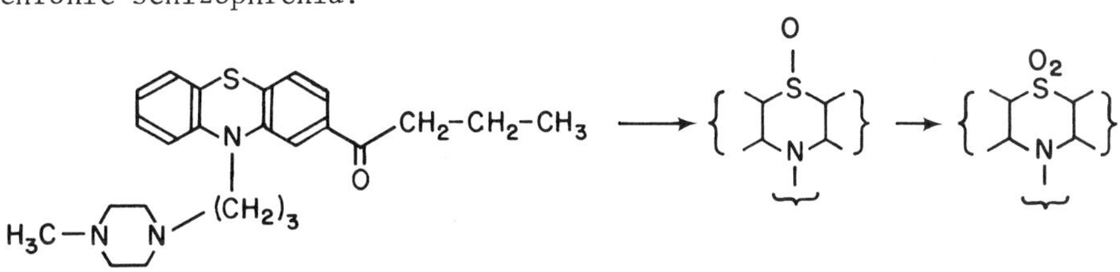

69 70 71

g. N-Oxidation.

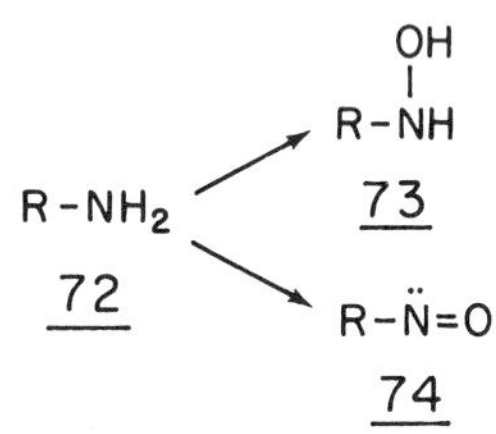

N-hydroxylation is an unusual metabolic reaction in that the product is more reactive, and in some cases more toxic, than the parent compound. However, it may be that hydroxylation followed by sulfate or glucuronyl conjugation occurs with greater facility and hence body elimination is enhanced. It is known that there is tissue specificity for N- and O-glucuronide conjugation and N-hydroxylation may introduce a group that will now serve as a substrate for glucuronyl transferase activity without translocation to other tissues.

Although not indicated above, secondary and tertiary amines undergo N-oxidation to the corresponding secondary hydroxyl amines and N-oxides:

75 76

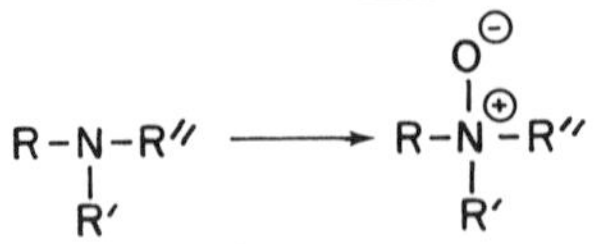

77 78

A mechanism recently proposed by Beckett and Belanger [29,30], in studying the anorexic drug phentermine and its derivatives, suggests that N-hydroxylation, 73, and nitroso metabolites, 74, are derived from a common precursor resulting from metabolic N-oxidation of the primary amine. Although the oxygen is derived from O_2 in this metabolic process, the usual P_{450} inhibitors for carbon oxidation are not effective, leading these investigators to conclude that N-oxidation and C-oxidation involve separate metabolic pathways [31].

Beckett and Belanger [28] have shown N-hydroxylation to occur in guinea-pig and rabbit hepatic microsomal homogenates for the drug phentermine, 79.

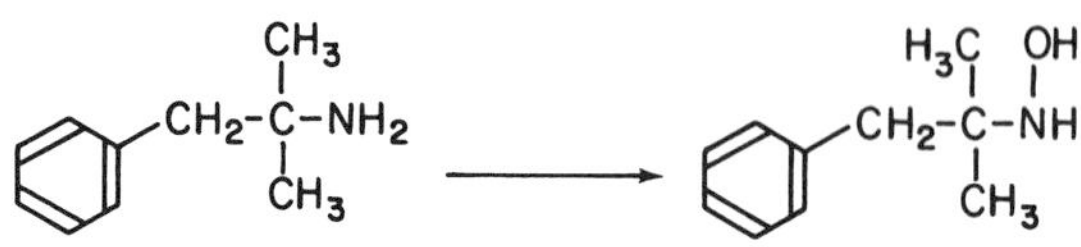

79 80

These workers identified, in addition to the N-hydroxylated metabolite 80, higher oxidative states of nitrogen--nitroso and nitro. Most interesting, however, was that control studies had indicated that the latter two materials were artifacts of base extraction. In addition, Sullivan et al. [25] have shown that the N-oxide detected in the urine of methadone-maintained subjects was not present when urine samples were immediately frozen and stored in this manner until analyzed. On the basis of these reports, oxidation states of nitrogen detected as metabolites should be subject to reevaluation and scrutiny.

Noguchi et al. [117] recently presented data indicating tertiary amine, 77, oxidation to the N-oxide, 78, for the antiinflammatory drug tiaramide, 81.

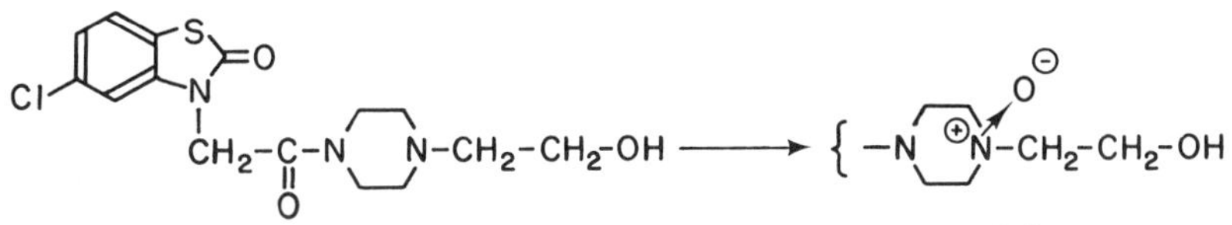

81 82

These authors had little trouble characterizing the N-oxide by solid-probe mass spectrometry, especially when combined with some simple reductive procedures that convert the metabolite back to the starting drug, 81. Tang et al. [343] have had difficulty characterizing the N-hydroxy metabolite 84 (which decomposed to amobarbital) by both electron-impact and chemical ionization mass spectrometry using the solid probe inlet. Similar results were obtained when the metabolite was prepared as the acetate or the trimethylsilyl derivative and gas chromatographed.

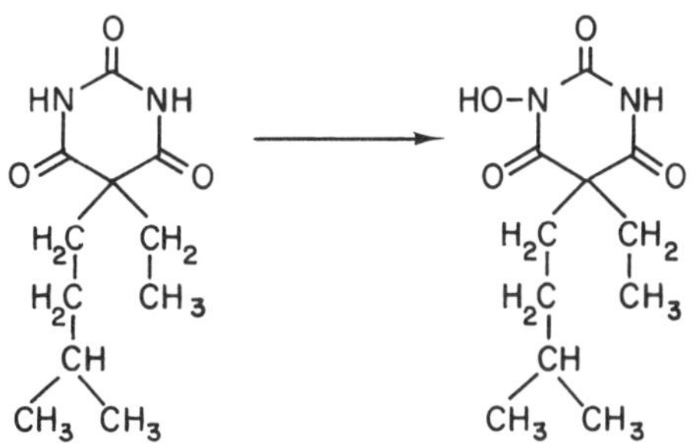

83 84

The problem was approached by on-column methylation with trimethylanilinium hydroxide, which gave the N-methyl derivative for the metabolite. Under these conditions, amobarbital gave the N,N-dimethyl product. The two products were easily resolved chromatographically and both products were identified by their mass spectra. On-column methylations are somewhat unreliable for quantitative work, however, and these authors avoided this complication by spiking the urine samples with ^{15}N-labeled N-hydroxy amobarbital.

h. Desulfuration.

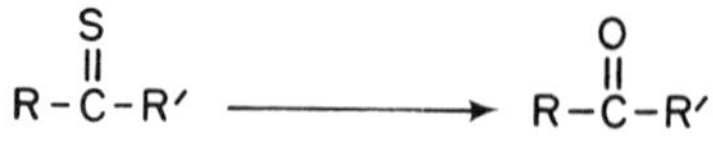

85 86

Desulfuration is essentially an oxygen replacement reaction acting on phosphorous and carbon compounds. The reaction is carried out by liver microsomes in the presence of molecular oxygen and NADPH. The desulfuration of phosphorothionates to phosphates has important implications in the metabolism of insecticides. Replacement of oxygen by sulfur has long been known in the metabolism of thiobarbiturates with conversion to the oxybarbiturate analog. Rat liver slices possess the capability of converting thiopental 87 to pentobarbital 88.

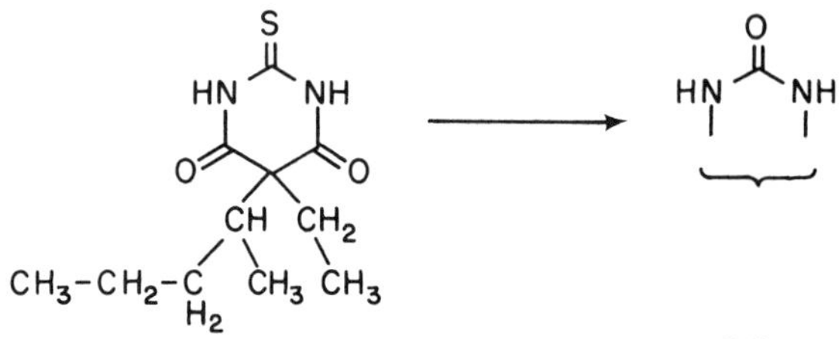

88

87

2. Reduction

89 90

Although drug metabolic routes may include enzymes that bring about reduction, the discussion here of the reductive capability of certain tissues from oxidative capacity is a separation not realized in most enzyme preparations. For example, the enzyme liver alcohol dehydrogenase can catalyze both the oxidation of carbinols to ketones and aldehydes and the reduction of these carbonyl compounds to carbinols.

Much has been written on these oxido-reductases, including substrate specificity, stereochemistry, and mechanism. In general, oxido-reductases in mammalian tissue act mainly as dehydrogenases. However, there have been reports of true reductases that require the coenzyme NADPH.

Another group of reductive enzymes active at yet another hetero-atom center are those that reduce nitro and azo compounds. Mammalian livers possess the ability to reduce many fully oxidized nitro compounds to the corresponding amines. As noted earlier in reference to nitrogen oxidation studies, the reactivity of these intermediates poses unique problems because of autooxidation and nonenzymatic reduction by reacting with biological reducing agents. Moreover, the facility with which enzyme systems carry out reduction and reoxidation of these intermediates makes it difficult to determine which mechanism or pathway is significant to the living organism.

There appears to be a soluble nitro-reductase mechanism exemplified by the xanthine oxidase system and two microsomal reductase pathways requiring cytochrome P-450 and NADPH-cytochrome c, respectively. It should be kept in mind, however, especially for those drugs involved in the enterohepatic cycle, that under anaerobic conditions that exist in the gut, bacterial flora may play an important role in nitro reduction.

Higuchi et al. [456] have shown that in both rats and dogs the metabolism of the new dihydropyridine diester cerebral vasodilator, 91, involves a complete reduction of the aromatic nitro group to the amine, 92.

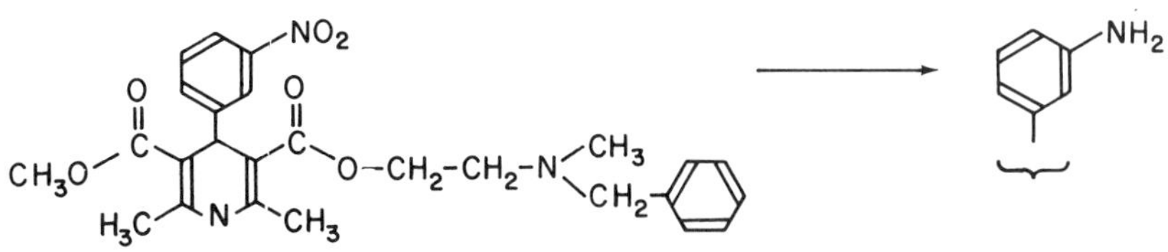

91 92

Beckett and Bélanger [28] have found, in addition to N-oxidation of the anorexic agent phentermine, 79, recent evidence of nitroso, 93, reduction by a 9,000xg microsomal fraction to the hydroxyl amine, 80.

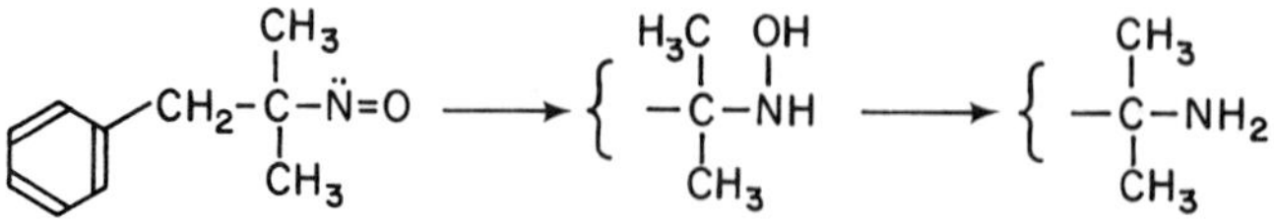

93 80 79

Other reductive metabolic transformations that can be considered with nitrogen compounds are those involved in reduction of the azo linkage. This metabolic reduction and cleavage received much attention when sulfonamide drug therapy was of great interest. The reduction of prontosil, 94, to the amine, sulfanilamide 95, an effective bacteriostatic agent, was one of the early examples of biotransformation in mammalian tissues of a foreign compound to an active agent.

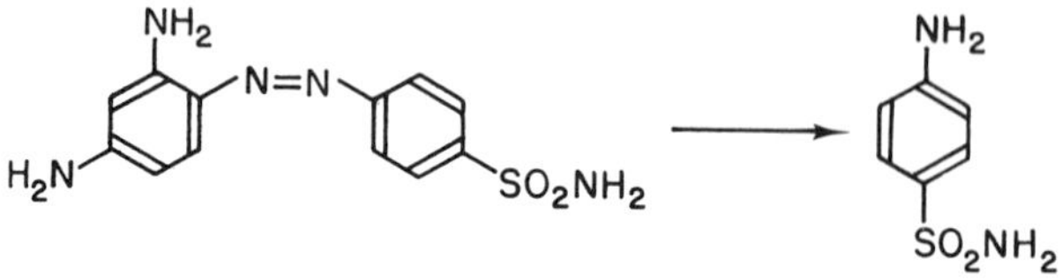

94 95

Another example of biotransformation in man is found in 2-[4'-di-(2-bromopropyl) amino-2'-methylphenylazo] benzoic acid 96. The product, 97, is a highly reactive alkylating agent thought to be responsible for the success of this drug, 96, in the treatment of primary hepatocellular carcinoma in man.

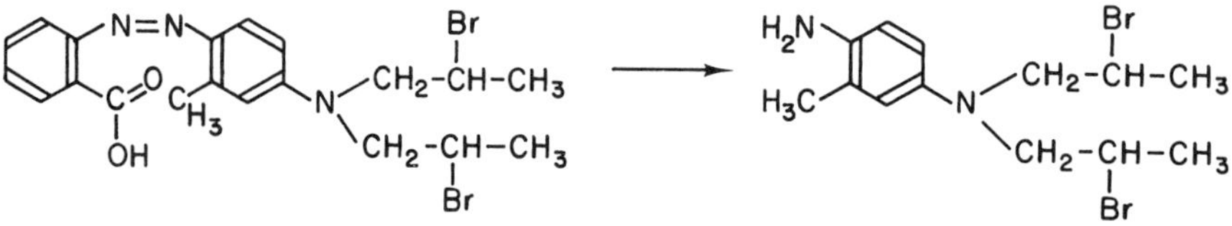

96 97

3. Hydrolysis

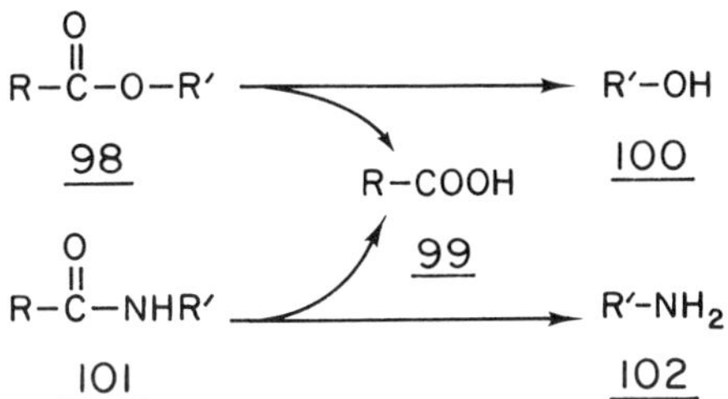

98 99 100 101 102

The hydrolysis reaction of esters, 98, and amides, 101, by tissue enzymes is widely distributed, occurring in serum, liver microsomes, and numerous other tissues. It is clear, however, that esterase activity varies considerably from one species to another and even from tissue to tissue within the same species. Meffin et al. [5] have discussed disposition kinetics of the adrenergic blocker acebutolol, 103, and its hydrolytic metabolites, 104.

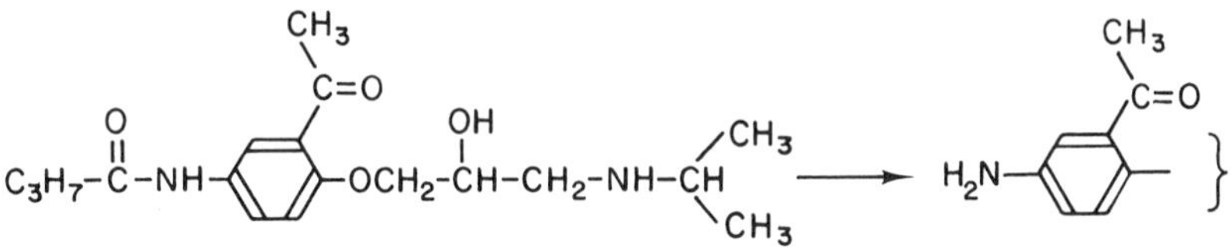

103 104

The enzymes involved in hydrolysis of the N-acyl group are classified more specifically as amidohydrolases and occur in homogenates of several rat tissues, particularly kidney and liver.

Human studies with the antithyroid drug carbimazole, 105, used for treating thyrotoxicosis, has been shown by Skellern et al [221] to undergo a hydrolytic cleavage to 3-methyl-2-thiohydantoin, 106.

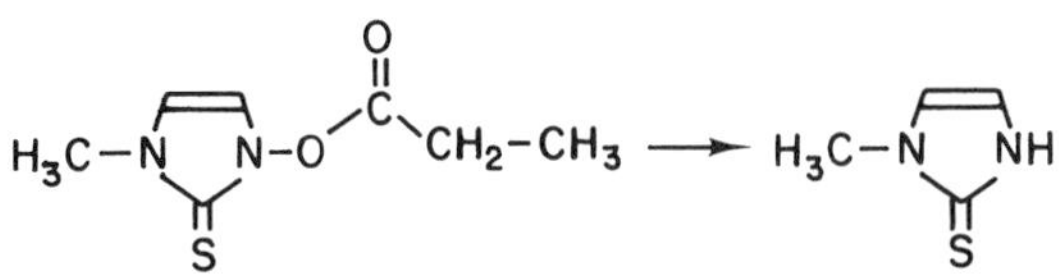

105 106

In an excellent example of the combined use of stable isotopes and gas chromatography-mass spectrometry, Due et al. [14] have studied the hydrolysis of the propionyl residue from propoxyphene, 3.

B. Phase II Transformations

The purpose of drug metabolism is to convert the compound into a form that is more water-soluble in order to aid in excretion. With the introduction of nucleophilic groups into the sample as discussed in Phase I reactions, there now exists a center for attachment of conjugating groups. In some cases, the Phase I reaction increases the water solubility sufficiently and conjugation does not occur.

1. *Sulfate Conjugation*

107 108

Sulfation is a major pathway in the metabolism of substances that contain phenolic hydroxyl groups or can be converted in vivo to compounds containing such groups. Less apparent are sulfate esters of alcoholic hydroxyl or primary amino groups. The conjugates of sulfate with nitrogen are of two types: the sulfanates, 109, and N-sulfates, 110.

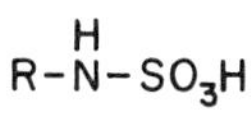

109

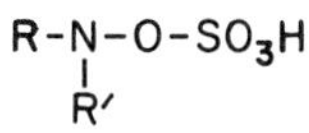

110

The sulfanate group is found in the mucopolysaccharide heparin and frequently, although incorrectly, called an N-sulfate. A notorious drug N-sulfate is that of N-hydroxy-2-acetamidofluorene 111, a compound that reacts readily with cellular nucleophilic components and fits the assumed role of ultimate carcinogen.

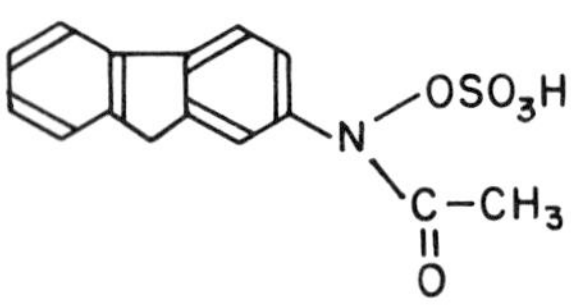

111

Sulfation markedly changes the physical properties of the drug, most important by aiding in its water solubility and hence excretion. However, sulfate conjugation with steroids appears to have important consequences in tissue distribution because of the enhanced ability of the conjugate to bind with protein.

The cofactor involved in enzymatic sulfation is 3-phosphoadenosine 5'-phosphosulfate (PAPA). Enzymes catalyzing this conjugation have not been isolated in a pure state. It must be assumed, however, that with the broad variation that exists in substrate, i.e., cerebrosides, mucopolysaccharides, steroids, and drugs to mention a few, there exists an equally diverse group of enzymes that serve as sulfotransferases.

Analysis of sulfate esters is complicated by the lability of the ester linkage and unfortunately this polar group cannot be blocked or derivatized for analysis by gas chromatography (GC) or mass spectrometry (MS). Identification of sulfate conjugates is usually based on strong ionic character coupled with sulfatase activity. However, when an accurate assessment of a sulfate conjugate is demanded because of a multiplicity of esterification points or the results obtained from chromatographic or enzymatic experiments are unequivocal, two alternative procedures discussed below may be considered.

One approach utilizes a combination of derivatization steps that makes use of the sulfate esters' lability upon trimethylsilylation (TMS). This technique provides convincing, albeit indirect, evidence for the point of esterification. An example of this approach was reported in a study by Bronaugh et al. [394] for identification of the 3-O-sulfate, 112, and 4-O-sulfate, 114, esters of dopamine. The sulfate esters are stable under conditions of acetylation and methylation. Thus, silylation of the products yields a trimethylsilyl group at the point of sulfate conjugation. The isomeric products show minor variation in ion intensity upon mass spectral analysis (Fig. 1) but even more convincing was the comparison with standards by chromatographic techniques of the original sulfate esters and derivatized products.

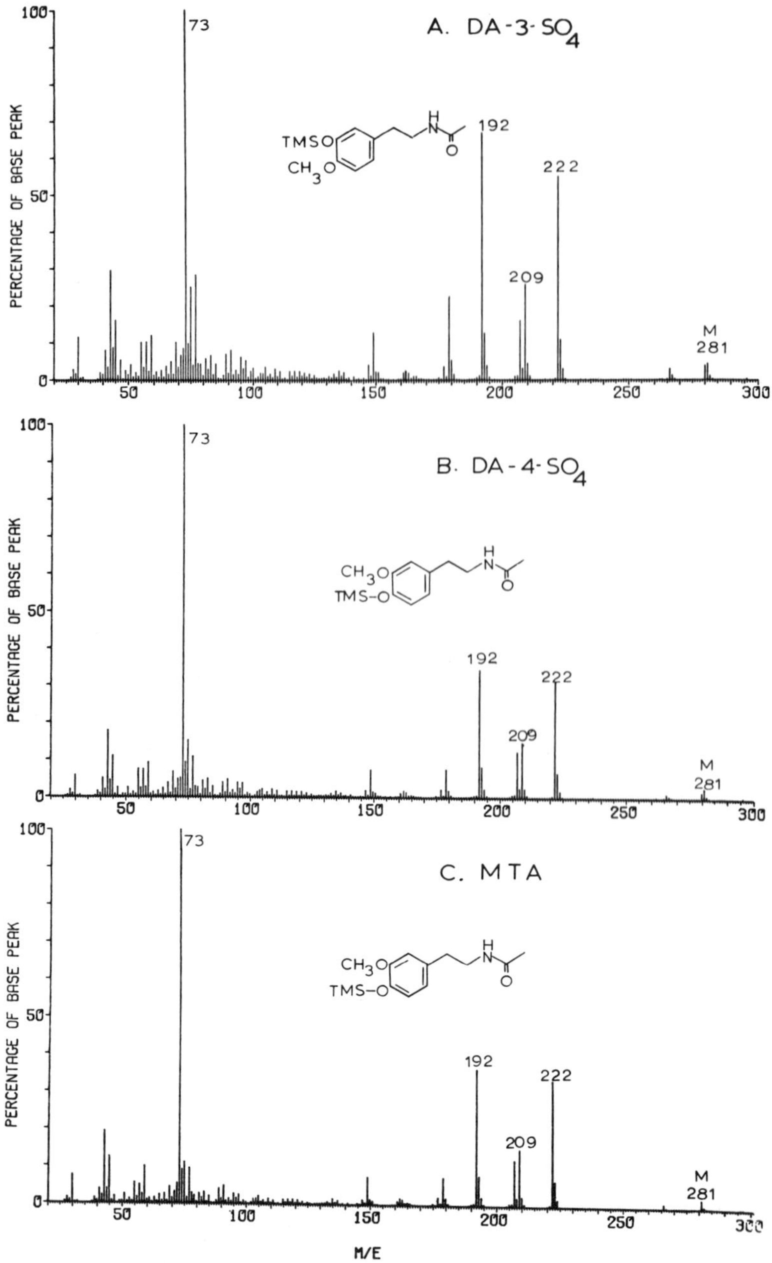

FIG. 1. Mass spectra of derivatives: (A) dopamine 3-O-sulfate 113; (B) dopamine 4-O-sulfate, 115; and (C) methoxytyramine 115. Spectra were obtained with a gas chromatograph-mass spectrometer using electron-impact ionization at 70 eV.

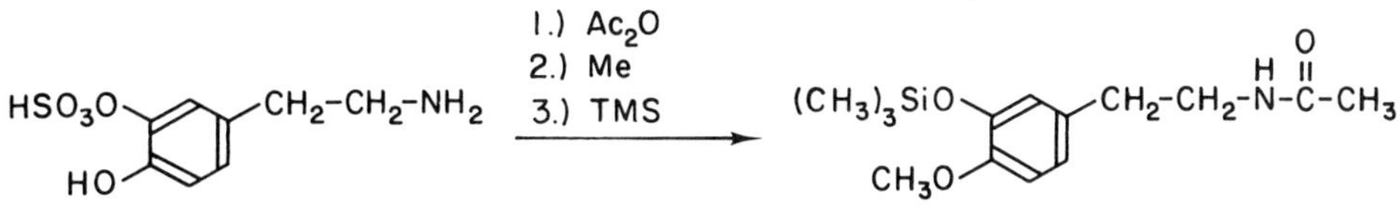

112 113

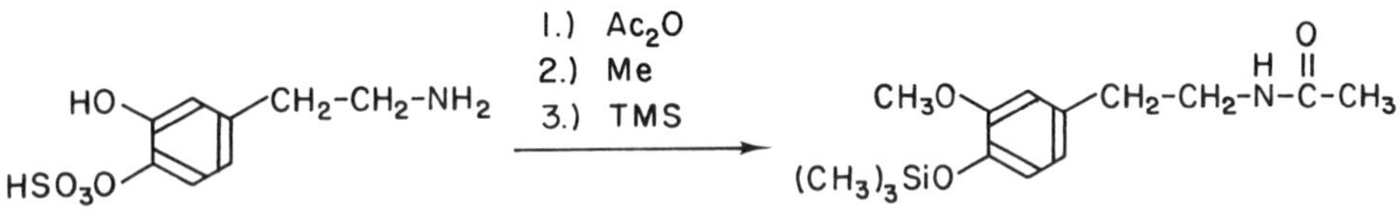

114 115

Field desorption mass spectrometry (FDMS) offers a unique approach for structural characterization of polar molecules. It allows mass spectral analysis by desorption and ionization without prior vaporization, and, in limited reports to date, organic esters of sulfuric acid appear amenable to this approach.

Games et al. [160] have analyzed three alkyl sulfate esters and cyclohexylphenyl-4-O-sulfate as their potassium salts as well as the sodium salt of 21-O-sulfuryl cortisone. Rather than electron expulsion, all ions were desorbed as cationized species (M+ cation). At lower emitter currents the base peak for cortisone was the doubly charged ion $[M+2Na]^{2+}$ whereas at higher emitter currents the singly charged cationized species predominated and ions due to the loss of the sulfate group with hydrogen transfer were observed. Schulten and Lehmann [170] studied the FDMS characteristics of three additional sulfate esters: α-napthyl, n-dodecyl, and dopamine-4-O-sulfate, in their potassium, sodium, and free acid forms respectively.

Field desorption mass spectrometry was used to aid in the characterization of the sulfate ester of 9-hydroxy ellipticine 117, one of two major metabolites from the antineoplastic drug ellipticine, 116 [121]. The spectrum, Fig. 2, that is characteristic for these sulfate esters shows prominent $[M+Na]^+$, $[M+H]^+$, and the disodium sulfate ester cation. Although it is impossible to eliminate all

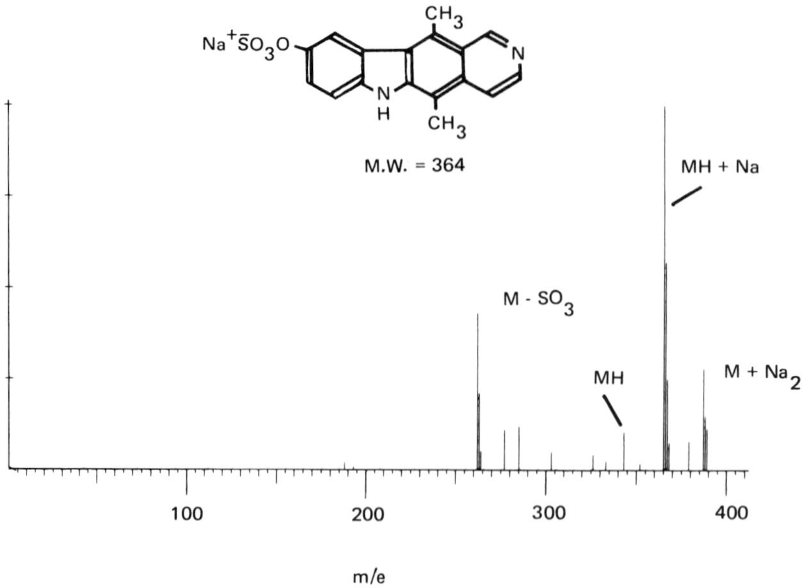

FIG. 2. Field desorption mass spectra of sulfate ester of 9-hydroxy ellipticine, 117. Emitter current, 25 mA.

salt, from these data and unpublished work in this laboratory, it appears that the salt form (Na or K) of the sulfate esters desorbs at lower emitter currents whereas the acid form frequently requires higher currents. Consequently, more fragmentation, and thus more structural detail, is detected at higher emitter currents.

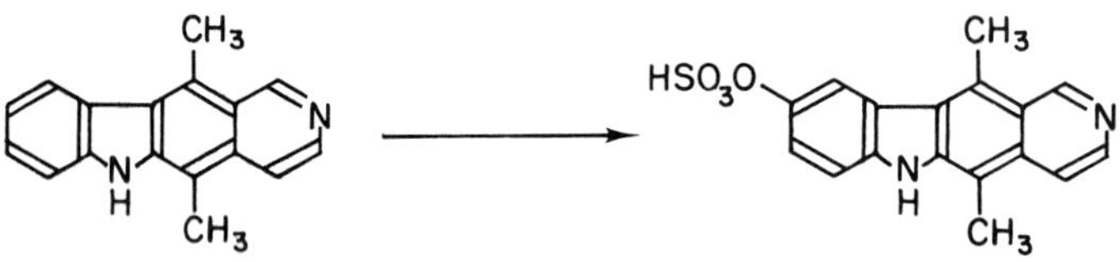

116 117

2. Glucuronide Conjugation

UDPGlcU + R-OH ⟶ 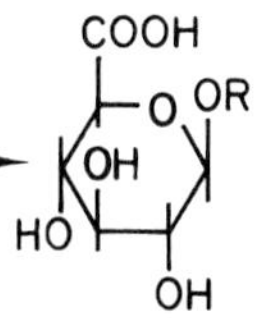

118   119   120

Glucuronide formation from a host of endogenous and foreign materials is probably the most common conjugation in mammals and has obvious pharmacological importance. Glucuronyltransferase enzymes are widely distributed in those tissues with external exposure; i.e., kidney cortex, liver, alimentary tract mucosa. The cellular location of this enzyme transferase is in association with the endoplasmic reticulum. Much like the sulfatransferases, the enzymes have yet to be purified, and preparations of UDP glucuronyltransferase have been known to yield erratic and surprising results in vitro. The aglycone group conjugated may be hydroxyl, carboxyl, amino, imino, or sulfhydryl and may in addition possess aliphatic, aromatic, or heteroaromatic character. It appears that the enzymes are inducible, for barbiturates increase glucuronide conjugation and excretion.

With many drug isolation procedures, solvent partition and chromatographic behavior of metabolites are suggestive of structure. Fractions thought to be glucuronides are, more often than should be, treated with glucuronidase and the aglycone isolated. This is especially the case with large-molecular-weight conjugates that cannot be handled by routine vapor phase analysis. Many structural evaluations, for completeness, demand better characterization, especially with those drugs having a multiplicity of functional groups and one or more heteroatoms in the aglycone.

The need to understand better the urine conjugates of marijuana-exposed subjects led Lyle et al. [273] to synthesize in vitro and characterize by gas-liquid chromatography and mass spectrometry the glucuronides of Δ^9-tetra-hydrocannabinol (Δ^9-THC), Δ^8-TCH, cannabinol, and cannabidiol. This may not always be fruitful for there clearly exist several UDP glucuronyl-transferases and activity toward one compound may not be found in the same tissue of another species.

Games et al. [160] have characterized the glucuronide of diethylstilbestrol, 121, by FDMS. Ions were detected for an assessment of molecular weight at *m/e* 489 $[M+Na]^+$ and the aglycone at *m/e* 268, the latter representing cleavage and hydrogen transfer to yield the parent phenol.

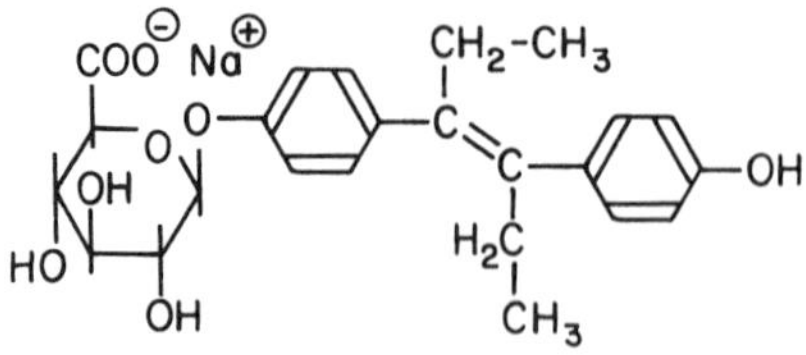

121

Although glucuronide formation is firmly established as a predominant conjugating mechanism for oxygen functions (R-OH, R-COOH), the case for N-glucuronides has been only recently documented. Smith et al. [77] in studying the metabolic fate of 5,5-diphenyldydantoin, 122, identified direct glucuronidation at the N-3 position. The glucuronide was characterized by GC-MS of the permethyl and perdeuteriomethyl derivatives.

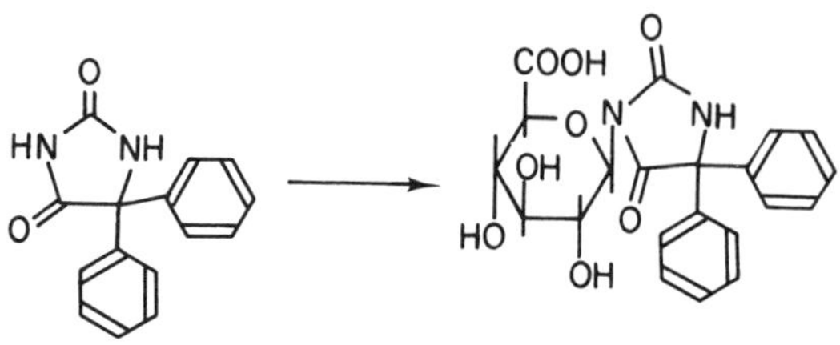

122 123

3. Glucoside Conjugation

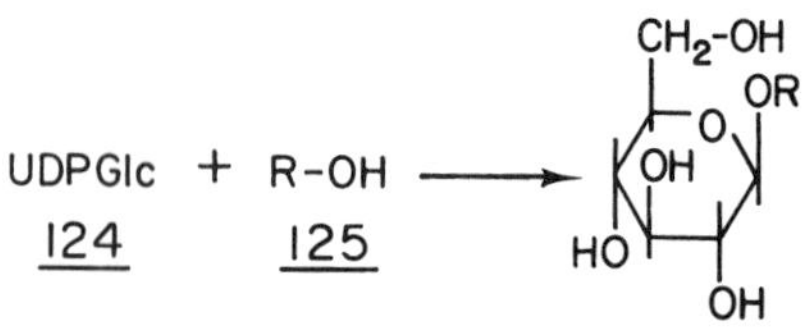

126

Glucose conjugation has been well documented in bacteria and insects, the conjugates being β-glucosides of the aglycone. However, a number of recent reports have shown the occurrence in mammalian tissues of glycosylated species of steroids, flavinoids, bilirubin, and diethylstilbesterol. Comparisons of glucose and glucuronic acid transfer from their respective nucleotides, in these studies have shown the latter to be several times more efficient. Glycosylation may represent an alternative conjugation mechanism in the absence of UDPGLcU, 118. Duggan et al. [201], however, have presented data showing the glucose conjugate to be a major polar metabolite for the xanthine oxidase inhibitor, 127, in dogs, rats, and monkeys.

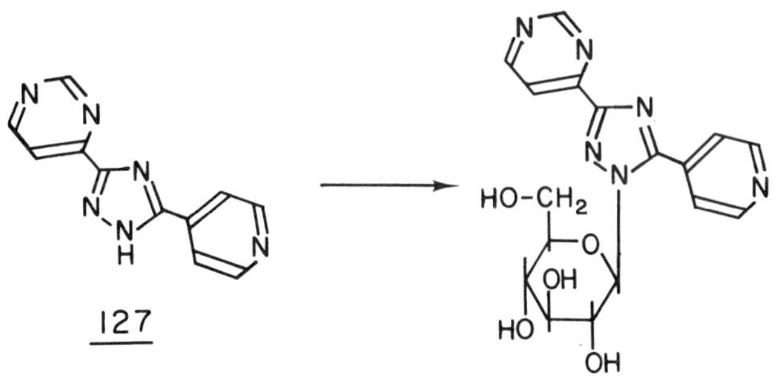

127 → 128

The structure 128 was assigned on the basis of nuclear magnetic resonance (NMR), ultraviolet absorbance and fluorescence spectra, and high-resolution mass spectral analysis.

4. Fatty Acid Conjugation

$$\underset{129}{\text{R-COOH}} + \underset{130}{\text{R}'\text{-OH}} \longrightarrow \underset{131}{\text{R-COOR}'}$$

Conjugation is usually thought of as a mechanism that increases drug polarity, effectively aiding in its excretion. Contrary to this, however, there have been a number of reports that indicate drug metabolites of less polarity than the parent drug. Such a case was recently reported by Leighty et al. [272], where Δ^9-THC and Δ^8-THC, 132, have been detected as hydroxylated, conjugated metabolites in which the conjugating groups are long-chain fatty acid esters, 130.

In order to produce metabolites in quantities sufficient for structural identification, the 11-hydroxy-Δ^9-THC and 11-hydroxy-Δ^8-THC <u>132</u> were incubated in rat liver microsomal preparations.

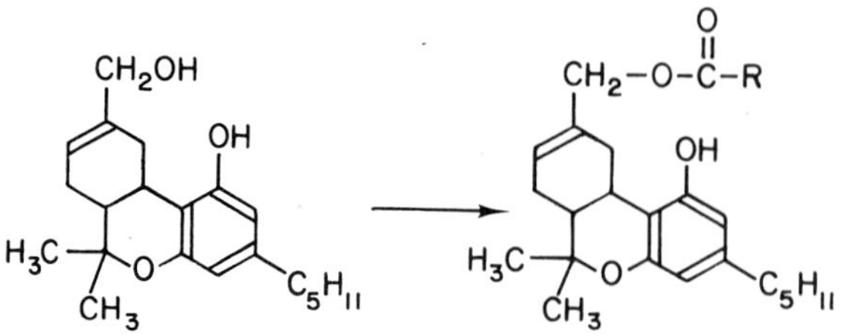

R = $C_{15}H_{31}$ or $C_{17}H_{35}$

<u>132</u> <u>133</u>

5. *Mercapturic Acid Conjugation*

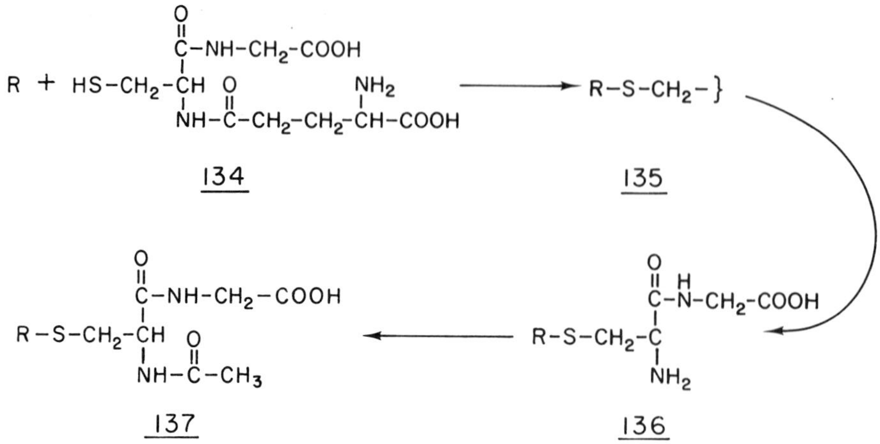

The conjugation of glutathionine, <u>134</u>, with exogenous compounds (perhaps after conversion to a more reactive form, e.g., an epoxide), followed by degradation to the dipeptide adduct, <u>136</u>, and mercapturic acid, <u>137</u>, is another prominent pathway for elimination of xenobiotics. The initial conjugations are catalyzed by glutathionine-5-transferases, although some nonenzymatic reactions are known. Cleavage of the glutamyl residue by the enzyme δ-glutamyltransferase is thought to be analogous to the catabolism of glutathione itself, and in most cases, the last step, acetylation to form the mercapturic acid, occurs before excretion from the body. Further oxidation of the sulfur to sulfoxide has been reported. The importance of conjugation and catabolism to mercapturic acid as a detoxification pathway is difficult to assess. It is known that the enzyme activities of this conjugation are low in man as compared with other species and it may be that this process of elimination represents vestiges of an archaic detoxification pathway for it does not require high-energy intermediates (e.g., ATP, UDP).

IV. FACTORS INVOLVED IN SAMPLE PREPARATION

The biological matrix in which drug metabolites are to be found and from which they must be distinguished and separated presents the first obstacle to the analyst. Mass spectrometry has an advantage in that it can provide a specific analysis of very small amounts of material, but also some disadvantage in that it will also detect a very wide range of endogeneous constituents and extraneous contaminants that may also be present in the sample.

A. Sample Selection

The sample to be analyzed may be any of several body fluids or tissues available and several factors have to be considered in making the selection. If something is already known about the distribution of drug and metabolites in the body compartments, the preferred sample may be obvious ab initio. Particularly in the case of new drugs, however, the determination of the suitable sample is the first stage of the analytical problem. Accessibility is also an important factor: the human subject may be more willing and able to participate in a study that requires donation of saliva or urine than of blood or tissue. In animal studies, quantification of drug or metabolite in a specific organ may require sacrifice of the animal and thus provide only a single data point for the experiment. For study of the transfer of drugs and metabolites from mother to child, cord blood or milk is the appropriate sample, although these are not otherwise considered. When possible, a noninvasive method is preferable and for this reason studies have been carried out to relate concentrations of drugs and metabolites in saliva to the serum levels. This technique should be seriously considered when multiple samples must be analyzed [295]. For the preparation of large quantities of metabolites in relatively "clean" media, in vitro syntheses can be useful. A rat liver microsomal enzyme system was utilized by Leighty et al. [272] for this purpose in the preparation of fatty acid conjugates, 138 and 139, of 11-hydroxy-Δ^9-tetrahydrocannabinol and 11-hydroxy-Δ^8-tetrahydrocannabinol. See Sec. II.4.

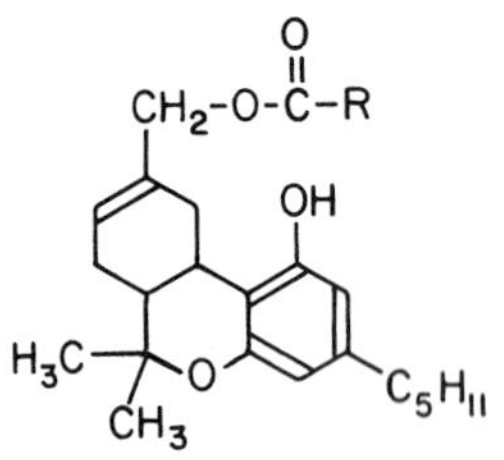

138

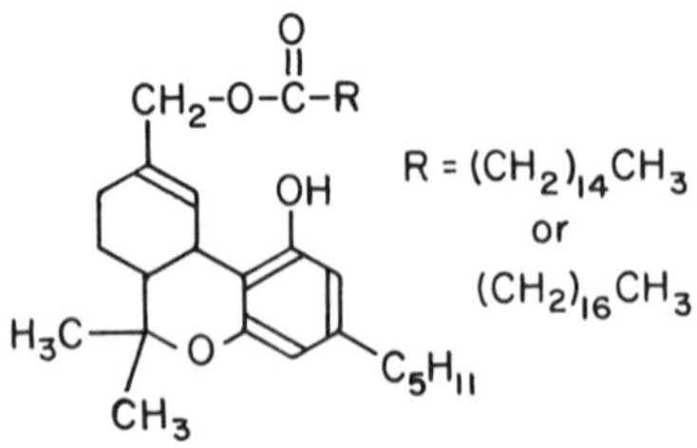

139

These synthetic compounds were later shown to be identical to cannabinoid metabolites isolated from the livers of rats after chronic interperitoneal injection of ^{14}C-labeled Δ^9-THC.

Purified enzymes may also be used to synthesize possible metabolites in vitro. Lyle and colleagues [273] have reported the synthesis of glucuronide conjugates of cannabinoids using a partially purified glucuronyltransferase enzyme immobilized on beaded sepharose. Δ^9-Tetrahydrocannabinol glucuronide, 137, and cannabidiol glucuronide, 140, were among the synthetic metabolites characterized. For structural studies, the glucuronides were analyzed as their per(trimethylsilyl) and poly(trimethylsilyl)methyl esters by both electron-impact and chemical ionization mass spectrometry.

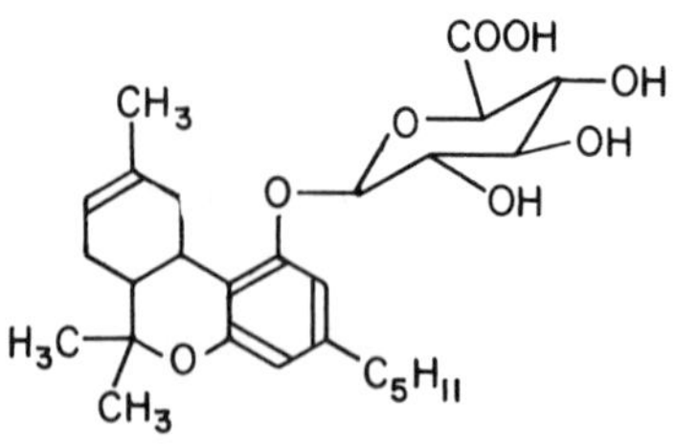

140

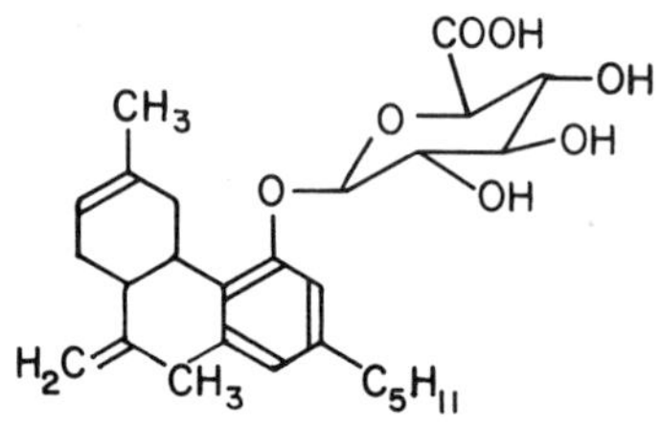

141

The human serum albumin used as a solubilizing agent for the cannabinoid aglycones during synthesis of the aglycones inhibited the enzyme activity between 50 and 75%, leading to long incubation times and low yields. Such interferences may impede, but need not deter, this type of synthesis, as successfully illustrated here.

In vitro syntheses have also been used by Rendic and co-workers [147] for the preparation of derivatives of chiral and achiral 1,4-benzodiazepines. They carried out an extensive study of the fragmentation pathways exhibited by these compounds. This type of information is extremely helpful for the elucidation of structures when new metabolites are encountered, and such studies have been carried out for many of the important drug types.

B. Sample Collection

Whenever possible, it is preferable to collect and store the samples, frozen without preservatives, in all-glass containers with Teflon-lined seals. Plasticizers, antioxidants, and other contaminants are ubiquitous and frequently soluble in organic solvents. Phthalates and adipates are the best known of these compounds and the most widely distributed, but others are also encountered. For example, in clinical studies involving blood samples, the use of Vacutainers[R] (Becton-Dickinson Co.) will result in the presence of tri-2-butoxyethyl phosphate 142 [289] (Fig. 3), particularly if the blood is stored or shipped in these containers. The use of polyvinyl chloride tubing in the sample-collection device sometimes leads to the presence of acetyltri-n-butyl citrate 143 (Fig. 4) in the samples [285].

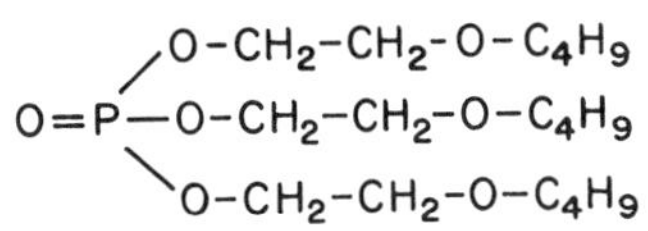

142

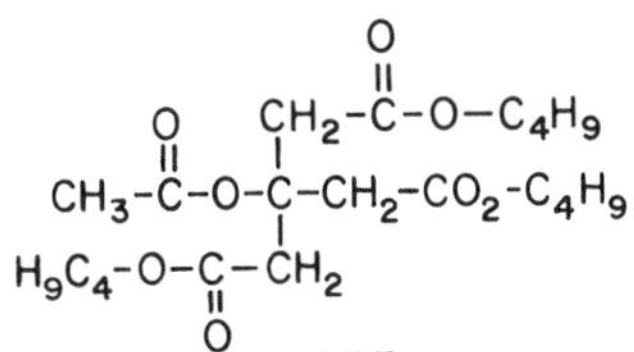

143

The possibility of this sort of contamination should be considered and its interference minimized in any experimental procedure. It is prudent that blanks of all workup procedures be prepared so that time

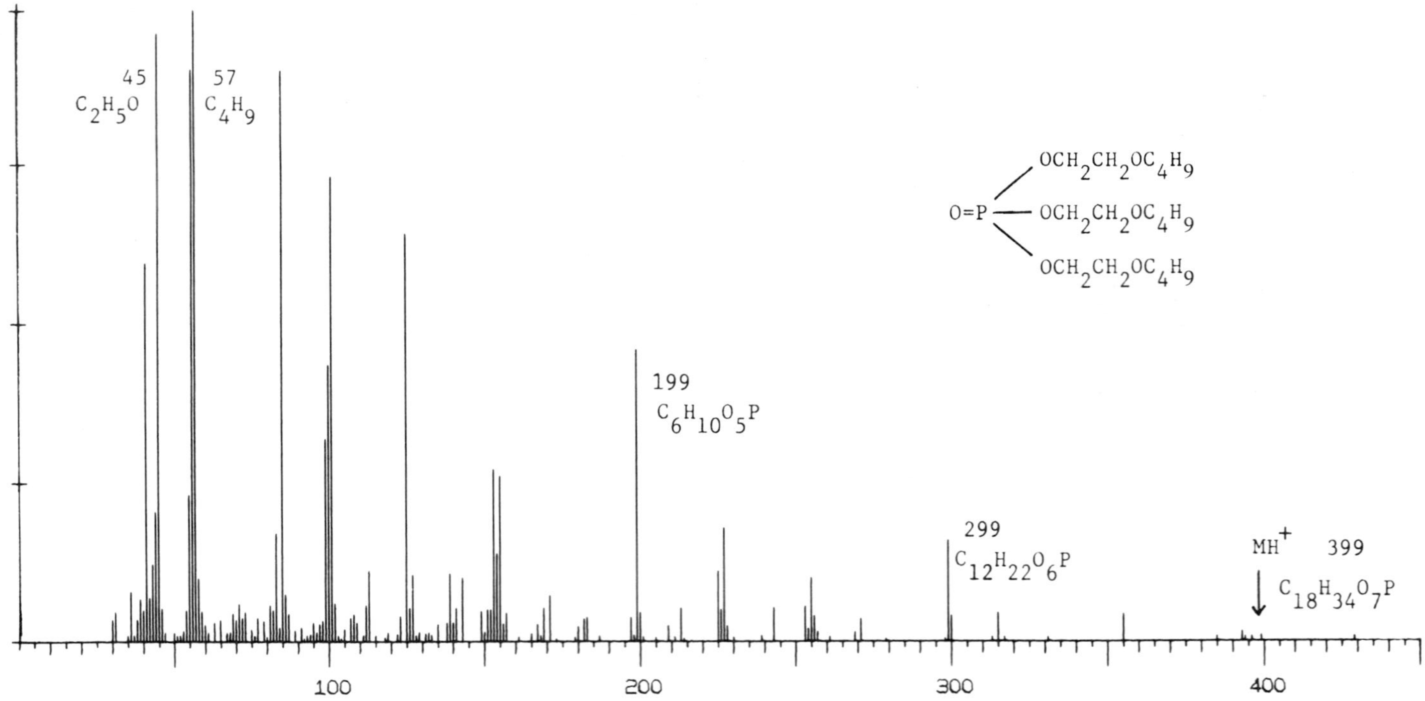

FIG. 3. Mass spectrum of tri-2-butoxyethyl phosphate, 142, the compound leached into blood samples from the stoppers of B-D Vacutainers[R].

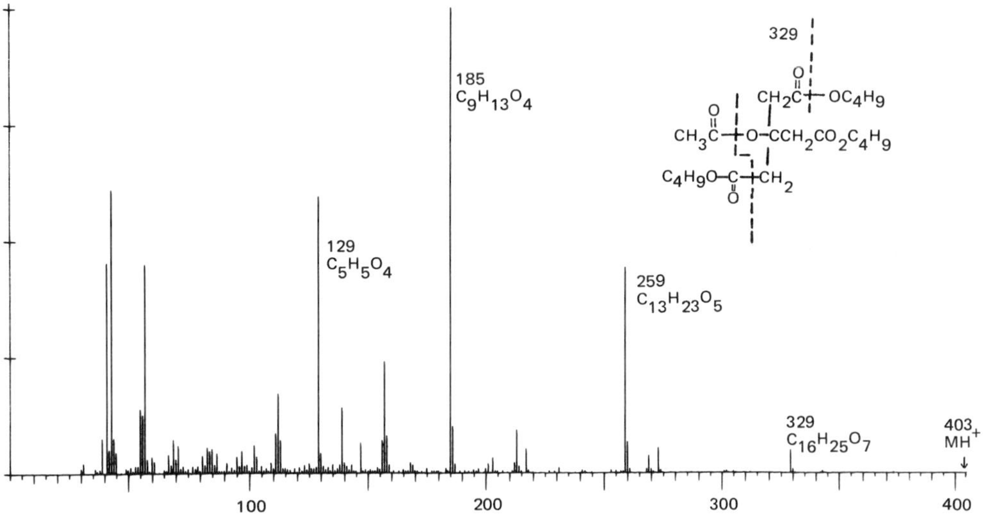

FIG. 4. Mass spectrum of acetyl tri-n-butyl citrate, 143, a plasticizer present in some types of polyvinyl chloride tubing, which may appear as a sample contaminant.

and effort are concentrated on identifying relevant substances. Extensive contamination of the sample may make the analysis impossible or unreliable.

The timing of the sample collection can affect the observed distribution of metabolites. Therefore, the details of the patient's history should be carefully noted. It is especially important that the time course of the sampling be recorded in distribution and excretion studies, but it is also important for qualitative studies, since different metabolites may predominate at different times after a dosage is administered or after continued exposure of the subject to the drug. The amount of drug administered can also make a difference in the metabolite pattern, particularly at very high dosages. In this situation, the ordinary metabolic routes may become saturated and the trivial pathways may give rise to measurable quantities of unusual metabolites. The patient's age and overall physical state will also influence the disposition of the drug. Horning and coworkers [293] have reported that women who had consumed caffeine in their diet excreted mono- and dimethylxanthines in larger quantities than they excreted caffeine immediately after childbirth, whereas their newborn infants excreted mainly caffeine itself.

C. Sample Isolation

For the purpose of toxicology in the case of overdoses, drugs and their metabolites may be extracted from physiological fluids by fairly simple and rapid techniques [288,297]. When it is the intention of the analyst, however, to isolate new drug metabolites or to quantify specific drugs or metabolites, more complex schemes are appropriate. That method must be chosen which minimizes interference from other materials and gives the most complete or most accurately quantified recovery.

Isolation begins with the homogenization, followed by filtration or centrifugation of the sample when these steps are necessary. Ethanol is useful to precipitate proteins from serum samples, but may result in loss of drugs or metabolites. Extraction of the sample should be carried out at a pH and with a solvent or combination of solvents that will permit the maximum recovery of the compounds of interest with minimal interference. Some compromises may have to be made, but it should be possible to separate the mixture into neutral, acidic, basic, and amphoteric components. It may be necessary to use several extractions at different pH to recover and separate the several metabolites of a single drug. Beckett [36] has used such a scheme to isolate the metabolites of mexiletine, 144. The detection limit for the quantification of 144a-144d were found to be 6-12 ng/ml and for 144e-144j, 40-50 ng/ml; 144f was not a metabolite itself, but was formed from 144 during the base extraction and was then determined to be appropriate for the analysis of 144d.

Improvement in the efficiency of solvent extraction can be accomplished not only by the use of buffers to optimize pH, but also by the use of salt/solvent pairs that minimize emulsion formation and provide pH control. This technique was introduced by Bastos and colleagues [283] and improvements have been suggested by Horning and her coworkers [296]. In a typical procedure, solid ammonium carbonate is

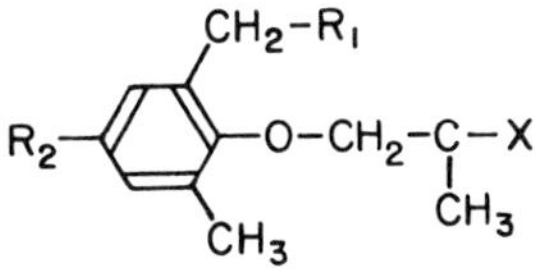

144 a-j

	R_1	R_2	X
144a	H	H	HNH_2 (mexiletine)
144b	H	H	$HNHCH_3$
144c	H	H	(=O)
144d	H	H	HOH
144e	H	H	HNHOH
144f	H	H	=NOH
144g	H	OH	HNH_2
144h	OH	H	HNH_2
144i	H	OH	HOH
144j	OH	H	HOH

added to urine to the saturation point. Ethyl acetate or methylene chloride is then added and the layers thoroughly mixed. After centrifugation, the organic layer is removed, the solvent is evaporated, and any necessary derivitization procedures are carried out prior to GC-MS analysis. Other combinations of salt/solvent pairs have also been found useful. Milberg et al. [405] reported the combination of sodium tetraborate and chloroform to be effective for the recovery of low levels of methylphenidate, 145, and its metabolite ritalinic acid, 146, from blood and urine.

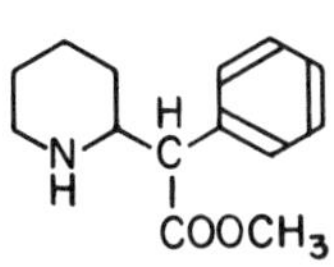

145

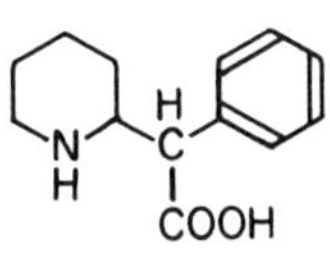

146

The compounds of interest can be derivatized in the extraction step. The terms "extractive alkylation" and "ion pair extraction" are used to describe this procedure. By means of a bulky quartenary ammonium cation, the drug or metabolite is extracted as an ion pair into the nonpolar solvent that contains the alkylating agent. Ehrsson [325] combined tetrahexylammonium hydrogen sulfate as the cation source and methyl iodide as the alkylating agent for the analysis of phenobarbital. Because of the strongly basic conditions developed in some of these techniques, degradation of substrate may occur. Determination of sample stability is therefore important when methods utilizing this convenient reaction are developed.

There is, of course, a change in the chemical properties of drugs due to metabolism. Oxidation and dealkylation lead to an increase in polarity and water solubility. Conjugation with glucuronates or sulfates will also lead to increased water solubility and these metabolic processes run counter to organic solvent extractive techniques. Therefore, adsorption can be an effective alternative means for removing these compounds from the aqueous phase and is frequently quicker and more quantitative than solvent extraction. The oldest and simplest technique of this type is adsorption onto charcoal, as described by Meola and Vanko [427]. In the first step, the organic compounds are adsorbed onto the solid surface and the aqueous phase is removed. The procedure is then reversed and a suitable solvent is used to remove the compounds of interest from the charcoal. Free acids may be collected if an anion-exchange resin is employed (DEAE-Sephadex) and a modification of this procedure is suitable for the analysis of certain drugs and their metabolites. This method has been extensively investigated by the Horning group [422]. According to the types of functional groups anticipated, the compounds recovered in the column effluent are derivatized (by methylation, trimethylsilylation, or hydroxyl-amine formation) and subjected to GC-MS analysis. For collection of sulfate and glucuronate conjugates, the nonionic resins (XAD-2, XAD-4) have proved effective [425,433,435]. These adsorbent resins first proved of value in purifications preceding thin-layer chromatography (TLC) analyses [433,435]. As is not uncommon, sample preparation schemes developed for other analytical techniques have proved to be suitable (with minor modifications) for mass spectrometric work as well. When first introduced, these resins were unsatisfactory for GC-MS work because of the extensive cleanup necessary to reduce background. The resins currently available have been substantially improved in this respect, and little preliminary washing is now necessary.

Hydrolysis of conjugates may be carried out enzymatically (β-glucuronidase, α-sulfatase) or by mild acid hydrolysis. The enzyme procedure is generally slower but leads to cleaner extracts and causes less degradation of sample. For quantitative determination, the free drug or metabolite concentration must be determined before and after the hydrolysis step with the difference attributed to the conjugate moieties [336,461]. It is also possible to quantify the intact conjugate [296].

Endogeneous compounds may interfere in an analysis or even react with the drugs of their metabolites. We have, for example, detected transmethylation between phosphatidyl choline (lecithin) and compounds that can serve as methyl acceptors (Fig. 5) [311]. In this instance, on-column methylation occurred to give the methyl derivatives of phenobarbital and diphenylhydantoin, as well as the methyl esters of the serum fatty acids. Selection of the proper solvent for the extraction can minimize this type of problem. In other cases, an additional preparative step may be necessary to eliminate the interference. Chang and colleagues [96] have resorted to a preliminary TLC procedure before GC-MS to remove the endogeneous acids when analyzing for the acidic metabolites of diphenhydramine. High-pressure liquid chromatography may also be useful and is being employed widely now, even for in vitro methods such as that of Leighty mentioned in the previous section. A less common procedure, but one that is highly useful when the equipment is available, is countercurrent distribution. In this manner, Abshagen and co-workers [364] carried out the isolation of a sulfur-containing metabolite, 148, of spironolactone, 147, from human urine. The new metabolite was assigned the structure 148 and this was confirmed by

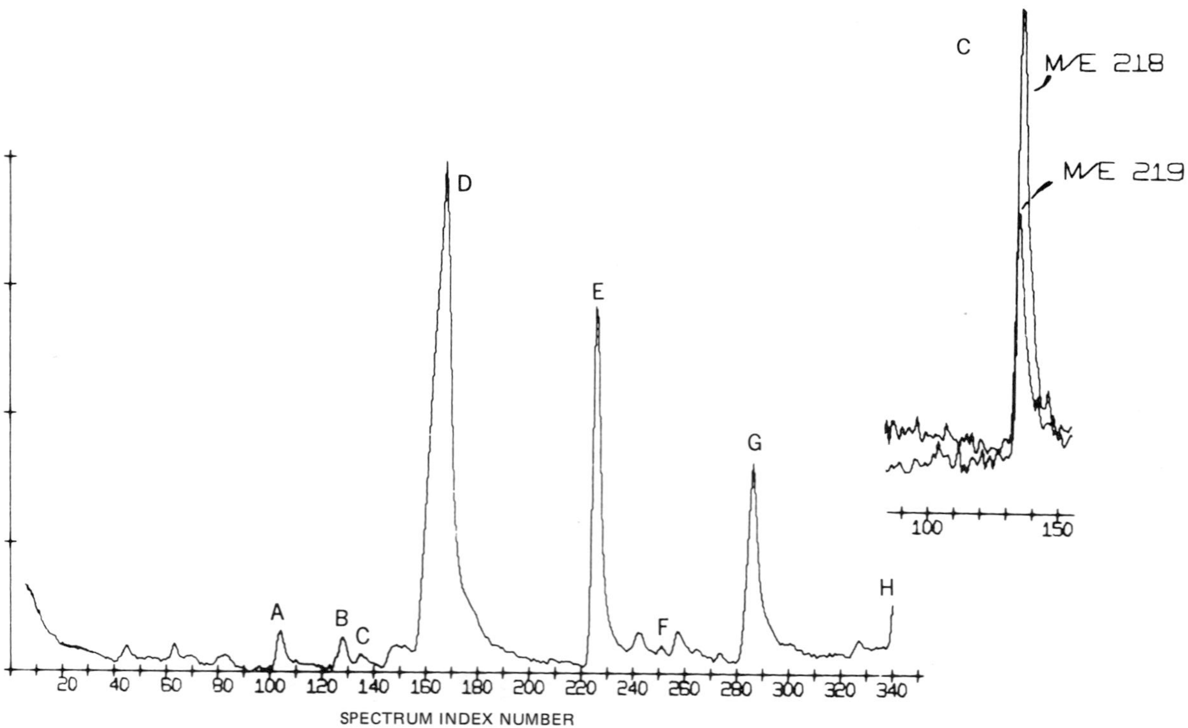

FIG. 5. Total ion plot of a serum extract obtained from dimethoxymethylphenobarbital-treated patient coinjected with phenobarbital-ethyl-$^{2}H_5$. A, B = phthalate plasticizers; C = N-methylphenobarbital ($^{2}H_0$ and $^{2}H_5$); D = phenobarbital; E = tri-2-butoxyethyl phosphate (impurity from B-D Vacutainer[R]); F = N-methyl phenyl-p-methylphenyl hydantoin; G = phenyl-p-methylphenyl hydantoin; H = cholesterol. The mass chromatograms of ions *m/e* 218, *m/e* 219 for compound C are shown in the top right insert. These arise by loss of C_2H_4 from the molecular ions and are the base peaks of the spectra of C ($^{2}H_0$ and $^{2}H_5$).

its TLC, MS, and NMR identity with the oxidation product of the known methylsulfinyl compound.

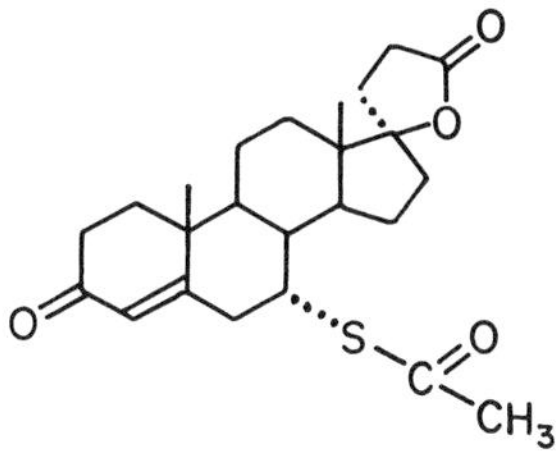

147

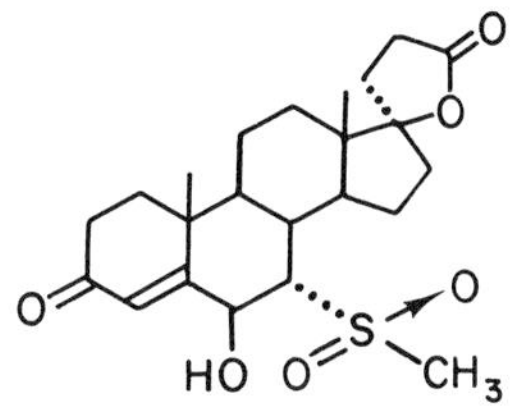

148

V. IDENTIFICATION

A. GC-MS Using Electron Impact

The success of the mass spectrometer as an analytical instrument can be accounted for, in large part, by the development of adjunct techniques that have greatly expanded its utility. For drug metabolism studies, the most practical instrumental combination is the gas chromatograph and the low-resolution mass spectrometer, utilizing an electron-impact ion source. More than just serving as a means of component separation, the gas chromatograph also provides a measure of quantification, class separations by column selection, and retention time information, all of which help to characterize the eluted compounds.

The choice of suitable chromatographic conditions (temperature range, isothermal or programmed heating, capillary or packed column, liquid phase, and solid support) is determined by the complexity of the separation and the type of information sought in the experiment. For qualitative analysis of an unknown mixture, a broad GC-temperature range needs to be examined and a wide mass range must be scanned on the MS, e.g., 020-500 amu (atomic mass units). For observation of a series of related compounds, a narrow GC-temperature range or isothermal conditions can be employed and a short mass range or a few selected ions may be monitored on the spectrometer. The latter conditions permit longer analysis time per unit mass and provide greater sensitivity.

The successful interfacing of the GC with the repetitive scanning MS has offered a quantum jump in the analytical understanding of complex biological mixtures. Coupled with this advancement was the burdensome need to peruse mountains of data. In order to solve the problem, the last decade has seen a proliferation of research groups and manufacturers who have developed hardware and software to handle, process, and manage this wealth of information. This imposing analytical package, GC-MS-computer, has now established a central position in all well-equipped analytical laboratories and not only has brought forth rapid mechanisms for assimilating data, but has introduced new insights for understanding complex mixtures. One of these is the mass chromatogram, a plot of one ion's intensity vs repetitive MS-scan number, a presentation that aids in detecting specific structural features and component homology of the complex mixture. A mass chromatogram of a given component will increase, reach a maximum, and decrease over a number of repetitive MS scans and the point of maximum intensity (identical for all the ions originating from that single component) defines the retention index for that material. Since all ions (mass chromatograms) from a single component will have maxima at identical retention index (which is not shared with trailing, overlapping, and background bleed peaks), a reconstruction of a mass spectrum from mass chromatograms at that retention index generates, in principle, a clean spectrum. This concept of mass spectra reconstruction was introduced by Biller and Biemann [414] as an alternative to simple background subtraction and has the advantage that total ion plots compiled from reconstructed spectra effectively, albeit artificially, increase GC-column resolution.

The possibilities afforded by this experimental approach are illustrated in Figs. 6-18. For reasons of simplicity, many of the strategies of analysis are illustrated with data from a single GC-MS run. The sample, a urine extract obtained from a woman who had ingested an overdose of an unknown number of drugs, was analyzed by a GC-MS-computer system. Subsequently, a routine computer program, using a library of mass spectra for comparison, quickly identified the major components, including some metabolites. The results, printed on a line printer (Fig. 6), represent a portion of the

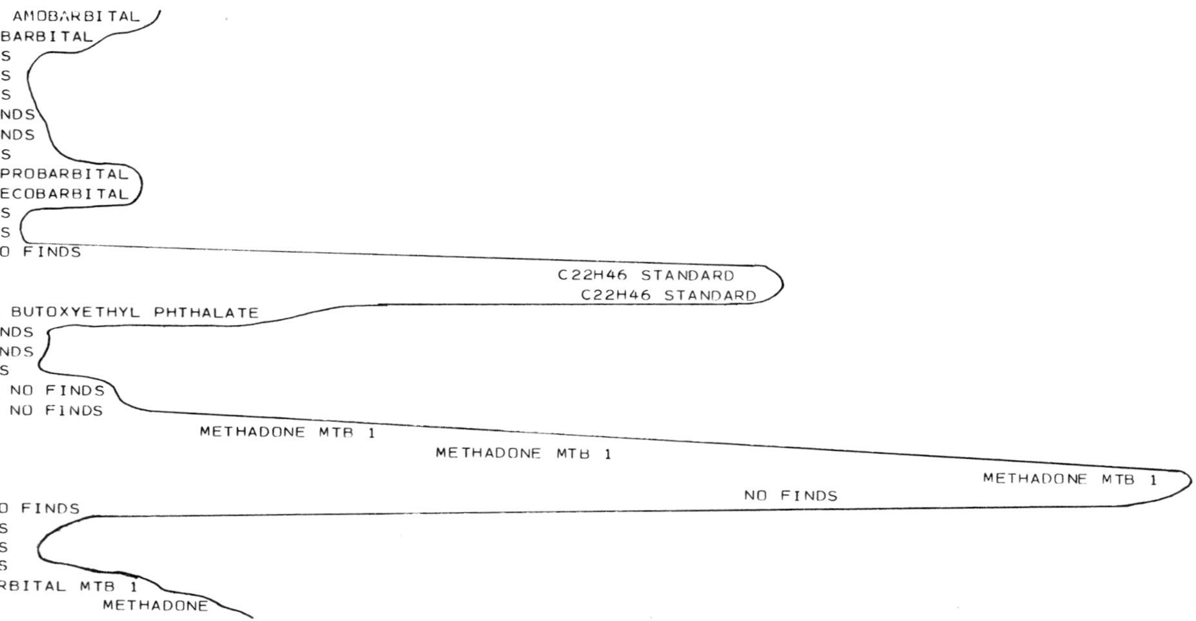

scan no.	sim. index	i.d. no.	search result
100	0.416	25	AMOBARBITAL
101	0.255	25	AMOBARBITAL
102			NO FINDS
103			NO FINDS
104			NO FINDS
105			NO FINDS
106			NO FINDS
107			NO FINDS
108	0.240	526	APROBARBITAL
109	0.325	148	SECOBARBITAL
110			NO FINDS
111			NO FINDS
112			NO FINDS
113	0.641	276	C22H46 STANDARD
114	0.553	276	C22H46 STANDARD
115	0.099	243	BUTYL BUTOXYETHYL PHTHALATE
116			NO FINDS
117			NO FINDS
118			NO FINDS
119			NO FINDS
120			NO FINDS
121	0.081	236	METHADONE MTB 1
122	0.148	236	METHADONE MTB 1
123	0.396	236	METHADONE MTB 1
124			NO FINDS
125			NO FINDS
126			NO FINDS
127			NO FINDS
128			NO FINDS
129	0.201	267	AMOBARBITAL MTB 1
130	0.151	143	METHADONE

FIG. 6. Portion of the library search results for scans 100 to 130 of the GC-MS analysis of urine from a patient described in the text, who was comatose after ingestion of an unknown number of drugs.

MOST PROBABLE COMPOUNDS

0.7878E 06	C22H46 STANDARD
0.4083E 06	C14H30 STANDARD
0.2691E 06	METHADONE MTB 1
0.2121E 06	C32H66 STANDARD
0.1054E 06	AMOBARBITAL
0.3769E 05	SECOBARBITAL
0.3313E 05	CHOLESTEROL
0.3271E 05	DIAZEPAM
0.1384E 05	METHADONE

FIG. 7. Summary of drug library search results for the major constituents of the urine and serum extracts from a patient with multiple drug ingestion.

search results, showing the library "best fit" for each scan and printed along the contours of the total ion plot. Figure 7 is a summation listing of the major component "finds" of the mixture in decreasing order of concentration.

Coinjection of the sample with a series of standard compounds (here $C_{14}H_{30}$, $C_{22}H_{46}$, $C_{32}H_{66}$) provides internal markers from which gas chromatographic retention indexes are automatically calculated [429]. These values are utilized in two ways. First, in the mass spectral library search routines, the retention index is used as a presearch in the computer programs so that only compounds of similar retention index are considered and compared. This retention index presearch minimizes computer time and, more important, increases confidence in the result. This approach is justified when it can reasonably be assumed that most or all of the compounds likely to be encountered in a set of samples are included in the reference collection. For well-defined projects, this is frequently the case. In addition to studies of drugs and peptide derivatives, retention index criteria have also been applied to studies involving urinary acids and steroids [308,432]. A second use of the retention index value is for its inherent information about molecular structure, as will be discussed below.

The complexity of the chromatogram (Fig. 8A) and the presence of several components whose polarity led to tailing of their chromatographic peaks suggested that reconstruction of the data [414] would be useful. This program generated a new total ion plot (Fig. 8B) and a new set of mass spectra that contain only masses whose mass chromatograms exhibited maxima at that scan number. The increase in effective resolution is clear by comparison of Figs. 8A and 8B. The cleanup of spectra obscured by overlapping peaks made possible the identification of several more drug metabolites. Scan 181 before and after reconstruction is shown in Figs. 9A and 9B. This cleanup is useful when a spectrum is to be interpreted manually and it also makes possible more effective use of library search programs. The search algorithms are primarily designed to deal with spectra of pure compounds and increasingly complex and time-consuming routines are necessary if they are to be used for adulterated spectra. The results of a search on this scan after cleanup are shown in Fig. 10. The mass spectrum corresponds to 149, the desmethylated metabolite of diazepam, 150.

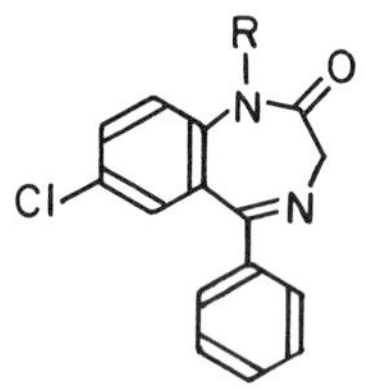

149 R=H

150 R=CH_3

Only a single scan (4 sec) later (scan 182), a metabolite of another drug eluted (Fig. 11A). Its structure could be assigned from the reconstructed spectrum. Although the spectrum of this compound was not in the reference collection, the library search (Fig. 12) provided information as to the compound type. The compounds that were retrieved are all highly aromatic nitrogen-containing compounds and two of these are derivatives of other

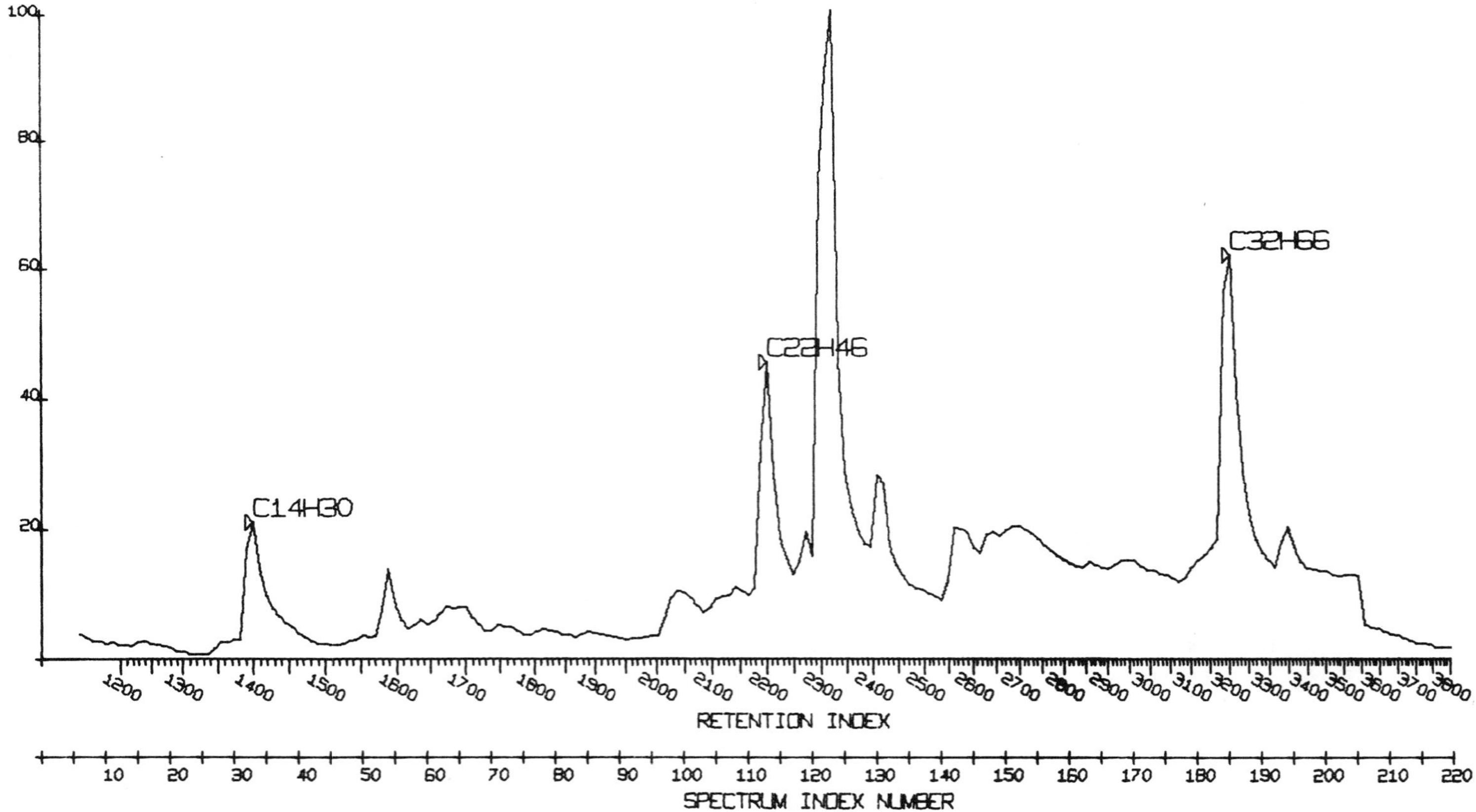

FIG. 8A. Total ionization plot for GC-MS analysis of urine extract for a patient with multiple drug overdose. GC conditions: 3% OV-17 on Gas Chrom Q, 6' glass column, programmed at 16°/min from 80° to 330°.

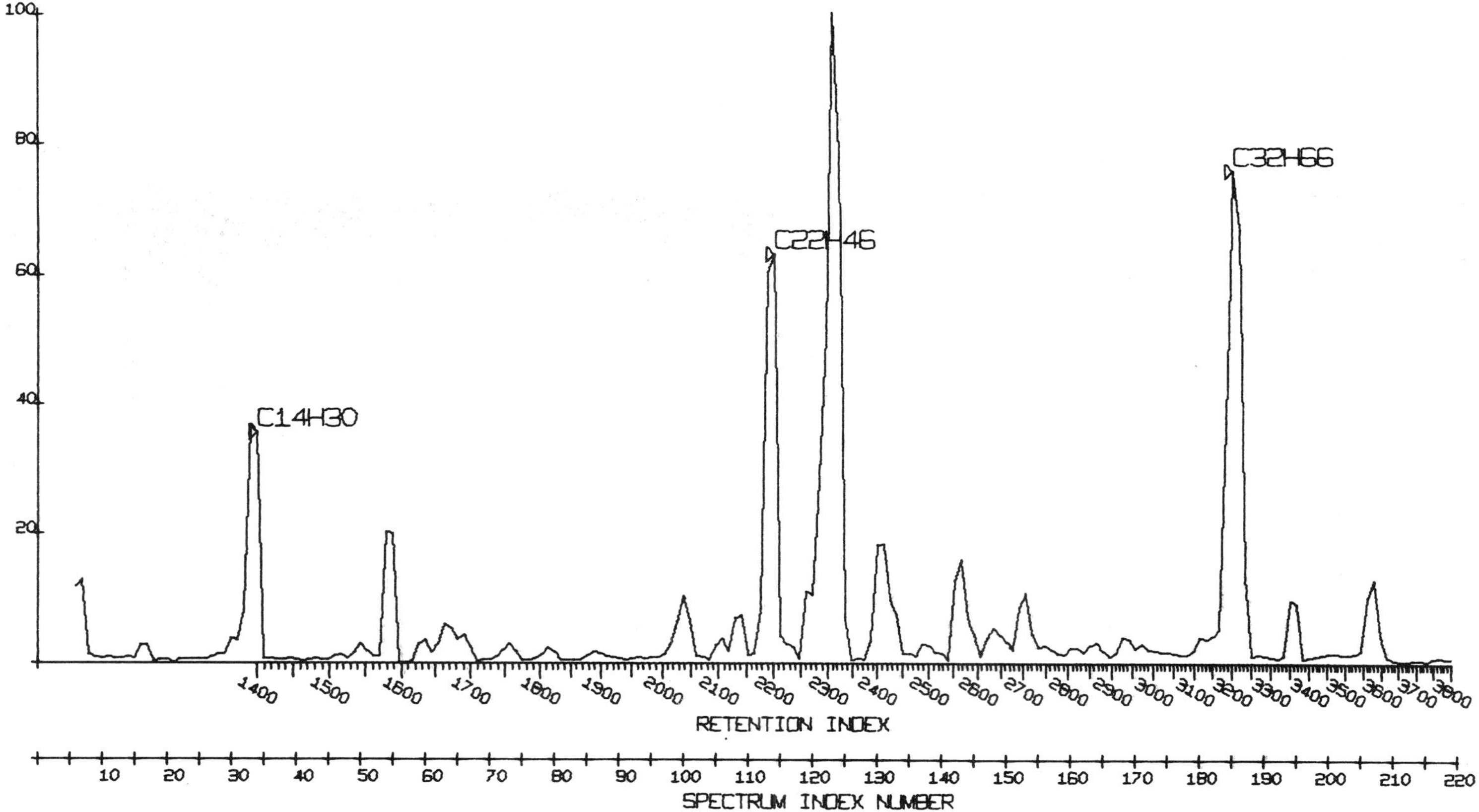

FIG. 8B. Reconstructed total ionization plot for same GC-MS run as (A). See text for details.

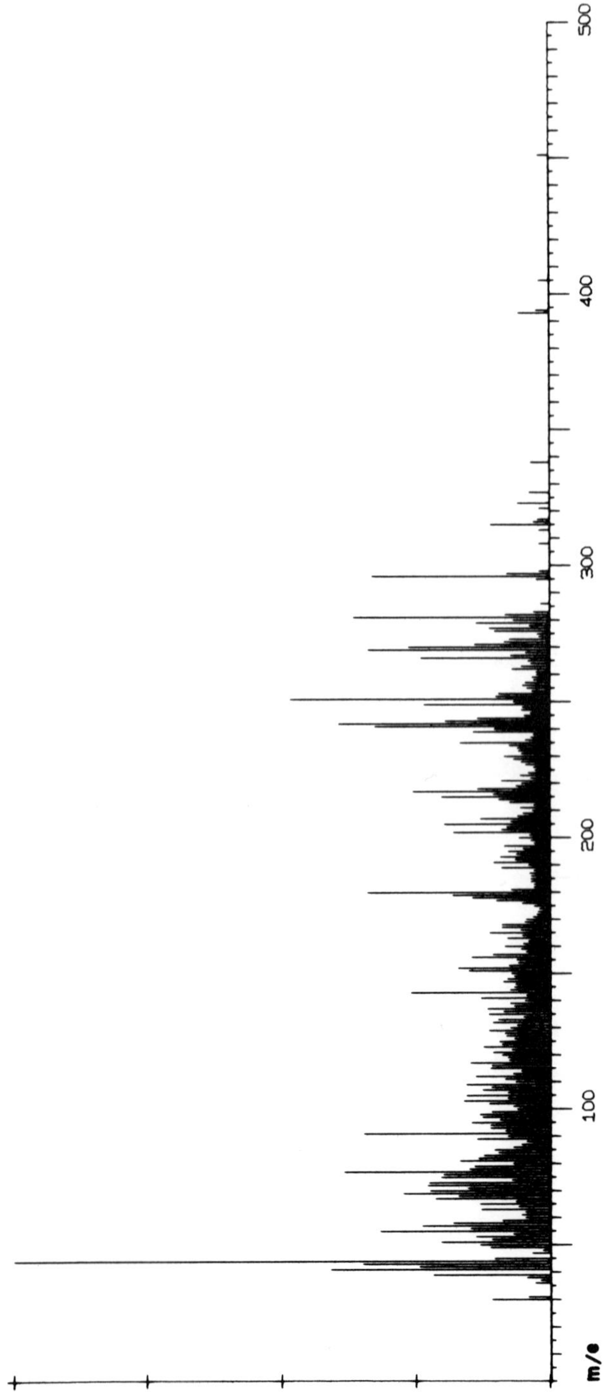

FIG. 9A. Scan 181 of GC-MS analysis in Fig. 8 before reconstruction.

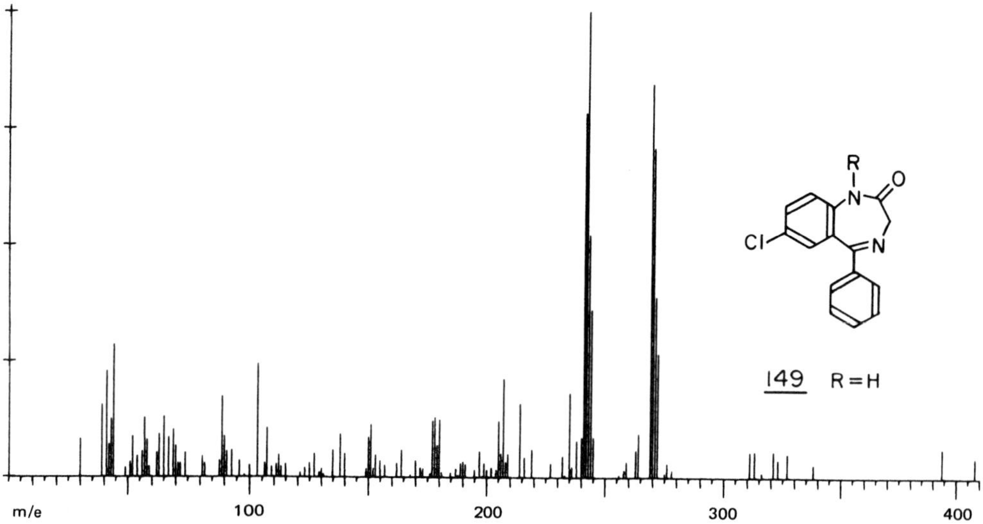

FIG. 9B. Same scan after reconstruction

RESULTS	181	RUN NO. =	2681
	ID.		SIM.
DIAZEPAM MTB 1	448		0.149
MEDAZEPAM	242		0.099

FIG. 10. Drug library search result for scan 181, Fig. 9B.

metabolites of the drug methaqualone 151. The spectrum also was markedly similar to that of the drug methaqualone, for which a spectrum (Fig. 11B) was obtained in the GC-MS of the accompanying serum extract. The compound eluting here was one of its metabolites, 152, which has both a hydroxy and a methoxy group on the phenyl ring, and has a molecular weight of 296.

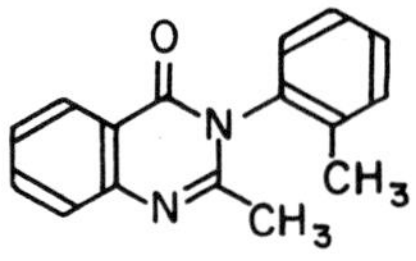

151

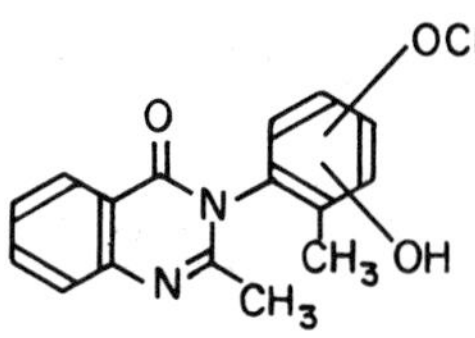

152

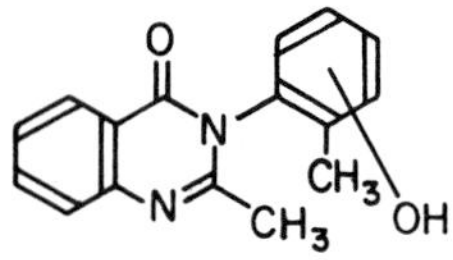

153

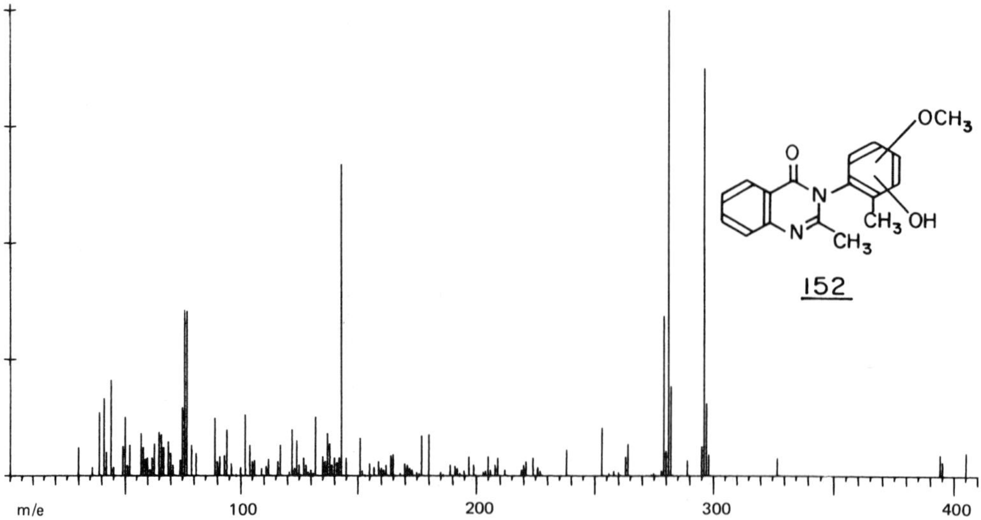

152

FIG. 11A. Mass spectrum of methaqualone metabolite, 152, from urine extract of same patient, scan 182 of the GC-MS run in Fig. 8, after reconstruction.

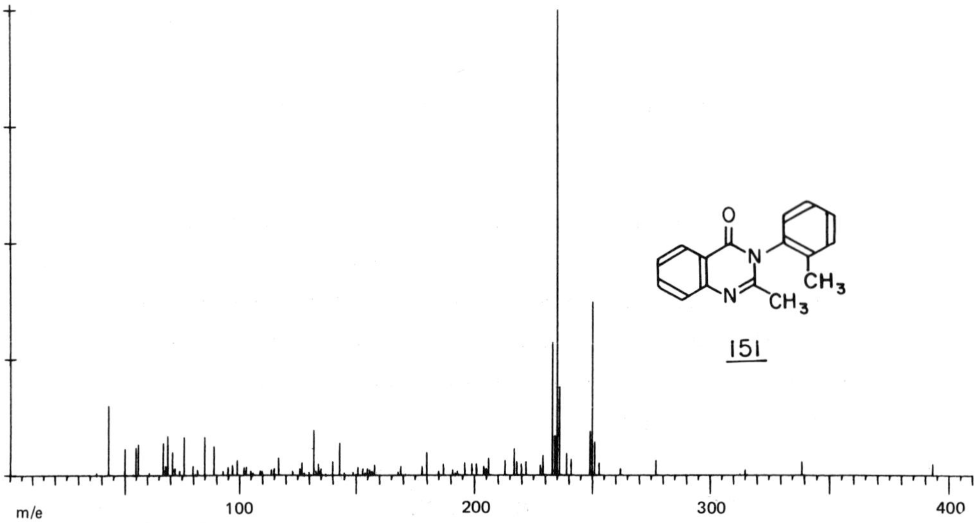

151

FIG. 11B. Mass spectrum of methaqualone, 151, from serum sample of patient who had ingested multiple overdose.

RESULTS	182	RUN NO. = 2681
	ID.	SIM.
DOXYLAMINE (GC)	241	0.055
METHAQUALONE MTB 4 TMS	312	0.053
THEBAINE	195	0.049
N,N-DIMETHYLTRYPTAMINE	181	0.044
THIORIDAZINE MTB 2	218	0.036
PERICYAZINE	135	0.033
2,5-DIMETHOXY-4-METHYLAMPHETAMINE	180	0.032
PROPOXYPHENE	572	0.031
METHAQUALONE MTB 3 TMS	311	0.031
DOXYLAMINE	85	0.030

FIG. 12. Drug library search result for scan 182, Fig. 11B. Retention index presearch has been bypassed. The authentic compound was not in the collection and related compounds were retrieved, but with low similarity indexes.

The corresponding hydroxy substituted compound 153 had eluted earlier. The mass chromatograms of *m/e* 296 and 266 (Fig. 13) point out their relative retention times.

The relationship between amobarbital, 154, and its hydroxylated metabolite, 155 is clear both from their mass spectra (Figs. 14A and 14B respectively) and the mass chromatogram of *m/e* 156 (Fig. 15), which arises by a McLafferty rearrangement and characterizes the ethyl barbiturate ring.

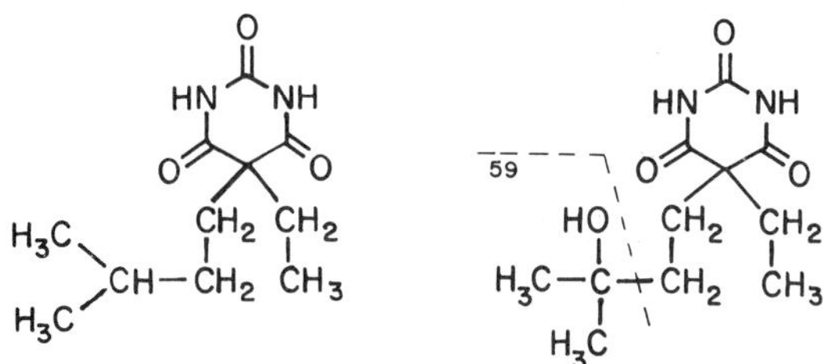

154 155

The shift in retention index of 340 (see Fig. 8) is typical for oxidation on the penultimate carbon of an aliphatic chain [429], and is a smaller shift than that observed for the aromatic hydroxylation on methaqualone above.

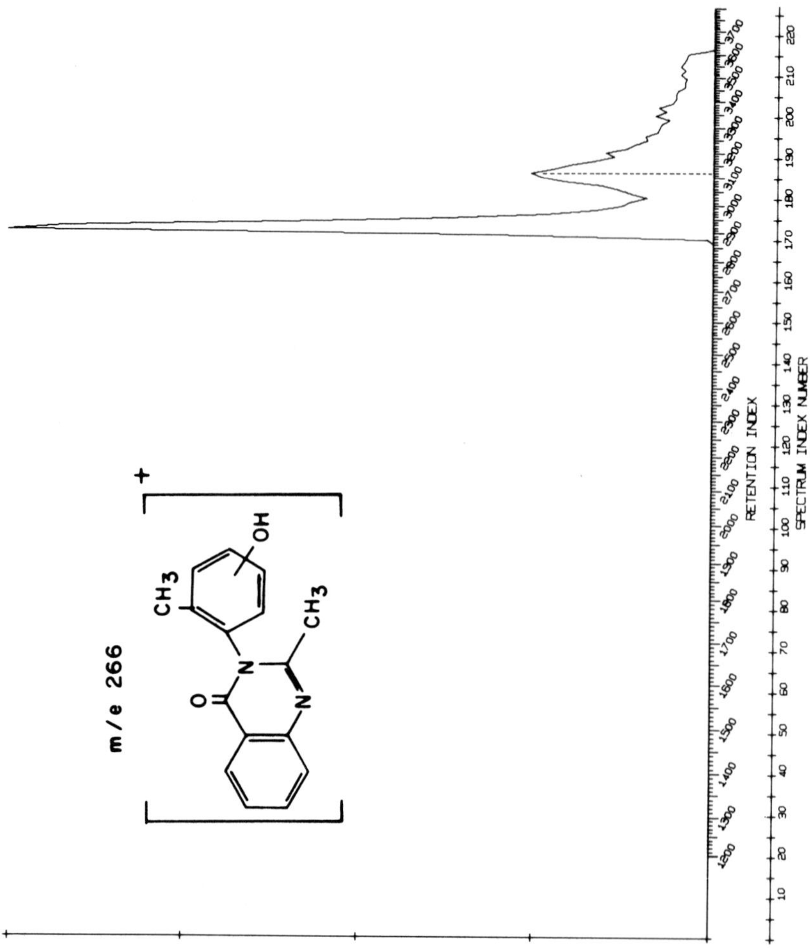
m/e 266
RETENTION INDEX
SPECTRUM INDEX NUMBER

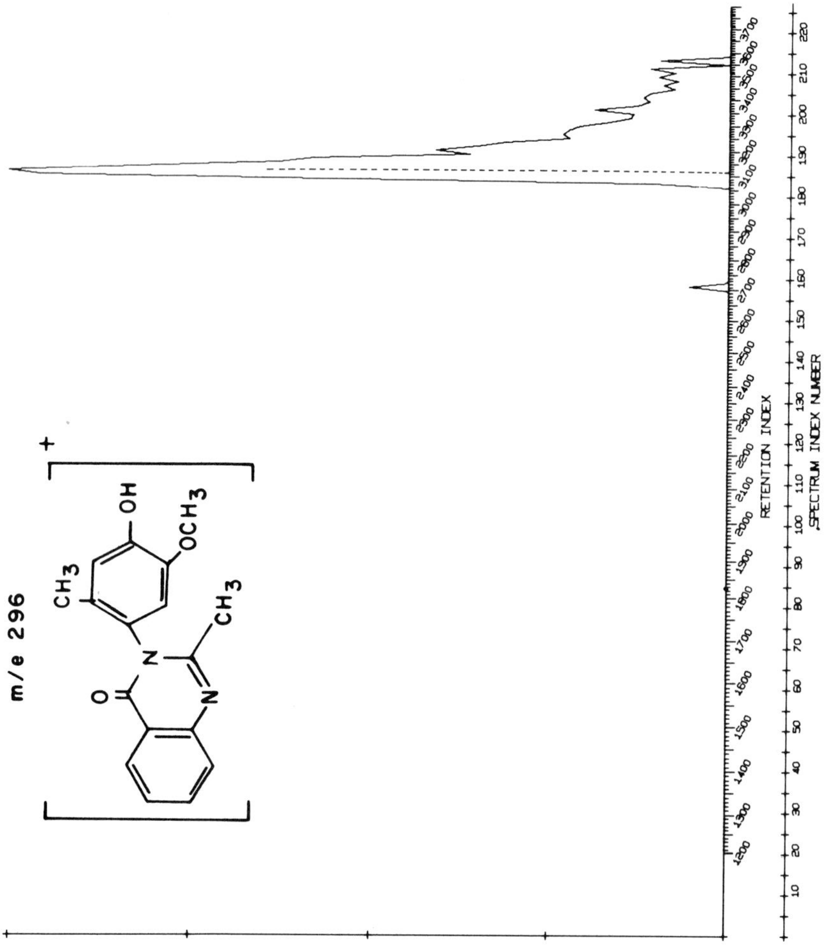

FIG. 13. Mass chromatograms of the molecular ions of compounds 152 and 153 in the GC-MS analysis of urine extract from comatose patient.

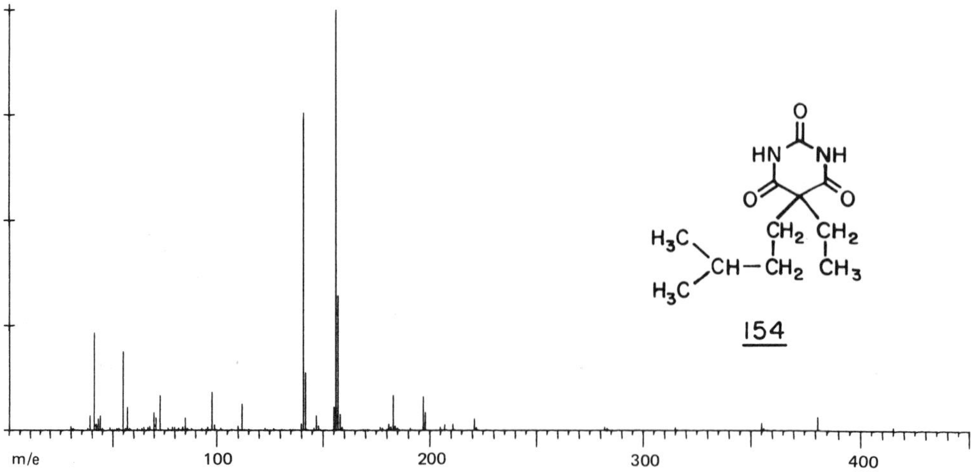

FIG. 14A. Mass spectrum of amobarbital, 154, scan 100 of GC-MS analysis of urine extract.

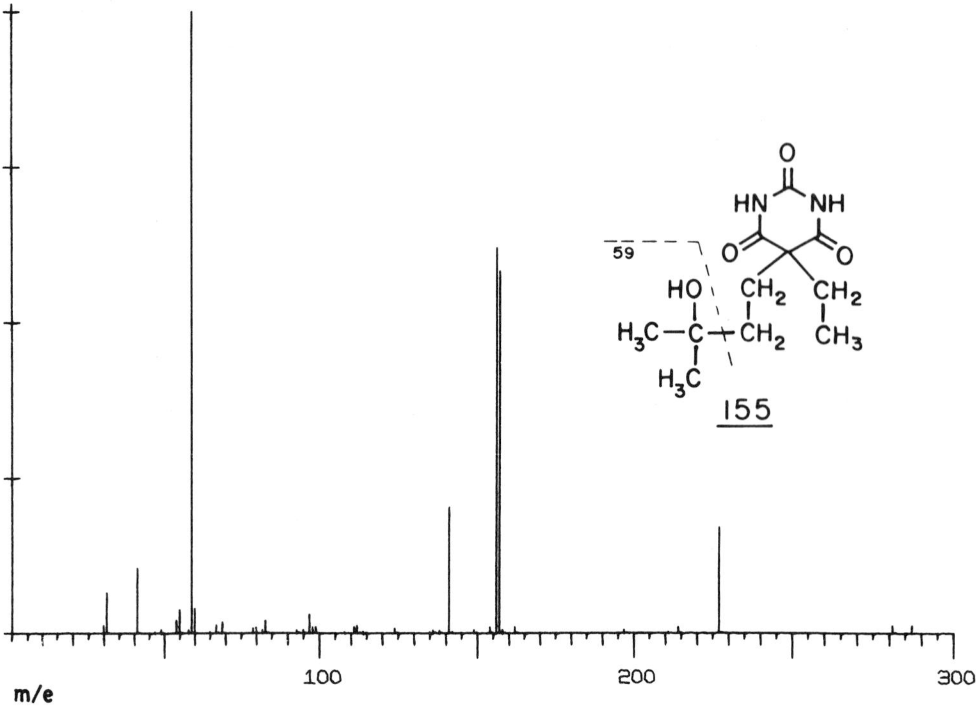

FIG. 14B. Mass spectrum of amobarbital metabolite, 155, scan 129 of GC-MS run in Fig. 8B.

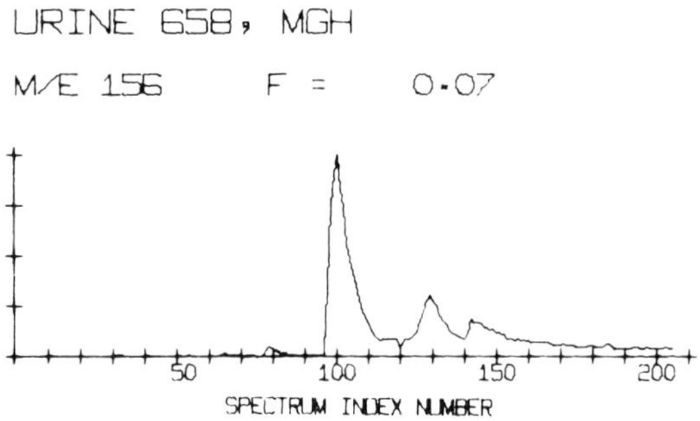

FIG. 15. Mass chromatogram of *m/e* 156 showing maxima at the retention times of amobarbital, 154 (scan 100), and its metabolite, 155 (scan 129).

It was also possible to identify the N-desmethyl metabolite, 156, of amitriptylene. It was clear from comparison of this spectrum (Fig. 16A) with that of the parent drug, 157 (Fig. 16B), that desmethylation of the molecule has occurred.

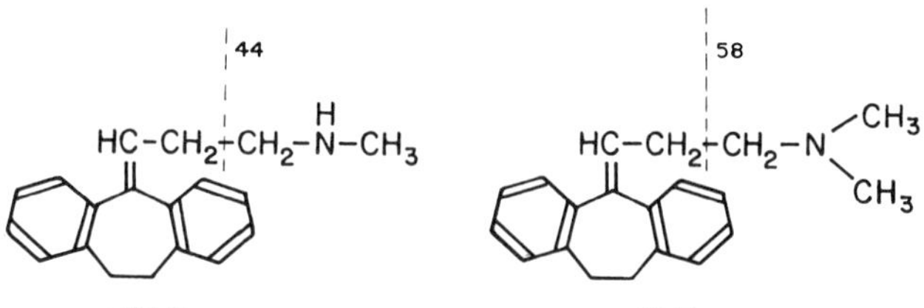

156 157

Thus, with GC-MS-computer techniques and deductive processes, it has been possible to identify in a single extract six drugs and at least nine of their metabolites (as well as a number of endogenous compounds). In addition, not only are compounds identified quickly, but previously unencountered compounds can be assigned structures.

Stable isotope-labeled species can be useful as marker compounds to follow biotransformation pathways. A simple illustration of the appearance of a spectrum that contains both labeled and unlabeled species is given in Fig. 17B for ^{18}O-labeled benzoic acid. Since the two oxygens in this molecule are equivalent, it is to be expected that an equal distribution of ^{18}O in these two positions is observed. It is apparent, therefore, that the analyst can easily ascertain

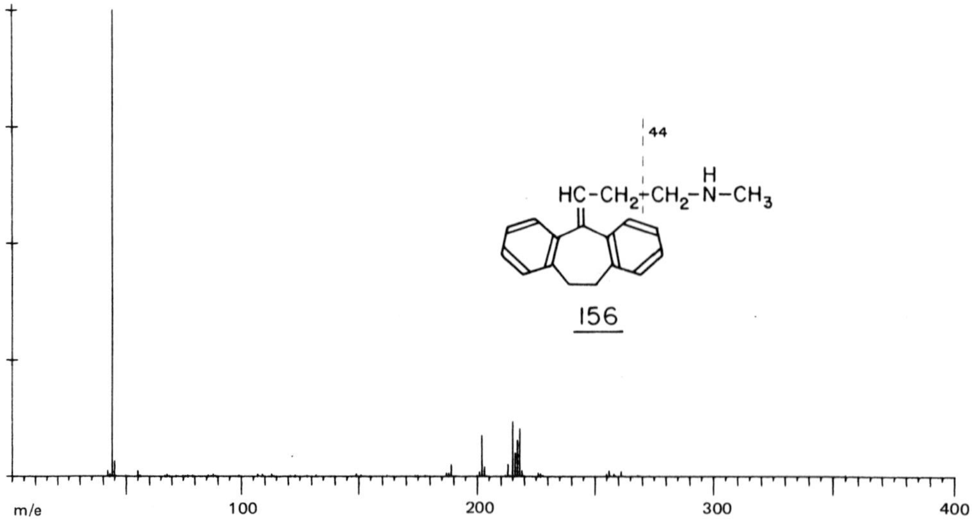

FIG. 16A. Reconstructed mass spectrum, scan 153, showing desmethylamitriptylene, 156, in urine extract of comatose patient.

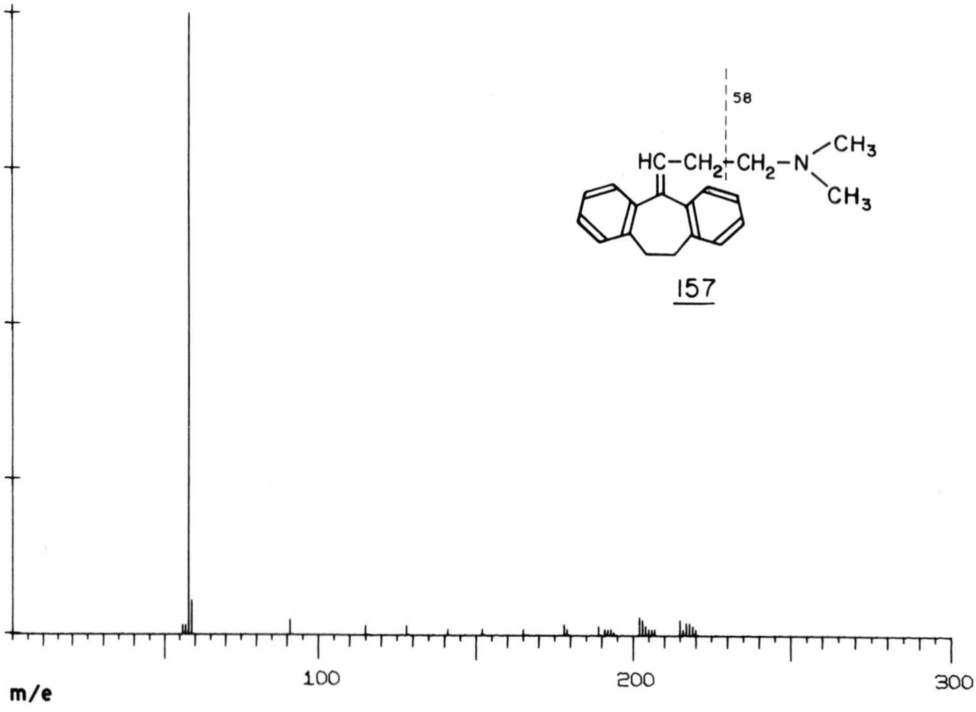

FIG. 16B. Mass spectrum of amitriptylene, 157, from the Massachusetts Institute of Technology drug library reference collection.

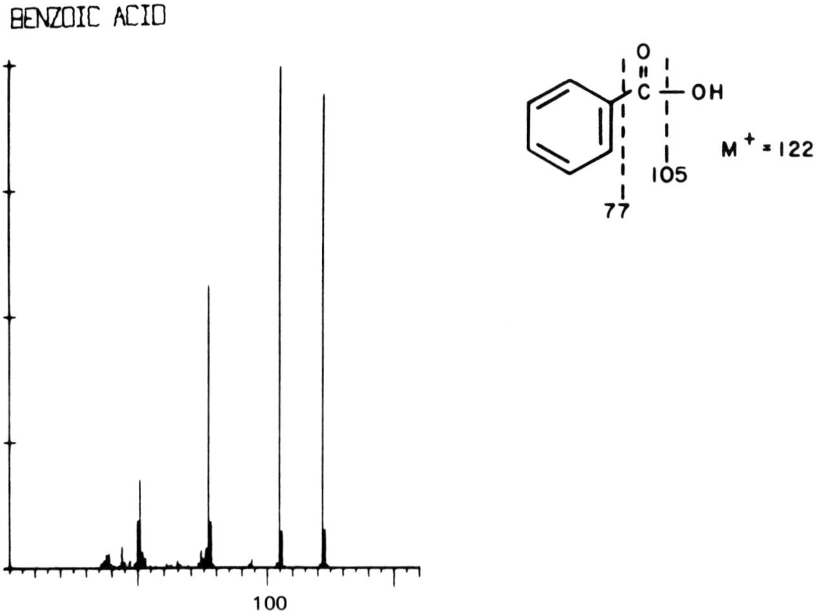

FIG. 17A. Mass spectrum of benzoic acid.

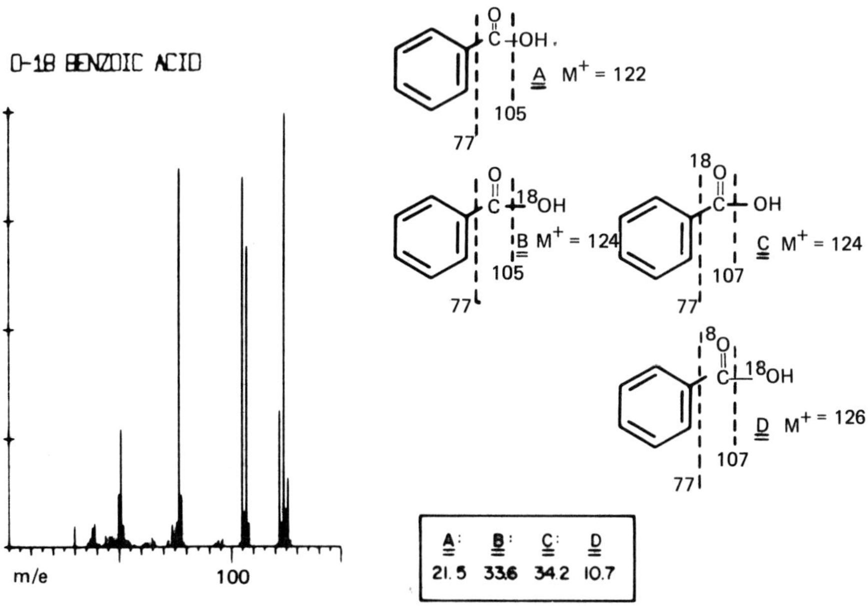

FIG. 17B. Mass spectrum of a mixture of unlabeled benzoic acid, and its ^{18}O- and $^{18}O_2$-labeled analogs.

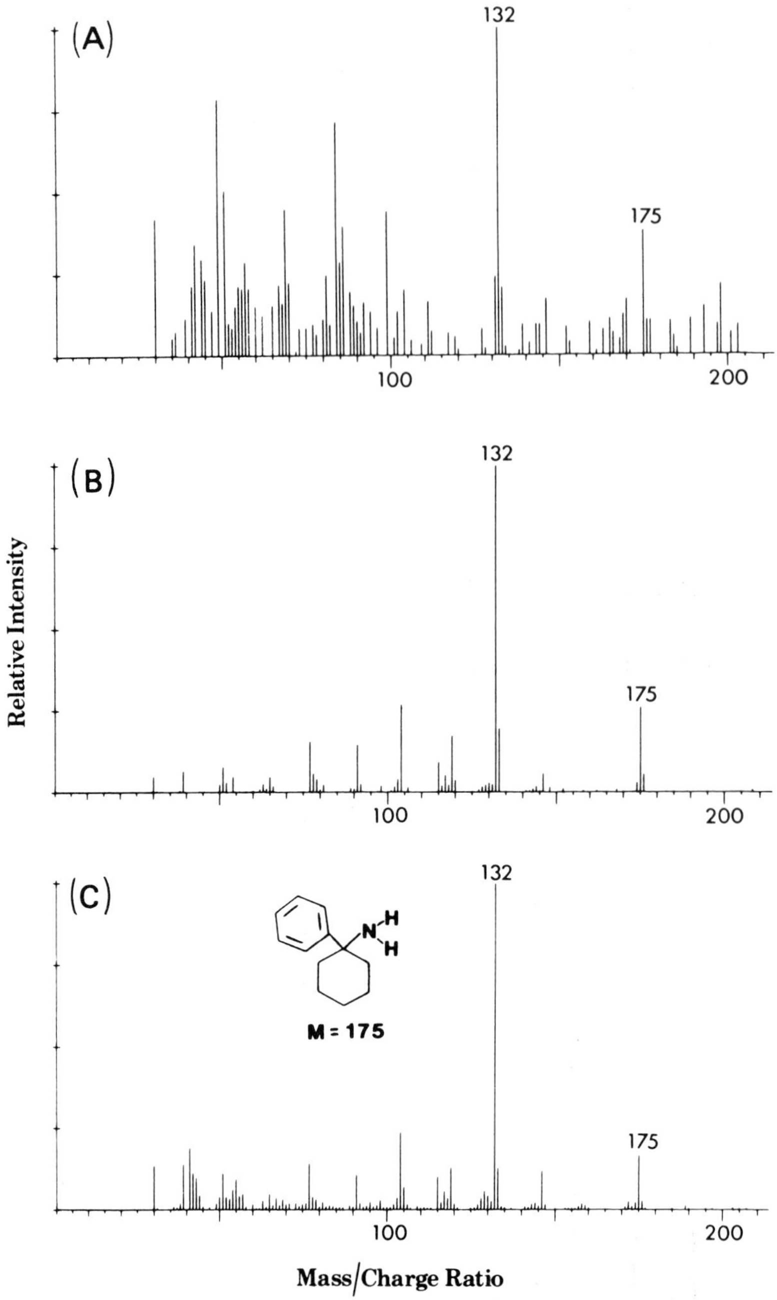

FIG. 18. (A) Mass spectrum of a phencyclidine metabolite detected in human urine. (B) Mass spectrum of the same metabolite detected in mouse urine. (C) Mass spectrum of 1-phenylcyclohexyl amine, 160, purchased from Aldrich Chemical Co.

which mass spectra are due to compounds with a partial label and even determine the position of the heavy atom. If the drug of interest has present in it an atom which has several important isotopes, this type of analysis can be carried out using the naturally occurring form of the drug.

The introduction of a stable isotope into a drug and the subsequent administration of the labeled and unlabeled drug to laboratory animals can make apparent the relationship between the drug and its metabolites by observation of double peaks in spectra of drug-derived species. Once characterized in this way, the same compounds can be searched for in human samples. In this manner, metabolites of the street drug phencyclidine, 158, have been identified first in rat urine then in human urine samples. A mixture of the $^2H_0$ and $^2H_5$ compound 159 was administered and the metabolite 160 was among those identified. Spectra of the authentic compound and the metabolite observed in mouse and human urine are shown in Fig. 18.

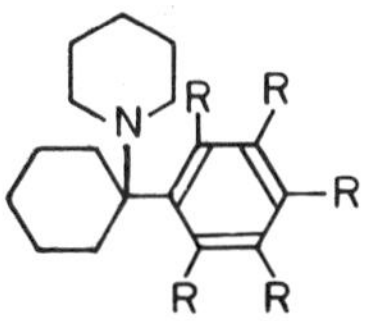

158 R=H
159 R=D

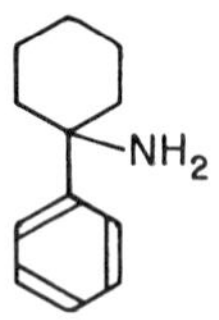

160

B. The Other Techniques

Although gas chromatographic mass spectrometry, utilizing electron-impact ionization and achieving unit mass resolution, is the predominant experimental approach in drug metabolism studies, it is by no means the only weapon in the mass spectrometrist's arsenal. Some of these techniques have been discussed in the preceding section. This section is a survey of the available methods, with brief discussions and representative applications.

1. Pyrolysis GC-MS

Ohya and Sano have developed a method for the determination of carpronium chloride, 161, in biological samples by monitoring specific ions of the lactone, 162, generated by a pyrolysis step preceding gas chromatography [431]. Deuterium-stable isotope analogs were produced for quantification and for determination of the degree of hydrogen scrambling under different pyrolysis conditions. This latter study was of course mandatory in order to assess the effects of the pyrolytically induced scrambling on the reliability of the quantitative data. In qualitative studies, the scrambling patterns are a key to understanding pathways of pyrolytic decomposition.

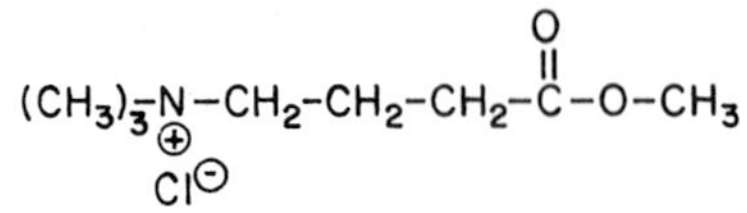

161

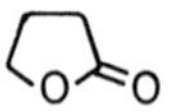

162

2. *Chemical Ionization*

Chemical ionization (CI) has become a standard technique in many mass spectrometry laboratories. Its use in drug metabolism studies falls into two areas: the determination of the molecular weight of compounds that do not exhibit a molecular ion under electron impact and the quantification of drugs and metabolites at low levels where the use of CI limits fragmentation to a few principal ions and thus provides greater sensitivity. In addition, quantifications of these primary ions (usually at higher mass) frequently decrease interference from extraneous compounds. The type and amount of reagent gas greatly affects ionization processes and it is worthwhile to investigate several choices before a new method is accepted. For qualitative studies, the chemical ionization spectrum of a partially purified biological sample can be useful as a preliminary screening technique to obviate a complex workup procedure. On the other hand, CI studies may be utilized to make fine distinctions among rather similar compounds. Gilbert and Brooks [400] have used GC-chemical ionization to characterize diastereomeric and enantiomeric ephedrines. Amino alcohols of the ephedrine group were examined by conversion to the TMS ethers of the diastereomeric amides, 163, formed with (R)-α-phenylbutyric acid. Resolution of diastereomers and enantiomers was achieved by GC and the isobutane chemical ionization spectra exhibited an abundant MH+. Moreover, these diastereomers could be differentiated by their MS fragmentation patterns.

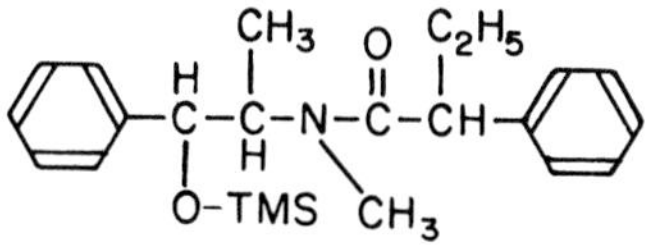

163

Chemical ionization has made important contributions in quantitative work as illustrated in Sec. VI.B in the analyses of morphine, 194; propantheline bromide, 196; and tolbutamide, 200.

3. *Atmospheric Pressure Ionization*

Atmospheric pressure ionization has been described by the Horning group [421] and commercial sources of this type are now available. The advantage of this type of source for drug metabolism studies is that it requires very little sample preparation and hence multiple analyses can be carried out quite rapidly. Sensitivities at the picogram and even femtogram level have been reported, making this technique particularly useful for monitoring therapeutic levels of anticonvulsants.

4. *Field Ionization*

Field ionization has had some use in drug metabolism studies. It has advantages similar to those of chemical ionization, in terms of minimal fragmentation, which contributes to greater relative sensitivity for the molecular species and decreased interference from other compounds. This instrumentation is not so widely available as that for chemical ionization. The corrollary to this is not surprising: it remains both more expensive and less convenient to operate. McReynolds and co-workers [336] have reported its use for the determination of picomole quantities of methaqualone, 164, and 6-hydroxy-methaqualone, 165, in human urine over a period of 11 days after the ingestion of a single 250-mg tablet. Extensive cleanup of the sample by several preparative TLC steps was necessary to separate the drug and metabolite from extraneous compounds and other monohydroxylated metabolites.

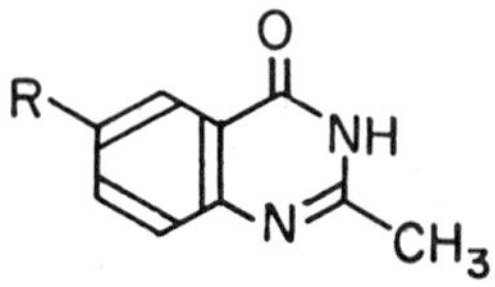

164 R=H
165 R=OH

The samples were introduced into the source via the solid probe and quantification was achieved using a stable isotope-labeled internal standard. The authors indicated a precision of approximately ± 30 pg/ml throughout the concentration range measured (apparently 200 pg/ml to 5 μg/ml in urine).

5. *Field Desorption*

Field desorption (FD) had become an important new tool for drug metabolism studies because, unlike the other ionization techniques, it does not require that the sample be volatile, since ionization takes place on the surface of a specially prepared emitter. Many drugs and their metabolites are salts or other types of highly polar or fragile molecules that undergo thermal decomposition at temperatures required for volatilization and consequently molecular ions are not observed by the other ionization techniques. For a discussion of the theories, instrumentation, and results of FD, the reader is referred to Chapter 5 of Part A of this text. Typical of the compounds that are amenable to FD are cymarin (strophanthin-K) 166, for which Schulten and Games [168] have reported M^+, MH^+, and structurally significant fragments, and the peroxy derivatives, 167 and 168, of the anticancer drugs ifosfamide and trofosfamide whose field desorption spectra were provided by Przybylski et al. [166].

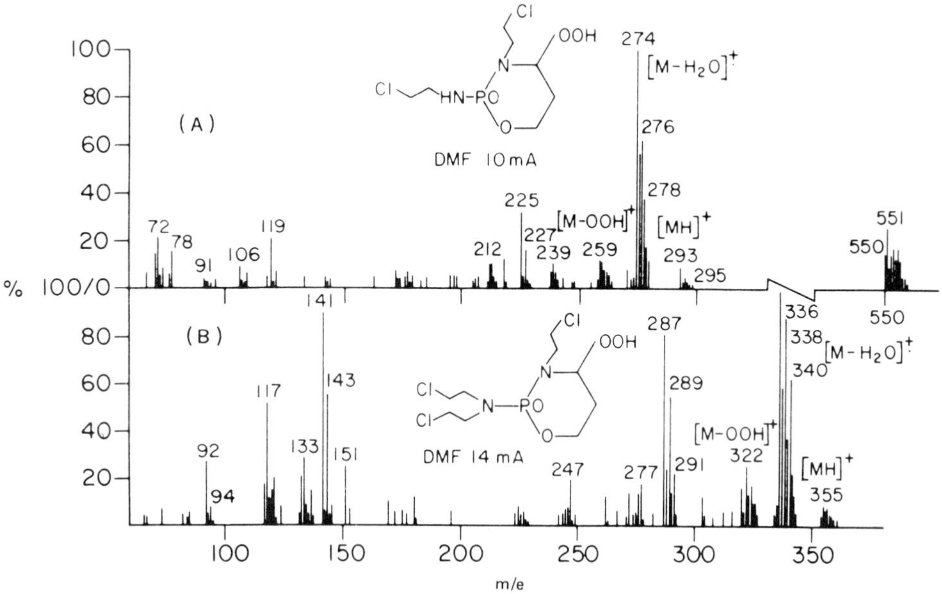

FIG. 19. (A) FD mass spectrum of 4-hydroperoxyifosfamide, 167. (B) FD mass spectrum of 4-hydroperoxytrofosfamide, 168.

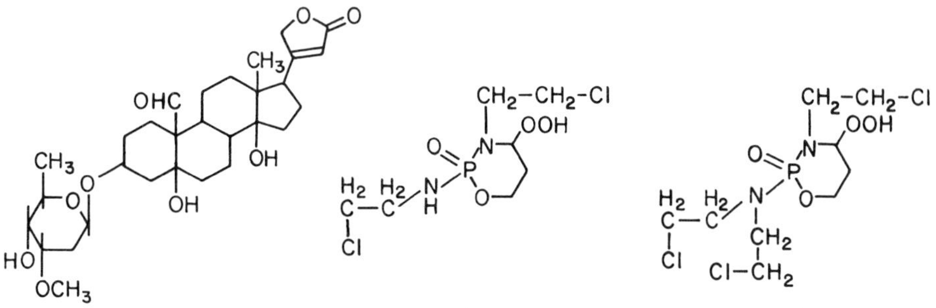

166 167 168

Other applications were discussed earlier, in the determination of compounds 112 and 117.

6. *New Ionization Methods*

As other new ionization techniques become available for more general use, they will surely be applied to drug metabolism studies. Among these are negative ion [424] and Californium-252 plasma desorption [164,171]. The latter has been used for very fragile molecules such as the neurotoxin tetrodotoxin, 169, for which even FD has been unable to provide a molecular ion.

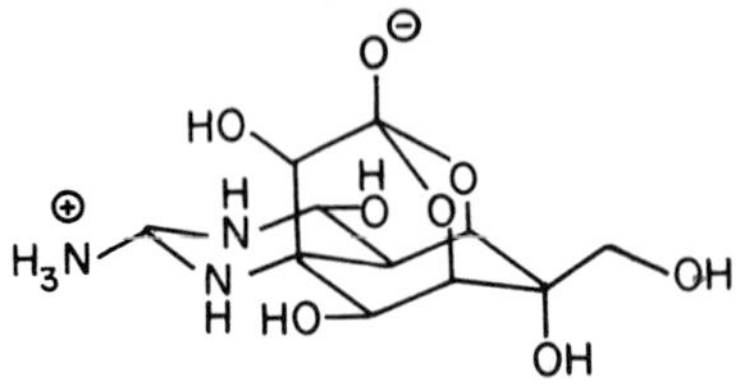

169

7. *High-Resolution Mass Spectral Analysis*

The discussion has so far been directed toward analysis by low-resolution mass spectrometry. However, at certain points in the solution of many problems, the ability to measure exact masses with an accuracy of a few parts per million is necessary. High-resolution data are useful to establish the elemental composition of compounds that are available in quantities too small for analysis by chemical techniques or not sufficiently pure or stable to give accurate results by other techniques. For delineation of the fragmentation pathways that lead to the observed mass spectrum, the knowledge of ion compositions from high-resolution data makes possible the distinction among alternative pathways. The shift of certain peaks in a mass spectrum of a drug metabolite with respect to the parent compound can be confidently related to structural changes if the origins of the fragments in the spectrum of the parent have been well established. Quantitative studies using single- or multiple-ion monitoring can be carried out at high resolution and thus gain a measure of specificity by eliminating interference even from other ions of different composition at the same nominal mass as that being observed.

High-resolution data have been used by so many of the authors whose work is described in the illustrations cited elsewhere in this chapter that no separate examples are provided here. It is regarded as a prerequisite to making structural assignments whenever there is a possibility of ambiguity.

VI. QUANTIFICATION

Mass spectrometric techniques were first employed in drug metabolism studies for qualitative analysis. The utility of mass spectral measurements for quantification is now making this application of equal or greater importance. Many of the considerations in sample handling already discussed are equally appropriate in quantitative work but, in addition, special care must be taken to assure the reliability of the measurements, especially when extremely small samples are analyzed.

Successful quantitative analysis with mass spectrometry instrumentation demands the general concern that experimental parameters need to be optimized. Within the limits of practicality, the sample cleanup procedure should remove interfering substances and minimize noninterfering compounds. Pragmatically, the established procedure is dictated at least partially by the number of samples to be analyzed and cleanup is pursued whenever the presence of extraneous materials causes deterioration of the experimental system or extends analysis time. Sample recovery should be maximized to achieve sensitivity and to assure that the sample is truly representative of the whole. The addition of internal standards early in the workup is useful both to monitor recovery and to compensate for losses. The preparation of standard curves on a daily basis is recommended. Maximum sensitivity is achieved with single-ion monitoring but a need for distinguishing among related compounds or removing background interference may make the monitoring of several ions preferable. The advantages of both are available if the selection of ions to be monitored can be changed during the course of a single run. A consideration of GC peak shape and the cycle time of the detection system is necessary in the selection of scanning parameters when more than one ion or a mass range is to be monitored. The overall condition of the mass spectrometer has much to do with the reliability of the data. Lee and Millard [359] have illustrated these general observations by carrying out a determination of dimethyl allobarbital under the conditions indicated in Table 1.

Extensive reviews have appeared that cover in detail the practical considerations peculiar to quantitative analyses with MS instrumentation and the reader is referred to these. Summaries of procedures employed in their laboratories and a discussion of the evolution of the current protocols have been presented by the Horning [290,356] and Sweeley [290] groups. A variety of quantitative applications of mass spectrometry has been provided by Watson [310]. The review by Blake, Crespi, and Katz [348] summarizes the results of many quantitative studies employing deuterated drugs.

The choice of an internal standard is a critical factor in the development of a quantitative method. The standard should have chemical behavior similar to the compound to be measured. Preferably, it may be an isomer or homolog or an isotopically labeled analog of the compound of interest. If a labeled analog is chosen, it has the advantage of serving as a carrier during isolation and chromatography so that the error introduced by sample loss is minimized. Alternatively, both a labeled carrier and a homolog may be added and this method is preferred by some workers, as will be discussed below. Labeled analogs of many drugs are now available commercially or from the National Institute on Drug Abuse, which has contracted for their synthesis. The National Institute of General Medical Sciences has arranged for the synthesis of some therapeutic drugs. There is a good possibility, therefore, that a suitable standard can be acquired for all but the newest drugs.

A. Statistical Analysis

Statistical analysis on quantitative measurements provides an evaluation of different approaches. Claeys, Markey, and Maenhaut [286,290] have applied variance analysis to evaluate precision in selected ion monitoring using stable isotope variants or chemical analogs as internal standards. The relative contributions of long- and short-term effects of instrument stability were considered, and, using this technique, the variance due to sample manipulation could be distinguished from that due to GC-MS instrumentation. The workers analyzed imipramine, 170, and desipramine, 171, using the $^2H_4$-labeled internal standards 172 and 173.

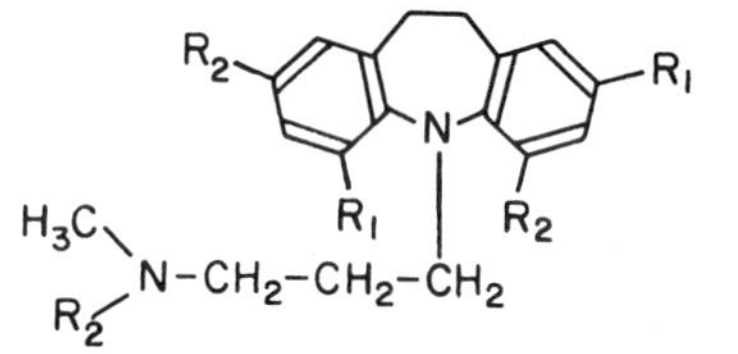

170 R_1=H , R_2=H
171 R_1=H , R_2=CH_3
172 R_1=D , R_2=H
173 R_1=D , R_2=CH_3

In addition, $^2H_4$ imipramine 173, maprotiline 174, and promazine 175 were evaluated as internal standards for the analysis of desipramine.

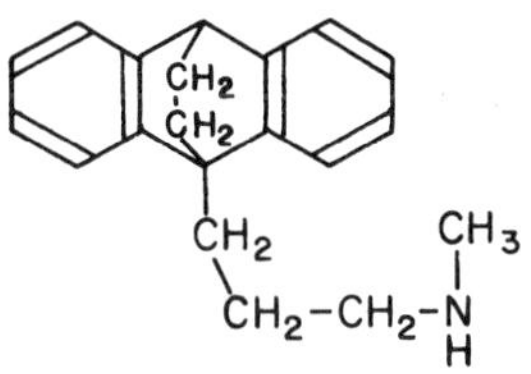

174

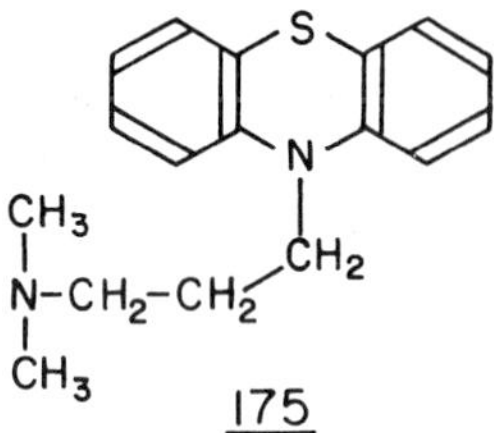

175

The samples were analyzed by methane CI using 171, 173, 175, and trifluoroacetyl (TFA) derivatives of 170, 171, and 174. MH^+ ions were monitored. In order to determine long-term stability, ten sets of two injections each were determined over a period of 5 hr. The peak ratios of desipramine/internal standard of a representative solution of 76.5 ng desipramine/ml plasma were presented. For $^2H_4$ imipramine as internal standard the desipramine/internal standard ratio determined varied from 0.588 to 1.637. With neither standard was a clear time trend apparent except that the first few measurements gave markedly lower results than the later ones. For $^2H_4$ desipramine/desipramine the reactions were found with a much narrower spread, 0.366 to 0.379. The results with maprotiline were also better (0.195 to 0.222) but not quite as satisfactory as the $^2H_4$ desipramine analysis. These two latter internal standards were evaluated over the likely concentration range. The relative standard deviations of the determinations using maprotiline as internal standard were three to ten times greater than those for the $^{2}H_4$ desipramine standard. Since the instrumental parameters were found to be the same for the two standards, the difference was attributed to the chemical dissimilarity of maprotiline and desipramine. On the basis of these and related experiments, the authors came to the following conclusions [286].

1. Stable isotope-labeled internal standards appear to produce the lowest variance factors due to long-term, instrumental, and manipulation errors. The criteria for unweighted linear regression are generally met by stable isotope variants.
2. In the selection of a suitable analog to be the internal standard for selected ion monitors (SIM), long-term variability should first be measured. Structural analogy is a critical factor in the choice of internal standards.
3. When chemical manipulations have to be performed, synthetic analogs may meet the criteria for linear regression analysis. However, the error dependency with respect to concentration should be measured to determine if weighting factors are necessary.

Lee and Millard have expressed a preference for use of two analogs during quantification experiments on the basis of their evaluation of instrumental parameters [301,359]. They compared 177 and 178 as internal standards in the determination of 176.

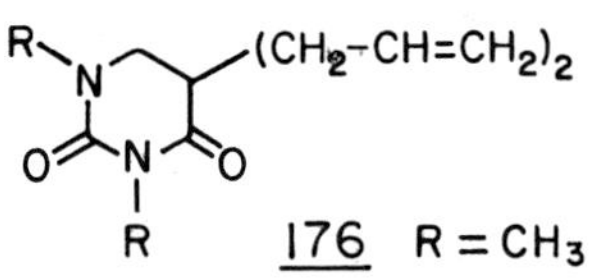

176 $R = CH_3$
177 $R = CH_2D$, CHD_2, or CD_3

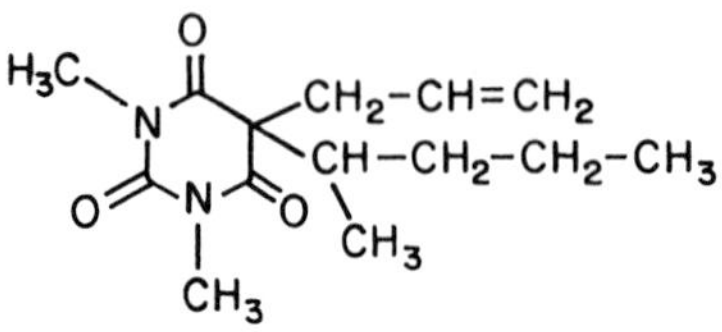

178

These authors assume that the chemical difference between compound and homolog can be adequately compensated for by measurement of standard mixtures that cover the whole concentration range of the drug or metabolite in the biological medium. Table 1 shows the results of their experiment.

TABLE 1

Coefficients of Variation for the Determination of Dimethylallobarbitone, 176, by Coinjection with Deuteriated Dimethylallobarbitone, 177, and Dimethylquinalbarbitone, 178; five injections of each solution

Amount of 176 injected (ng)		0.04	0.1	0.4	1.0	2.0	10.0	50.0
Multiple-ion monitoring[b] with 177 as standard	(1)[a]	—	—	—	23.7	14.5	5.7	3.2
	(2)[a]	19.8	15.3	1.9	1.3	—	—	—
Multiple-ion monitoring[b] with 178 as standard	(1)[a]	—	—	—	13.2	8.3	3.6	3.3
	(2)[a]	23.1	6.5	4.0	0.9	—	—	—
Single-ion monitoring of *m/e* 195	(1)[a]	—	—	9.9	5.6	4.2	3.0	3.0
	(2)[a]	11.5	2.0	0.8	0.6	—	—	—

[a](1) Before and (2) after extensive overhaul of instrument, including the fitting of a new electron multiplier.
[b]Multiple-ion monitoring used *m/e* 195 and *m/e* 199.

They interpret their results as favoring use of a homolog:

> This work shows that small quantities of a compound are most accurately determined by adding a labeled version of the compound to act as a carrier, plus an internal standard which gives an ion in common with the compound being measured so that single ion monitoring may be employed.

Schoeller [362] has reviewed the use of linear regression analysis for the reduction of isotope dilution data and surveyed errors made in performing the regressions. The data often violate the basic assumptions about the relationships among the variables, and neglect of these violations can lead to erroneous or misleading calibrations. Some of these errors are unique to stable isotope-labeled standards; others are general sources of error in measurements using internal standards. Four specific sources of error are considered.

The first error can occur in the dilution and preparation step if the uncertainty introduced by variance in the mole ratios approaches the precision of the isotope ratio measurement. If this error is significant, "replicate observations of the isotope ratio of a single dilution do not increase the precision of the calibration; instead it is necessary to analyze replicate dilutions to obtain greater precision."

Memory effects in the mass spectrometer can prevent the isotope ratio measured for a given standard mixture from having its requisite independence from the other standard mixtures. Since the analyst has good control over this parameter, its effect can be eliminated or minimized as a possible source of error.

The presence of isotopic impurities in either the unlabeled compound or the labeled internal standard can introduce nonlinearity into the extremes of the calibration curves. If the curvature is small compared with the error in isotope measurement, it can be ignored. One way to extend the linear portion of the data is to plot the ratios using the isotopically purer substance as the denominator, Fig. 20. A second approach is to calculate the isotope dilutions and make the appropriate corrections in the measured values. The atom % excess is then calculated according to the formula:

$$\text{Atom \% excess} = [R'/(R' + 1)] \times 100 \tag{1}$$

where R' = isotope ratio corrected for natural abundance contributions by subtraction of the isotope ratio of the unlabeled standard from the isotope ratio of the diluted standard

This problem can be minimized if a sufficient mass shift is introduced into the internal standard. Then the contribution from natural abundance to the mass peak to be monitored for the internal standard is negligible. If the labeled compound is prepared in sufficient purity that its content of unlabeled or partially labeled species is vanishingly small, the problem is also minimized. Pickup and McPherson [290,361] have discussed this problem also and the equation they have developed for treatment of data of this type appears here as Eq. (2).

Inaccuracy can also be introduced by assuming that the standard deviations of each population of isotope measurements are equal. Schoeller cites this as the most frequently violated assumption in stable isotope quantifications. The result is manifested as an inaccuracy near the origin of the isotope dilution calibration curve.

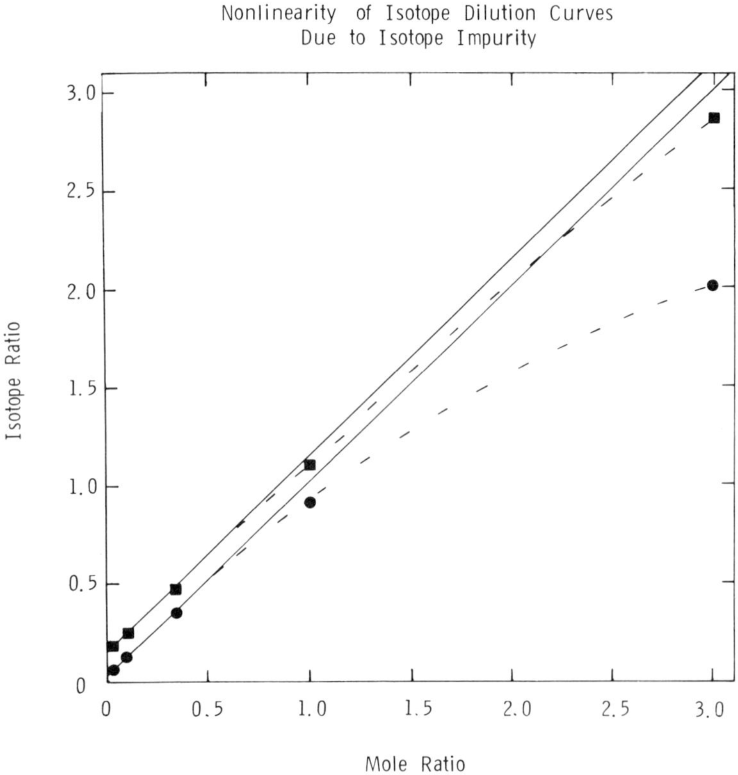

FIG. 20. Theoretical relationship for a mixture of natural abundance standard with 15% of the isotope label and a labeled standard with 2.5% unlabeled impurity. The lower curve is the plot of the ratio of the ion current "observed" for the labeled species and that of the unlabeled species. The upper curve is the inverse. The dotted lines are the lines described by the linear portion of the data.

The greater precision in the determination permits the use of a weighted linear regression. The effect of the weighting on the slope and intercept is small but it brings about greater accuracy in the determination of small ratios. When measurements are made over a wide concentration range, the values of the smaller ratios can approach the standard deviation of the larger ratios. Schoeller has shown that, for these cases, this correction becomes very significant. He has presented the necessary formulas for performing this type of calculation. The weighting factor should be the inverse of the variance. Finally, he has presented an example of testing the validity of calibration data to assess when it is appropriate to prepare a new calibration curve.

Pickup and McPherson [290,361] have developed equations to describe isotope clusters and calibration curves which are based on probability functions. Equation (2) is the result of their calculations.

$$R = \frac{(x/y)(p_a/M) + (q_a/M')}{(x/y)(p_b/M) + (q_b/M')} = \frac{[\text{Unlabeled compound}]}{[\text{Labeled internal standard}]} \tag{2}$$

where x = weight of unlabeled compound

y = weight of labeled compound

M = molecular weight of unlabeled compound

M' = molecular weight of labeled compound

p_a and q_a = probability of occurrence of the isotope form representing unlabeled material in unlabeled and labeled material respectively

p_b and q_b = probability of occurrence of the isotope form representing labeled material in unlabeled and labeled material respectively

Although, in theory, use of the equation should permit the analyst to construct a calibration graph using known abundances rather than experimental results, the variability in natural abundances among samples from different sources and the accuracy with which enrichment values are known present practical difficulties. The calculations can, be used, however, to predict the potential utility of a proposed procedure or to assay the true enrichment of labeled internal standards. In addition, comparison of an experimentally determined calibration curve with the theoretical one provides guidance in the evaluation of a new assay and can be helpful in determining the reasons for divergence of the data from the theoretically derived curve. For compounds containing only C, H, N, and O the theory predicts that an increase of 4 amu in the isotopically labeled internal standard should be sufficient to provide linear calibration graphs. For smaller increments, the equation can be used to calculate the deviation from linearity which is to be expected.

If careful attention is paid to all the factors that affect the precision and accuracy of isotopic measurements, reliable measurements can be made even at very low enrichments. For example, Von Unruh, Hauber, Schoeller, and Hayes [363] have reported the limit of detection in the differential comparison of CO_2 derived from drugs and metabolites in human urine as 5×10^{-5} gram excess ^{13}C/gram material. In most drug metabolism studies, this precision in isotope ratio determination itself is not required, but much the same demands are made when the concentrations of natural and labeled materials are to be measured at very low levels of both.

B. Stable Isotopes as Internal Standards

For the quantification of drugs and their metabolites by mass spectrometry, stable isotopes have been the most widely used internal standards. In the space available, it is impossible to cover even a fraction of the published methods, but a few selected papers serve to exemplify current applications and they will be presented below. Many more applications can be found in the reference section.

In an elaborate study, Horning and co-workers [295] carried out pharmacokinetic studies involving plasma, urine, breast milk, saliva, and liver homogenates using stable isotope-labeled internal standards and chemical ionization GC-MS. Regression analysis of timed biological samples was used to obtain the concentration at zero time and the slope of the concentration curve. The half-lives, the apparent volume of distribution, the elimination constant, and the plasma clearance rates were then calculated. The unlabeled drugs in the study were administered orally and their stable isotope-labeled analogs were added to the biological fluids before the start of the isolation procedure. Included in the study were 2,4,5-^{13}C-phenobarbital <u>179</u>, 2,4,5-^{13}C-diphenylhydantoin <u>180</u>, 2,4,5-^{13}C-pentobarbital <u>181</u>, N-C^2H_3-antipyrine <u>182</u>, and N-C^2H_3-caffeine <u>183</u>.

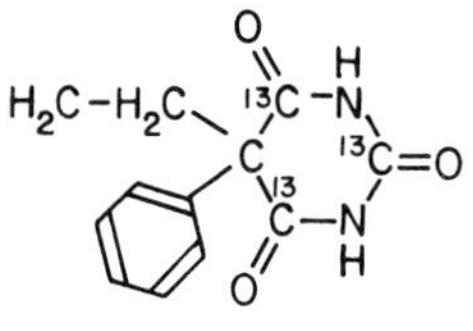

179

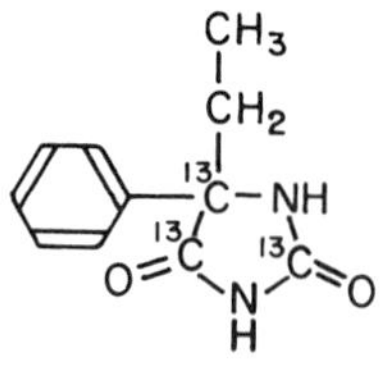

180

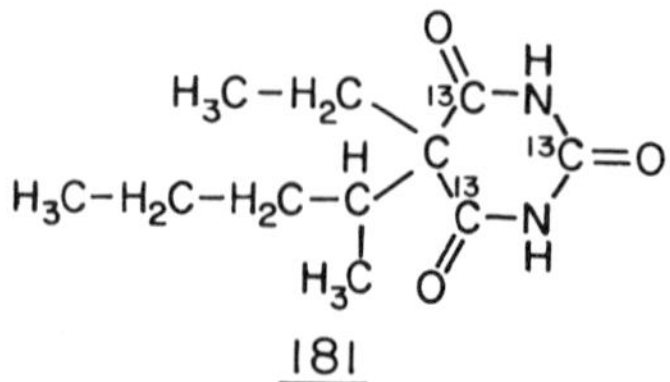

181

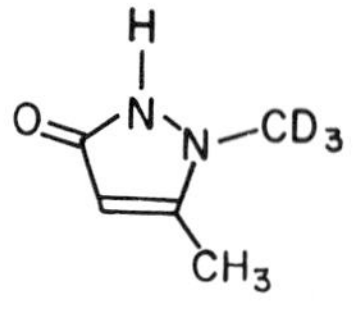

182

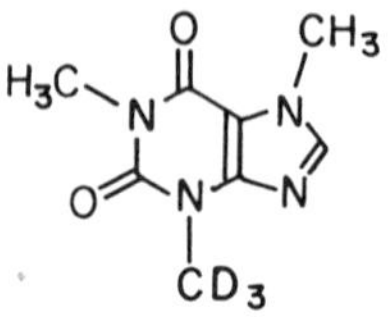

183

The results of these studies included determination of the half-life of caffeine and of other parameters for mephobarbital and antipyrine and some of their metabolites after oral doses. An analysis of paired samples of breast milk and plasma showed that the half-life of diphenylhydantoin is longer in breast milk than in plasma. This is consistent with previous results for other drugs and may be the result of slower metabolism. The excretion of mephobarbital, 184, and its metabolite, phenobarbital, 185, in the urine of a newborn whose mother had received daily doses of the drug during pregnancy enabled the authors to quantify these compounds during the first 22 days of the infant's withdrawal from the drug. Labeled pentobarbital, 181, was used as the internal standard for the determination of secobarbital, 185, and its hydroxylated metabolite, 184, during an in vitro study using rat liver homogenates. For all these analyses, the stable isotope internal standards were found to give satisfactory results, both in terms of the reliability of the data and in the speed of the analysis.

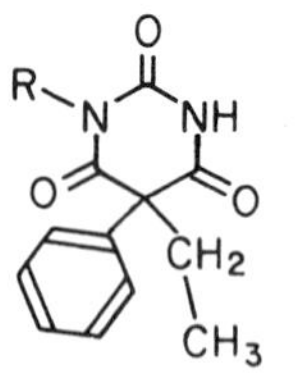

184 R = CH

185 R = H

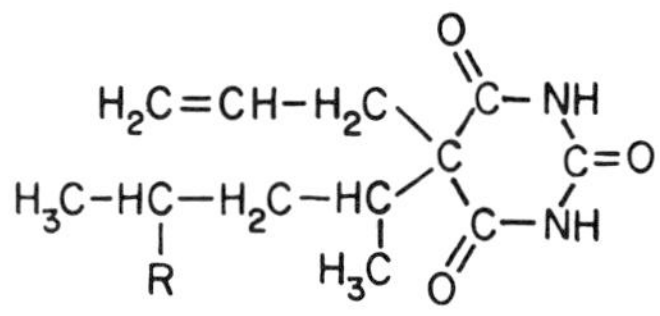

186 R = H

187 R = OH

The use of stable isotope-labeled compounds also makes possible the investigation of steady-state kinetics. For this purpose, two different labeled analogs are employed. After the subject has reached an equilibrium drug level through maintenance doses of the unlabeled drug, a "pulse dose" of the labeled analog is administered and then administration of the unlabeled drug is continued in the normal fashion. Quantification is achieved using a second labeled analog, which is added to the biological fluid before the extraction is begun. This approach has two advantages: it permits the determination of half-life and other parameters during the steady state, thus reflecting the usual administration protocol for drugs prescribed on a maintenance regimen; and it does not require the withdrawal of a drug whose therapeutic efficacy for the patient has been established. In a study involving human subjects, Sullivan and co-workers [252] reported the determination of steady-state pharmacokinetics of methadone, 188.

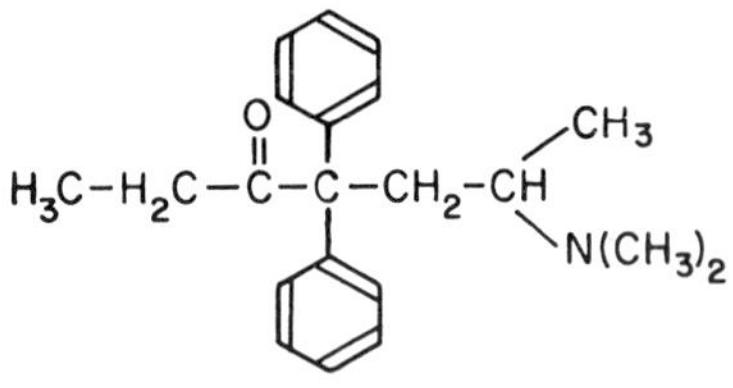

188

They used the $[M-15]^+$ (*m/e* 293) ion, which has an abundance of 1.2% of the base peak *m/e* 72 in the electron-impact mass spectrum and were able to make measurements down to 16 pmol/ml (ca. 50 ng/ml). (Their sensitivity limit for the GC/FID method was 50 pmol/ml.) Interestingly, they observed an elevation of unlabeled methadone upon administartion of the $^2H_3$-labeled methadone and attributed this effect to the displacement of the unlabeled compound from readily available tissue-binding sites. A similar study measured the kinetics of propoxyphene, 189, in doses where the $^2H_2$-labeled propoxyphene 190 was used for the pulse dose. In this work, $^2H_7$-labeled propoxyphene 191 was used as the internal standard for the quantification of the plasma levels.

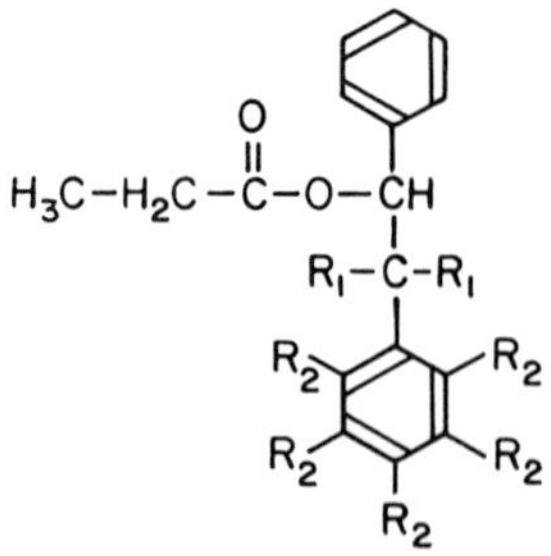

189 $R_1 = H$, $R_2 = H$
190 $R_1 = D$, $R_2 = H$
191 $R_1 = D$, $R_2 = D$

The same initial rise in the plasma level of the unlabeled compound was observed. The decline of the two curves showed a longer half-life for the unlabeled species. The authors concluded that "several pools of the drug exist [in the dogs] at the steady state and the distribution equilibrium of $^2H_2$-propoxyphene is not complete even at 48 hr after a single pulse dose. Plasma levels of propoxyphene are sustained by pools generated by multiple dosing, whereas plasma levels of the labeled drug are dependent upon elimination and exchange distribution into less accessible ('deep') pools of drug." Further experiments were carried out to ascertain that the effect was not merely an isotope effect upon distribution and/or metabolism. The ions chosen for monitoring in these studies were *m/e* 208, 210, and 215, which arise by loss of 131 amu from the molecular ions and have an abundance of 15% vs the base peak *m/e* 58 in the electron impact spectrum. The limit of sensitivity was found to be 8 pmol/ml (ca. 25 ng/ml) plasma with a coefficient of variation of less than 3.5%. The group had previously reported the lower limit of sensitivity for FID measurements for this compound to be 74 pmol/ml serum.

Selected ion monitoring was used by Baty et al [13] for the measurement of acetanilide, 191, in plasma and the results compared with those obtained using an ultraviolet spectrophotometric method. For the GC-MS analysis, the trimethylsilyl derivatives of 188 and 189 were prepared and the $[M-15]^+$ ions were monitored for the unlab-

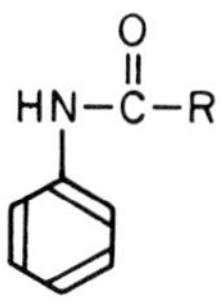

192 R = CH_3
193 R = CD_3

eled and labeled species (*m/e* 192 and 195 respectively). For the ultraviolet (UV) method, the plasma was hydrolyzed with acid to convert the acetanilide to aniline, and this compound was diazotized and coupled with N-1-naphthylethylenediamine. The absorbance was read at 550 nm. The values determined by the GC-MS method were consistently lower than the UV measurements. It was shown that this discrepancy was due to interference of acetanilide metabolites in the determination of the drug by UV. Although the sensitivity of the two methods was comparable, the mass spectral method was preferred because of its greater specificity.

Exchange of the stable isotope label could occur during metabolism and chemical workup and even during measurement. To prevent such exchange, the analyst must place the label at a more inert site in the molecule or must modify the analytical procedure to prevent the possibility of exchange. Clarke and Foltz [237] described the quantitative analysis of morphine, 194, in urine by GC-CIMS using methane as the reagent gas. Their internal standard was the NC^2H_3 analog 195. Bis-trimethylsilyl ether derivatives were prepared and two masses of each spectrum monitored.

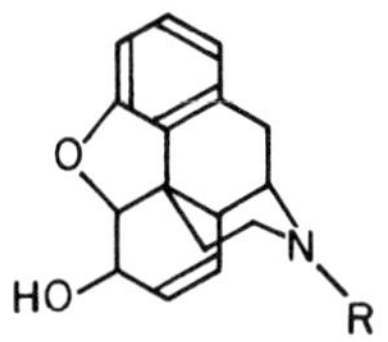

194 R = CH
195 R = CD_3

The standard curves that they obtained were linear over the range 5 to 1,000 ng/ml. No exchange of label was noted.

When Ford and co-workers [352] used methane for the CI reagent gas in the GC-MS analysis of the quarternary amine propantheline bromide 196, using the deuterated analog 197 as internal standard, scrambling was observed. This drug undergoes thermal decomposition on the GC injection port to give the two tertiary amines, 198 and 199.

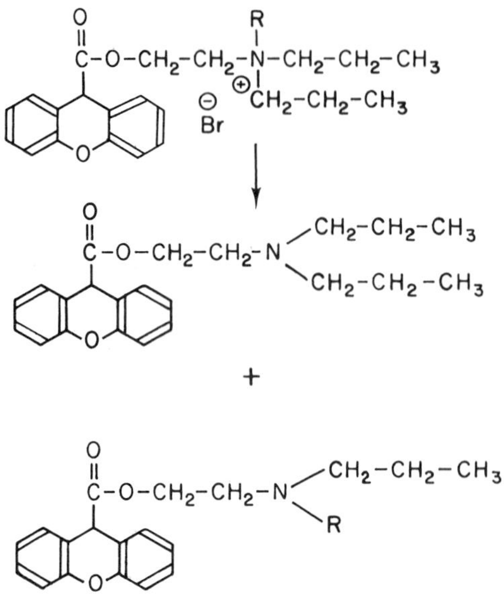

196 $R = CH_3$

197 $R = CD_3$

198

199

Using methane as the reagent gas, the authors reported 7.1 ± 0.857 exchange of the C^2H_3 label with the CH_5^+ present in the ion source and the limit for the quantification in plasma was about 20 ng/ml. When the same procedure was followed but ammonia was employed as the reagent gas, no such exchange was observed and the sensitivity limit was about 5 ng/ml.

When compounds do not exhibit molecular ions in their electron-impact spectra or when the molecular ions are of very low abundance, quantification can frequently be achieved using chemical ionization mass spectrometry. The sample may be introduced either through the gas chromatograph or by means of the direct insertion probe. Matin and Kinght [175] have described both approaches in the measurement of tolbutamide, 200, and its metabolites 1-p-hydroxy-methylbenzenes sulfonyl-3-butyl urea, <u>201</u>, and 1-p-carboxyphenylsulfonyl-3-butyl urea, <u>202</u>, as their N-1 methyl derivatives using deuterium-labeled analogs of each.

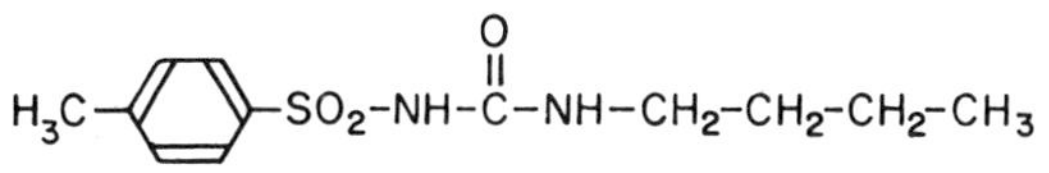

<u>200</u>

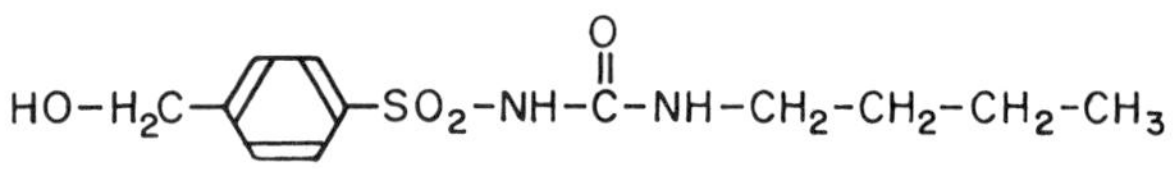

<u>201</u>

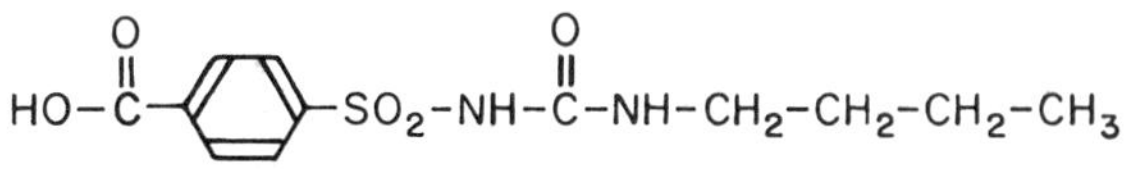

<u>202</u>

The drug and its metabolites all undergo thermal decomposition during GC-MS analysis, but the conditions were sufficiently reproducible that reliable data were obtained. Decomposition was not detected with the direct insertion probe method and MH^+ ions were observed. Quantification of the compounds in plasma after administration of a single dose was carried down to the 200-ng/ml level, and this was not the limit of the sensitivity. The standard deviation of measurement of tolbutamide was given as 3.6 and 3.8%, respectively, for the solid probe and the gas chromatographic inlet methods, when the concentration in plasma was 2.5 ng/ml. An earlier GC method of Aggarwal and Sunshine [172] quantified the products of injection port decomposition and was reported to have a useful analytical range of 0.5 to 100 μg/ml with a coefficient of variation of 2.5%. The products of the thermal decomposition had been identified by their retention times and electron-impact spectra, but the quantitative method depended on GC peak areas.

C. Other Internal Standards; External Standards

Internal standards prepared with stable isotopes are the most widely used analogs for quantitative work but a variety of substances, including isomers, homologs, and chemically similar compounds, have proved useful. A few representative analyses will be discussed below, as well as ones that make use of an external standard.

Bertilsson and Palmer [393] reported an analysis of two dopamine metabolites, homovanillic acid (HVA, 203) and iso-homovanillic acid (iso-HVA, 205), which allows determination of the relative and the total amount of the pair.

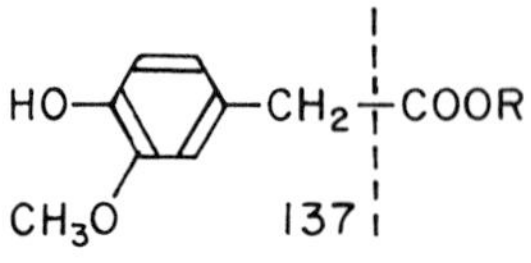

203 R = H

204 R = C_2H_5

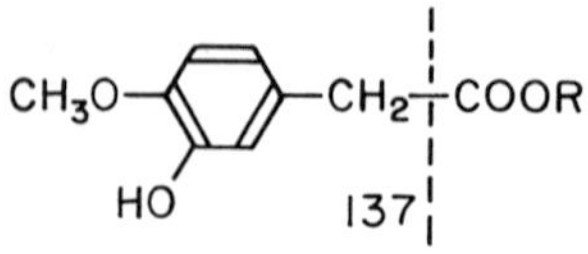

205 R = H

206 R = C_2H_5

Under the chromatographic conditions employed, the heptafluorobutyryl methyl ester derivatives of the two compounds were not resolved. However, the authors were able to quantify the metabolites by construction of a calibration curve that related the relative intensities of *m/e* 393 and 392, which are abundant in the mass spectra of both derivatives 207 and 208.

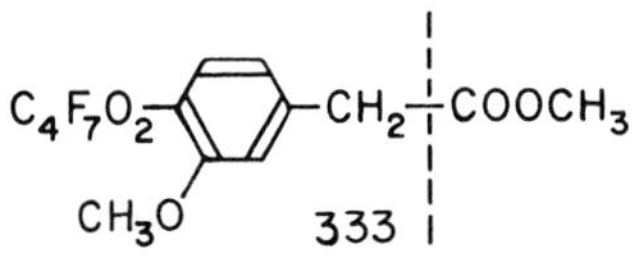

207

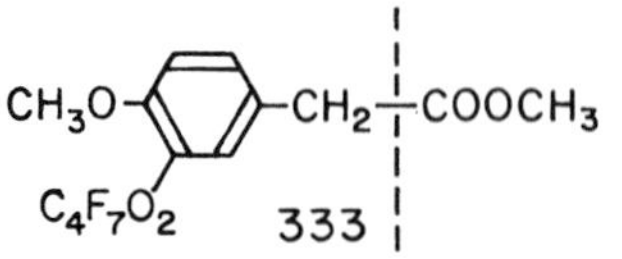

208

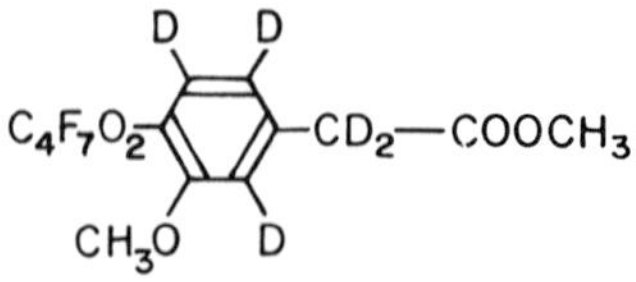

209

For HVA, the ratio of *m/e* 333/392 was determined to be 1.45 ± 0.01. For iso-HVA, it was 4.98. Since the ratios of the two ions in the spectra of the isomers are not equal, they found it possible to use standard mixtures to calculate the relative amounts of the two species in an extract that contained both. In this manner, they showed that, for a series of cerebrospinal fluid (CSF) samples from 12 patients, the range of ratios was 1.38 to 1.50, the mean being 1.46 with a standard deviation of 0.03. The iso-HVA then always constituted less than 2% of the HVA content. This value was significantly lower than the 30% previously reported from paper chromatographic studies (which lack specificity). The total HVA plus iso-HVA was calculated by comparing the intensity of *m/e* 392 with *m/e* 397, the molecular ion of the derivative 209 formed from the stable isotope-labeled internal standard. The sample was shown to contain less than 100 ng of HVA and iso-HVA in 2 ml of CSF.

Narasimhachari [428] approached the same analytical problem with a method that used iso-HVA ethyl ester, 206, or 4-methoxymandelic acid TMS ether ethyl ester, 210, as the "internal" standard. This method would be more aptly referred to as a standard addition method. For determination of the ratio of iso-HVA to HVA, the peak height of the *m/e* 137 ion was monitored as the chromatographically resolved isomeric ethyl esters eluted. Coinjection of a known amount of iso-HVA ethyl ester during a second chromatographic run made possible the calculation of the total content of the isomers. The calibration curve was constructed over a relatively narrow range (10-100 ng of each) and sample sizes had to be adjusted to fall within the range. Samples analyzed included lyophilized CSF, urine, and plasma. Because of the complexity of the extracts it was found necessary to carry out preparative TLC before the GC-MS analysis of urine or plasma samples. For three human CSF samples, the total content of the isomers was found to be 300, 120, and 400 ng/ml. For these, the iso-HVA level was determined to be 4.0, 4.2, and 3.2% that of the HVA. When HVA and iso-HVA were converted to their TMS ether ethyl esters, they cochromatographed. 4-Methoxymandelic acid TMS ether ethyl ester, 210, was coinjected and used for quantification. The results for the first two samples above were then 350 and 150 ng/ml respectively.

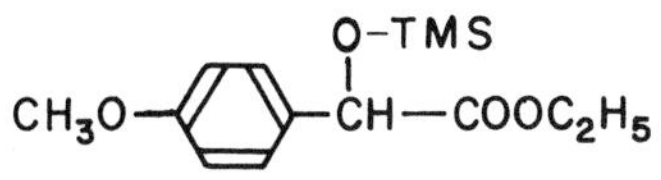

210

Because of the small number of values given for comparison, it is not possible to evaluate the statistical integrity of the method. Although it is ostensibly an improvement over the Bertilsson and Palmer method, the advantages of this method are unclear and open to question.

Strong and Atkinson [47] made use of a homolog trimecaine, 213, to quantify lidocaine, 211, and its active metabolite monoethylglycinexylidide, 212. Two masses were monitored for each compound as the chromatographically resolved species were eluted. The ions monitored were *m/e* 48, 86, and 120. For 211 the range measured was 12-200 ng (0.3-5 μg/ml); for 212 it was 20-400 ng (0.5-10 μg/ml) with standard deviations of 7.4 and 3.1% respectively.

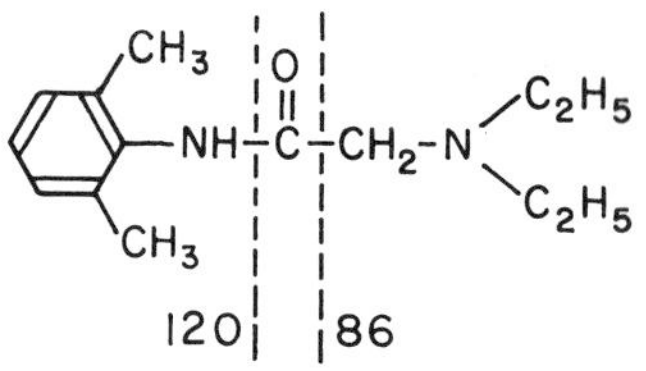

211

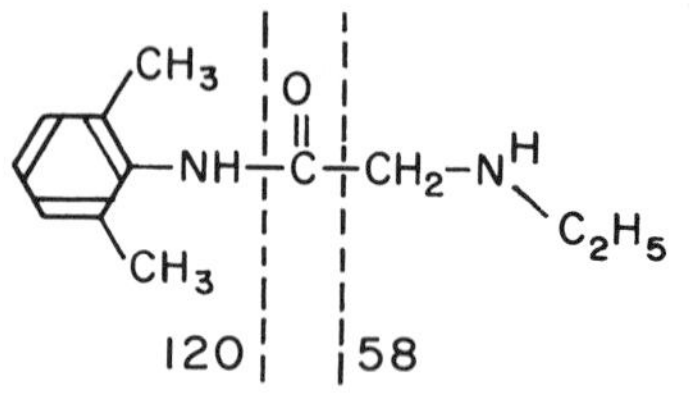

212

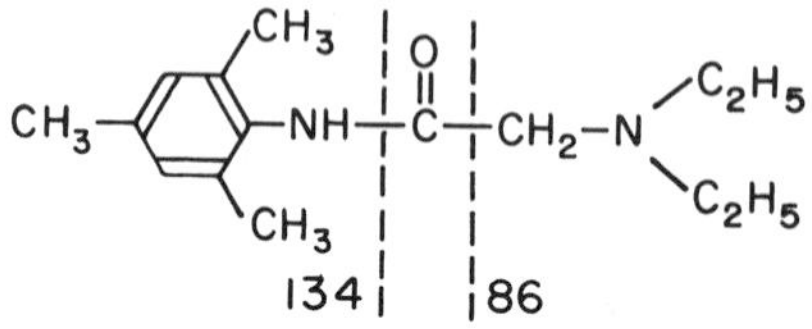

213

A judicious choice of ions to be monitored may permit the simultaneous measurement of more than one compound of interest, and even a nonbiological isomer of the compound of interest may be selected as the internal standard. Maume et al. [381] reported the use of epiestriol, 216, as the added internal standard for the analysis of the heptafluorobutyryl derivatives 217, 215, 219, respectively, of estriol, 214, and estradiol, 218, in blood, illustrated in Fig. 21. The ion monitored for the added standard *m/e* 449 arises by loss of two heptafluorobutyryl groups and one hydrogen. The *m/e* 451 ion is the normal abundance M + 2 isotope peak for the same fragment. This method requires 0.5 ml of plasma for quantification at the 200-ng/100-ml level. These workers were interested in the measurement of endogeneous compounds, but the approach is equally appropriate for analysis of drugs and their metabolites.

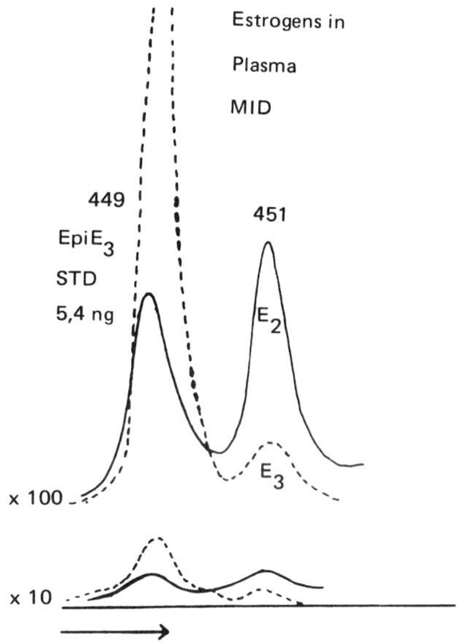

FIG. 21. Mass fragmentogram of estrogen in extract of 500 μl of plasma from a nonpregnant woman (15th day of ovarian cycle). Estriol (E3, 214), epi-estriol (epi-E3, 216), and estradiol (E2, 218) were analyzed as their heptafluorobutyrate derivatives, 215, 217, 219. Internal standard epi-E3 (5.4 ng) and E3 were recorded by using the fragment ion at *m/e* 451 (M - 213 = M - one HFB group): solid line.

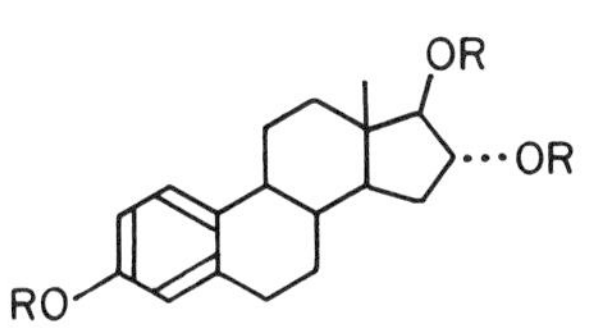

214 R = H
215 R = C_4F_7O

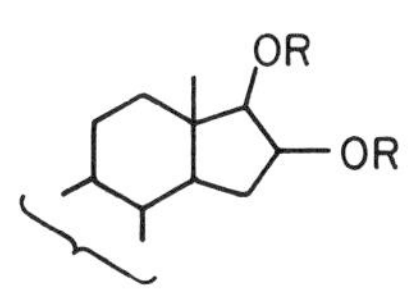

216 R = H
217 R = C_4F_7O

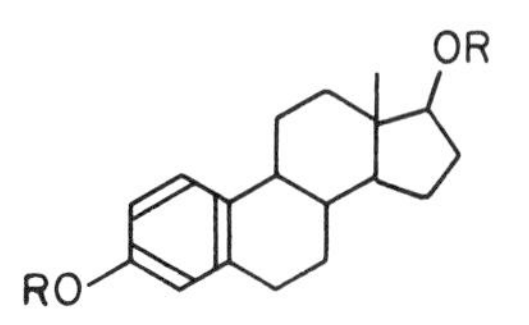

218 R = H
219 R = C_4F_7O

The use of chiral reagent for derivatization makes possible the determination of optical isomers. Matin, Wan, and Knight [403] have described the analysis of dextro and levo amphetamine, 220, in plasma. Following basic extraction with hexane and derivatization with the chemical reagent N-pentafluorobenzoyl-S-(-)-prolyl-1-imidazolide, the resulting diasteromers can be chromatographically resolved and the effluent can be quantified using methane CIMS.

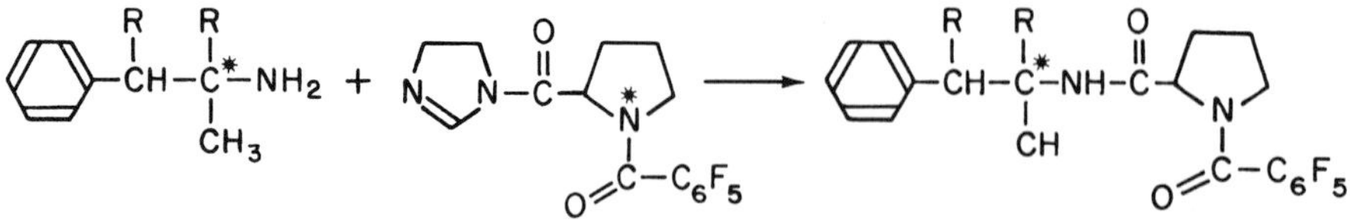

220 221 222

Linear calibration curves were obtained over the range 0-80 ng/ml plasma or saliva with a coefficient of variation of 3.5% at 50 ng/ml (N = 5). These workers established that saliva levels were considerably higher than blood levels and sufficiently parallel to blood levels that salivary measurements were suitable for determining amphetamine intake.

There are a few reports of quantitative analysis by GC-MS that make use of external standards to construct a calibration curve. If careful attention is paid to the reproducibility of all steps of the extraction and measurement, this approach can also provide good accuracy even at low levels. Milberg et al. [405] have used this method for the analysis of low levels of methylphenidate, 223, and its metabolite ritalinic acid, 224, in blood and urine.

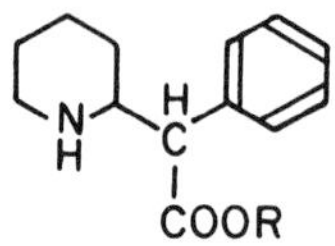

223 R=CH

224 R=H

Methylphenidate was isolated by extraction into chloroform. Ritalinic acid was isolated by salting out into isopropyl alcohol followed by methylation and subsequent solvent extraction. Using 10 ml of urine or 5 ml of plasma, the sensitivity limits for 223 were 2 ng/ml urine and 1.5 ng/ml plasma. For 224 the recovery was only 30% compared with 223 and the detection limits were 50 ng/ml urine and 56 ng/ml plasma. For both compounds the calibration was linear to the upper concentrations examined, 100 ng/ml for 223 in plasma, 500 ng/ml for 224 in plasma and for both compounds in urine. Milberg and his coworkers concluded that this GC-MS approach provided a rapid, reliable method that was quantitative over the concentration ranges likely to be encountered.

When Simon and co-workers [290] wished to quantify alclofenac, 225, and its metabolites, 228 and 231, in human serum, they chose compounds that were chemically similar, 229, 230, and 233, for internal standards. Careful selection of the peaks to be monitored permitted accurate quantification of derivatives of all three compounds in serum following a single 500-mg dose of alclofenac. The

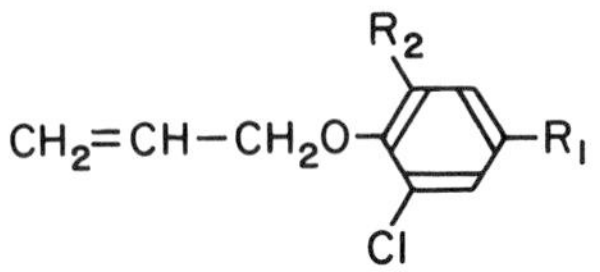

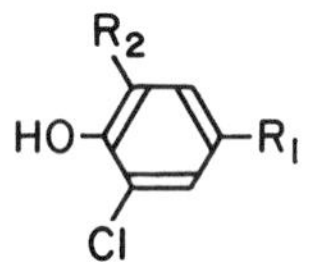

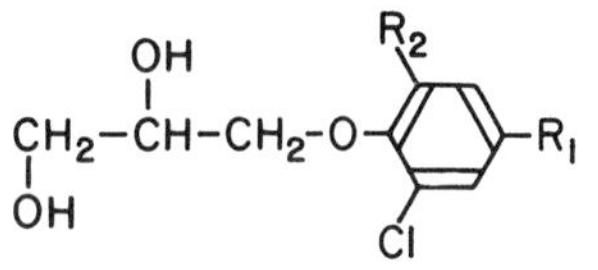

225	228	231
	$R_1 = CH_2\text{-}COOH$	$R_2 = H$
226	229	232
	$R_1 = CH_2\text{-}COOH$	$R_2 = \text{-O-}CH_2\text{-}CH{=}CH_2$
227	230	233
	$R_1 = COOH$	$R_2 = H$

two metabolites that had previously been reported in urine extracts were found to be present in serum at levels only 1% that of the drug. The limits of sensitivity for the metabolites, 228 and 231, were found to be 200 and 100 ng/ml plasma while the parent drug was found at concentrations of 1-50 μg/ml. The error in the measurements was estimated at 3% for the drug and 5% for the metabolites in the concentration ranges encountered. These results constituted a great improvement in sensitivity for these compounds over existing methodologies. The metabolites had not previously been detected in serum.

VII. CONCLUSION

Society's increasing demand for the modification of biological processes by means of drugs is a "Pandora's box" that may never close. This flood of chemical agents demands a concern for safeguards against untoward effects of drug and drug metabolite action. Better analytical methodology is needed to match the greater understanding of target and drug receptor knowledge, a factor that will certainly lead to new drugs of greater specificity and lower or closely controlled therapeutic dosages. To keep abreast with the analytical requirements, instrumentation of greater specificity and sensitivity must be utilized, and the GC-MS-computer appears to be the instrumental package of choice.

We have tried in this chapter to provide an overview of developments in mass spectrometry and its adjunct techniques and a survey of how they have been successfully applied to drug-related problems. It is apparent that significant insight into structural interpretation can be gained by intuitive guesses obtained from sample workup and a knowledge of drug metabolism. Thus, we have outlined major metabolic transformations and their effects on the physiochemical properties of the administered drugs. Examples in the recent literature illustrate how mass spectrometry and its related techniques can most effectively be used to characterize and quantify drugs and their metabolites. These existing methods do not define the limits of mass spectral analysis but serve as an indication of its wide-ranging possibilities.

ACKNOWLEDGMENT

We wish to thank Ms. Kathleen O'Brien for her excellent assistance in cataloging and organizing the references and Professor Klaus Biemann, in whose laboratory much of the work reported was carried out, for providing the important elements of encouragement and guidance. This work was supported by the National Institutes of Health (RR00317 and HD-066-45).

Permission has been received from the following publishers to reproduce the indicated figures and table that appear in the text: Fig. 1, Williams and Wilkins [394]; Fig. 5, Elsevier [311]; Figs. 18, 19, and 20, and Table 1, Heyden and Sons, Ltd. [281], [166], [362], and [359] respectively; and Fig. 21, the American Chemical Society [381].

REFERENCES

β-Adrenergic Blockers

1. M. S. Benedetti, D. A. Larue, C. V. Perrier, and L. Sitavanc, Pharmacokinetic and metabolic pathways of dimethophrine in man, *Arzneim Forsch, 27,* 158-161 (1977).

2. N. O. Bodin, Identification of the major urinary metabolite of alprenolol in man, dog, and rat, *Life Sci., 14,* 685-692 (1974).

3. H. Ehrsson, Simultaneous determination of (+) and (-) propranolol by gas chromatography-mass spectrometry using a deuterium labeling technique, *J. Pharm. Pharmacol., 28,* 662-663 (1976).

4. T. E. Gaffney, D. Knapp, T. Walle, P. J. Privitera, and D. Saelens, The importance of stable isotope labelled drugs in clinical pharmacology: with illustrations from studies on the disposition of propranolol and nortriptyline in man, *South Med. J., 67,* 990-1002 (1974).

5. P. F. Meffin, S. R. Harapat, and D. C. Harrison, Quantitation in plasma and urine of acebutolol and a major metabolite with preliminary observations on their disposition kinetics in man, *Res. Commun. Chem. Pathol. Pharmacol., 15,* 31-51 (1976).

6. K. K. Midha, A. H. Beckett, and A. Saunders, Identification of the major metabolites of propylhexedrine *in vivo* (in man and *in vitro* in guinea pig and rabbit), *Xenobiotica, 4,* 627-635 (1974).

7. H. Mori, M. Kido, N. Murakami, S. Morita, and H. Kohri, Metabolic fate of carteolol hydrochloride [5-(3-tert-butylamino-2-hydroxypropoxy)-3, 4-dihydrocarbostyril hydrochloride, OPC-1085], a new beta-blocker. V. Identification of metabolites in rat, dog and human (author's transl.), *J. Pharm. Soc. Jpn., 97,* 305-308 (1977).

8. D. A. Saelens, T. Walle, and P. J. Privitera, Quantitative determination of propranolol, propranolol glycol, and N-desisopropyl propranolol in brain tissue by electron capture gas chromatography, *J. Chromatog., 123,* 185-192 (1976).

9. W. G. Stillwell, and M. G. Horning, Metabolism of pronethalol in the rat, the guinea pig, and the mouse, *Res. Commun. Chem. Pathol. Pharmacol., 9,* 601-619 (1974).

10. D. J. Tocco, A. E. Duncan, F. A. Delauna, H. B. Hucker, V. F. Gruber, and W. J. VandenHeuval, Physiological disposition and metabolism of timolol in man and laboratory animals, *Drug Metabl. Dispos., 3,* 361-370 (1975).

11. T. Walle, J. I. Morrison, and G. L. Tindell, Isomeric ring hydroxylated metabolites of propranolol in rats, man, and dogs, *Res. Commun. Chem. Pathol. Pharmacol., 9,* 1-10 (1974).

12. T. Walle, GLC determination of propranolol, other beta-blocking drugs, and metabolites in biological fluids and tissues, *J. Pharm. Sci., 63,* 1885-1891 (1974).

Analgesics

13. J. D. Baty, J. Playfer, D. A. Price Evans, and J. Lamb, The measurement of acetanilide in plasma by spectrophotometric and selected ion monitoring methods, *Biomed. Mass Spectrom., 4,* 255-257 (1977).

14. S. L. Due, H. R. Sullivan, and R. E. McMahon, Propoxyphene: pathways of metabolism in man and laboratory animals, *Biomed. Mass Spectrom., 3,* 217-225 (1976).

15. W. A. Garland, K. C. Hsiao, E. J. Pantuck, and A. H. Conney, Quantitative determination of phenacetin and its metabolite acetaminophen by GLC-chemical ionization mass spectrometry, *J. Pharm. Sci., 66,* 340-344 (1977).

16. J. A. Hinson, S. D. Nelson, and J. R. Mitchell, Studies on the microsomal formation of arylating metabolites of acetaminiphen and phenacetin, *Mol. Pharmacol., 13,* 625-633 (1977).

17. G. Holzman, P. H. Kreutzer, and K. Roth, Metabolism of trans-tilidine: isolation and identification of the metabolite from human urine (author's transl.), *Arch. Pharm. (Weinheim), 308,* 169-178 (1975).

18. D. J. Jollow, S. S. Thorgeirsson, W. Z. Potter, M. Hashimoto, and J. R. Mitchell, Acetaminophen-induced hepatic necrosis. VI. Metabolic disposition of toxic and nontoxic doses of acetaminophen, *Pharmacology, 12,* 251-271 (1974).

19. R. E. McMahon, H. R. Sullivan, and S. L. Due, Proceedings: The combined use of gas chromatography-mass spectrometry (GC-MS) together with deuterium labelling in a study of the disposition of *d*-propoxyphene, *Acta. Pharm. Suec., 11,* 639-640 (1975).

20. J. E. Mrochek, S. Katz, W. H. Christie, and S. R. Dinsmore, Acetaminophen metabolism in man, as determined by high-resolution liquid chromatography, *Clin. Chem., 20,* 1086-1096 (1974).

21. J. F. Nash, I. F. Bennett, R. J. Bopp, M. K. Brunson, and H. R. Sullivan, Quantitation of propoxyphene and its major metabolites in heroin addict plasma after large dose administration of propoxyphene napsylate, *J. Pharm. Sci., 64,* 429-433 (1975).

22. M. Stafford, G. Kellerman, R. N. Stillwell, and M. G. Horning, Metabolism of antipyrine by the epoxide-diol pathway in the rat, guinea pig and human, *Res. Commun, Chem. Pathol. Pharmacol., 8,* 593-606 (1974).

23. H. R. Sullivan, S. L. Due, and R. E. McMahon, Metabolism of alpha-l-methadol. N-acetylation, a new metabolic pathway, *Res. Commun. Chem. Pathol. Pharmacol., 6,* 1072-1078 (1973).

24. H. R. Sullivan, J. L. Emmerson, F. K. Marshall, P. G. Wood, and R. E. McMahon, Quantitation of plasma levels of propoxyphene and norpropoxyphene by combined use of stable isotope labeling and selected ion monitoring, *Drug Metab. Dispos., 2,* 526-532 (1974).

25. Y. Tanaka and M. Sano, Analytical studies on mepirizole and its metabolites. II. Identification of the human urinary metabolites of mepirizole with stable isotope labeling, *Chem. Pharm. Bull. (Tokyo), 24,* 804-807 (1976).

26. Y. Tanaka and M. Sano, Analytical studies on mepirizole and its metabolites. IV. Quantitative determination of mepirizole in human serum by mass fragmentography, and isotope effect on drug metabolism, *Chem. Pharm. Bull. (Tokyo), 24,* 1305-1309 (1976).

27. E. Zietz and G. Spiteller, 2-Acetyl-2-methyl-1-phenyl-hydrazine: a frequently occurring artifact in processing biological material after ingestion of drugs containing antipyrine (author's transl.), *Arzneim Forsch, 27,* 355-356 (1977).

Anorexics

28. A. H. Beckett and P. M. Bélanger, The identification of three new metabolic products of phentermine after liver microsomal incubation, *Xenobiotica, 4,* 509-519 (1974).

29. A. H. Beckett and P. M. Bélanger, The mechanism of metabolic N-oxidation of phentermine and chlorophentermine to their hydroxy amino- and nitroso-compounds, *J. Pharm. Pharmac., 26,* 558-560 (1974).

30. A. H. Beckett and P. M. Bélanger, The microsomal N-oxidation of phentermine, *J. Pharm. Pharmac., 28,* 692-699 (1976).

31. A. H. Beckett and P. M. Bélanger, Metabolic incorporation of oxygen into primary and secondary aliphatic amines and the consequences in carbon-nitrogen bond cleavage, *J. Pharm. Pharmac., 27,* 547-552 (1975).

32. A. K. Cho, The identification of *p*-hydroxyphentermine as a urinary metabolite of phentermine, *Res. Commun. Chem. Pathol. Pharmacol., 7,* 67-78 (1974).

33. R. B. Franklin, L. G. Dring, and R. T. Williams, The metabolism of phenmetrazine in man and laboratory animals, *Drug Metab. Dispos., 5,* 223-233 (1977).

34. J. E. Sinsheimer, L. G. Dring, and R. T. Williams, Species differences in the metabolism of norephedrine in man, rabbit, and rat, *Biochem. J., 136,* 763-771 (1973).

35. B. Testa, Some chemical and stereochemical aspects of diethylpropion metabolism in man, *Acta Pharm. Suec., 10,* 441-454 (1973).

Antiarrhythmics, Cardiotonics

36. A. H. Beckett and E. C. Chidomere, The identification and analysis of mexiletine and its metabolic products in man, *J. Pharm. Pharmacol., 29,* 281-285 (1977).

37. W. H. Bulger, S. J. Stohs, and D. M. Wheeler, 6 Beta-hydroxy-3-epidigitoxigenin--the major metabolite of digitoxigenin by rabbit liver homogenates, *Biochem. Pharmacol., 23,* 921-929 (1974).

38. F. I. Carroll, D. Smith, and M. E. Wall, Carbon-13 magnetic resonance study. Structure of the metabolites of orally administered quinidine in humans, *J. Med. Chem., 17,* 985-987 (1974).

39. D. R. Clark, and S. M. Kalman, Dihydrodigoxin: a common metabolite of digoxin in man, *Drug Metab. Dispos., 2,* 148-150 (1974).

40. L. Dodion, J. M. deSuray, M. Deblecker, and A. Georges, Pharmacokinetic characteristics and biodegradation of aprindine in man, *Therapie, 29,* 221-232 (1974).

41. D. E. Drayer, M. M. Reidenberg, and R. W. Sevy, N-acetylprocainamide: an active metabolite of procainamide, *Proc. Soc. Exp. Biol. Med., 146,* 358-363 (1974).

42. W. A. Garland, W. F. Trager, and S. D. Nelson, Direct (nonchromatographic) quantification of drugs and their metabolites from human plasma utilizing chemical ionization mass spectrometry and stable isotope labeling: quinidien and lidocaine, *Biomed. Mass Spectrom., 1,* 124-129 (1974).

43. H. Greenwood, W. Snedden, R. P. Hayward, and J. Landon, The measurement of urinary digoxin and dihydrodigoxin by radioimmunoassay and by mass spectroscopy, *Clin. Chim. Acta, 62,* 213-224 (1975).

44. E. Hauser, U. Boffo, L. Meister, L. Sawlewicz, H. H. Linde, and K. Meyer, 3-amino-3-desoxy compounds from uzarigenin, oleandrigenin, gitoxigenin and digoxigenin. Partial synthesis experiments in the series of cardiac poisons. 6, *Helv. Chim. Acta, 56,* 2782-2795 (1973).

45. P. J. Meffin, S. R. Harapat, and D. C. Harrison, High-pressure liquid chromatographic analysis of drugs in biological fluids II: determination of an antiarrhythmic drug, tocainide, as its dansyl derivative using a fluorescence detector, *J. Pharm. Sci., 66,* 583-586 (1977).

46. P. J. Murphy, Metabolic pathways of aprindine, *Acta Cardiol. (Brux), 18,* 131-142 (1974).

47. J. M. Strong, and A. J. Atkinson, Jr., Simultaneous measurement of plasma concentration of lidocaine and its desethylated metabolite by mass fragmentography, *Anal. Chem., 44,* 2287-2290 (1972).

48. K. E. Wirth, J. C. Frolich, J. W. Hollifield, F. C. Falkner, B. S. Sweetman, and J. A. Oates, Metabolism of digitoxin in man and its modification by spironolactone, *Eur. J. Clin. Pharmacol.*, *9*, 345-354 (1976).

49. A. G. Zacchei, L. L. Weidner, G. H. Bessebaar, and L. S. B. Raftery, Physiological disposition and metabolic fate of a new antiarrhythmic agent, α, α-dimethyl-4-(α, α, β, β-tetrafluorophenethyl) benzylamine in the rat, dog, monkey, baboon, and man, *Drug Metab. Dispos.*, *4*, 387-401 (1976).

Antibacterials, Antifungals

50. C. J. Brooks and M. T. Gilbert, Studies of urinary metabolites of 2-(4-isobutylphenyl) propionic acid by gas-liquid chromatography-mass spectrometry, *J. Chromatogr.*, *99*, 541-551 (1974).

51. L. W. Brown, GLC determination of clindamycin and related compounds, *J. Pharm. Sci.*, *63*, 1597-1600 (1974).

52. K. K. Chan and P. A. Harris, A fluorometric determination of adriamycin and its metabolites in biological tissues, *Res. Commun. Chem. Pathol. Pharmacol.*, *6*, 447-463 (1973).

53. M. J. Cooper, M. W. Anders, and B. L. Mirkin, Ion-pair extraction and high-speed liquid chromatography of cephalothin and deacetylcephalothin in human serum and urine, *Drug Metab. Dispos.*, *1*, 659-662 (1973).

54. D. L. Hill and R. F. Struck, A microsomal amidase which cleaves chloramphenicol, *Biochem. Pharmacol.*, *22*, 127-129 (1973).

55. C. C. Lin, J. Magat, R. Chang, J. McGlotten, and S. Symchowicz, Absorption, metabolism and excretion of 14C-griseofulvin in man, *J. Pharmacol. Exp. Ther.*, *187*, 415-422 (1973).

56. A. M. Nadzan, K. L. Rinehart, Jr., and W. T. Sokolski, Hydroxynybomycin: isolation, structure and bioactivity, *J. Antibiot. (Tokyo)*, *30*, 523-524 (1977).

57. T. Nakagawa, M. Masada, and T. Uno, Gas chromatographic determination and gas chromatographic-mass spectrometric analysis of chloramphenicol, thiamphenicol, and their metabolites, *J. Chromatogr.*, *111*, 355-364 (1975).

58. J. C. Phillips, K. Hardy, R. Richards, R. C. Cottrell, and S. D. Gangolli, Studies of the metabolic fate of p-hydroxybenzoic acid in male and female cats, *Toxicology*, *7*, 257-263 (1977).

59. F. F. Sun, Metabolism of clindamycin. II. Urinary excretion products of clindamycin in rat and dog, *J. Pharm. Sci.*, *62*, 1657-1662 (1973).

60. S. Takanashi, and N. R. Bachur, Adriamycin metabolism in man. Evidence from urinary metabolites, *Drug Metab. Dispos.*, *4*, 79-87 (1976).

61. C. Y. Wang, C. W. Chiu, B. Kaiman, and G. T. Bryan, Identification of 2-methyl-4-(5-amino-2-furyl) thiazole as the reduced meta-olite of 2-methyl-4(5-nitro-2-furyl) thiazole, *Biochem. Pharmacol., 24,* 291-293 (1975).

Anticonvulsants

62. J. G. Allen, M. J. Blackburn, and S. M. Caldwell, The metabolism by man of an anticonvulsant oxadiazole. An unusual metabolic route for tertiary amides, *Xenobiotica, 1,* 3-12 (1971).

63. J. D. Alvin, and M. T. Bush, Synthesis of millimole amounts of 14C- and 2D-labelled primidone, phenylethylmalonamide, and phenobarbital for high-sensitivity detection in biologic materials, *Mikrochim. Acta, 1,* 685-696 (1975).

64. G. Belvedere, C. Pantarotto, and A. Frigerio, Carbamazepine-10, 11-epoxide metabolites in rat urine, *Res. Commun. Chem. Pathol. Pharmacol., 11,* 221-232 (1975).

65. K. H. Dudley, D. L. Bius, and C. D. Waldrop, Urinary metabolites of N-methyl-α-methyl-α-phenylsuccinimide (methsuximide) in the dog, *Drug Metab. Dispos., 2,* 113-122 (1974).

66. M. Eichelbaum, and L. Bertilsson, Determination of carbamazepine and its epoxide metabolite in plasma by high-speed liquid chromatography, *J. Chromatogr., 103,* 135-140 (1975).

67. B. Ferrandes, and P. Eymard, Metabolism of valproate sodium in rabbit, rat, dog and man, *Epilepsia, 18,* 169-182 (1977).

68. N. Gerber, R. A. Seibert, and R. M. Thompson, Identification of a catechol glucuronide metabolite of 5, 5-diphenylhydantoin (DPH) in rat bile by gas chromatography (GC) and mass spectrometry (MS), *Res. Commun. Chem. Pathol. Pharmacol., 6,* 499-511 (1973).

69. S. Goenechea, and E. Hecke-Seibicke, The metabolism of carbamazapine (author's transl.), *Z. Klin. Chem. Clin. Biochem., 10,* 112-113 (1972).

70. M. G. Horning, K. Lertratanangkoon, J. Nowlin, W. G. Stillwell, R. N. Stillwell, T. E. Zion, P. Kellaway, and R. M. Hill, Anticonvulsant drug monitoring by GC-MS-COM techniques, *J. Chromatogr. Sci., 12,* 630-635 (1974).

71. M. G. Horning, C. Stratton, J. Nowlin, D. J. Harvey, and R. M. Hill, Metabolism of 2-thyl-2-methylsuccinimide (ethosuximide) in the rat and human, *Drug Metabl. Dispos., 1,* 569-576 (1973).

72. T. Kuhara, and I. Matsumoto, Metabolism of branched medium chain length fatty acid, *Biomed. Mass Spectrom., 1,* 291-294 (1974).

73. K. K. Midha, I. J. McGilveray, and D. L. Wilson, Sensitive GLC procedure for the simultaneous determination of phenytoin and its major metabolite from plasma following single doses of phenytoin, *J. Pharm. Sci., 65,* 1240-1243 (1976).

74. I. A. Muni, C. H. Altshuler, and J. C. Neicheril, Identification of a blood metabolite of methsuximide by GLC-mass spectrometry, *J. Pharm. Sci., 62,* 1820-1823 (1973).

75. J. Naestoft, and N. E. Larsen, Quantitative determination of clonazepam and its metabolites in human plasma by gas chromatography, *J. Chromatogr., 93,* 113-122 (1974).

76. P. G. Preste, C. E. Westerman, N. P. Das, B. J. Wilder, and J. H. Duncan, Identification of 2-ethyl-2-methyl-3-hydroxysuccinimide as a major metabolite of ethosuximide in humans, *J. Pharm. Sci., 63,* 467-469 (1974).

77. R. G. Smith, G. D. Davis, Jr., R. K. Lynn, and N. Gerber, Hydantoin ring glucuronidation: characterization of new metabolite of 5,5-diphenylhydantoin in man and the rat, *Biomed. Mass Spectrom., 4,* 275-279 (1977).

78. J. M. Strong, T. Abe, E. L. Gibbs, and A. J. Atkinson, Jr., Plasma levels of methsuximide and N-demethylmethsuximide during methsuximide therapy, *Neurology (Minneap.), 24,* 250-255 (1974).

79. R. M. Thompson, J. Beghin, W. K. Fife, and N. Gerber, 5, 5-Bis(3-hydroxyphenyl) hydantoin, a minor metabolite of diphenylhydatoin (dilantin) in the rat and human, *Drug Metab. Dispos., 4,* 349-356 (1976).

Antidepressants

80. B. Alexanderson, Prediction of steady-state plasma levels of nortriptyline from single oral dose kinetics: a study in twins, *Eur. J. Clin. Pharmacol., 6,* 44-53 (1973).

81. G. Alfredsson, F. A. Wiesel, B. Fyro, and G. Sedvall, Mass fragmentographic analysis of clominpramine and its mono-demethylated metabolite in human plasma, *Psychopharmacology, 52,* 25-30 (1977).

82. G. Belvedere, A. Frigerio, C. Pantarotto, V. Rovei, M. Zanol, F. Paraavicine, and G. Pifferi, Mass spectrometric identification of 10, 11-dihydro-10, 11-dihydroxyprotriptyline in rat urine after administration of protriptyline, *Farmaco Sci., 30,* 904-916 (1975).

83. A. G. Bolt, G. Graham, and P. Wilson, Stereoselective demethylation of the enantiomers of nefopam, an experimental antidepressant and skeletal muscle relaxant, *Xenobiotica, 4,* 355-363 (1974).

84. M. Claeys, G. Muscettola, and S. P. Markey, Simultaneous measurement of imipramine and desiprimine by selected ion recording with deuterated internal standards, *Biomed. Mass Spectrom.*, *3*, 110-116 (1976).

85. M. Constatin, J. F. Pognat, and G. Streichenberger, Urinary metabolites of amixetrine in man and in the dog, *Arzneim Forsch.*, *26*, 80-87 (1976).

86. J. J. de Ridder, P. C. Koppens, and H. J. van Hal, Mass fragmentographic assay of nanogram amounts of the antidepressant drug mianserin hydrochloride (Org GB 94) in human plasma, *J. Chromatogr.*, *143*, 289-297 (1977).

87. A. Frigerio, C. Pantarotto, R. Franco, R. Gomeni, and P. L. Morselli, Quantitative determination of doxepin and desmethyldoxepin in rat plasma by means of gas-liquid chromatography-mass fragmentography, *J. Chromatogr.*, *130*, 354-360 (1977).

88. W. A. Garland, Quantitative determination of amitriptyline and its principal metabolite, nortriptyline, by GLC-chemical ionization mass spectrometry, *J. Pharm. Sci.*, *66*, 77-81 (1977).

89. J. A. Hoskins, The formation of meta-hydroxyphenylacetic acid through a direct ring hydroxylation reaction in humans, *Biochem. Biophys. Res. Commun.*, *77*, 50-56 (1977).

90. H. B. Hucker, A. J. Balletto, J. Demetriades, B. H. Arison, and A. G. Zacchei, Urinary metabolites of amitriptyline in the dog, *Drug Metab. Dispos.*, *5*, 132-142 (1977).

91. C. Pantarotto, G. Belvedere, and A. Frigerio, Quantitative determination of imipramine and desmethylimipramine in human plasma by mass fragmentography, *Z. Klin. Chem. Klin. Biochem.*, *12*, 224-225 (1974).

92. V. Rovei, G. Belvedere, C. Pantarotto, and A. Frigerio, Isolation of 10, 11-epoxide of protriptyline in rat urine after protriptyline administration, *J. Pharm. Sci.*, *65*, 810-815 (1976).

93. F. Sjoqvist, A pharmacokinetic approach to the treatment of depression, *Int. Pharmacopsychiatry*, *6*, 147-169 (1971).

94. J. M. Wilson, L. J. Williamson, and V. A. Raisys, Simultaneous measurement of secondary and tertiary tricyclic antidepressants by GC/MS chemical ionization mass fragmentography, *Clin. Chem.*, *23*, 1012-1017 (1977).

Antihistamines

95. N. K. Chandhuri, O. A. Servando, M. J. Manniello, R. C. Luders, D. K. Chao, and M. F. Bartlett, Metabolism of tripellennamine in man, *Drug Metab. Dispos*, *4*, 372-378 (1976).

96. T. Chang, R. A. Okerholm, and A. J. Glazko, Identification of diphenydramine (Benadryl) metabolites in human subjects, *Res. Commun. Chem. Pathol. Pharmacol., 9,* 391-404 (1974).

97. A. Frigerio, N. Sossi, G. Belvedere, C. Pantarotto, and S. Garattini, Identification of desmethylcyproheptadine-10, 11-epoxide and other cyproheptadine metabolites isolated from rat urine, *J. Pharm. Sci., 63,* 1536-1540 (1974).

98. K. L. Hintze, J. S. Wold, and L. J. Fischer, Disposition of cyproheptadine in rats, mice, and humans and identification of a stable epoxide metabolite, *Drug Metab. Dispos., 3,* 1-9 (1975).

99. C. C. Porter, B. H. Arison, V. F. Gruber, D. C. Titus, and W. J. VandenHeuval, Human metabolism of cyproheptadine, *Drug Metab. Dispos., 3,* 189-197 (1975).

100. G. S. Rao, G. Krishna, and J. R. Gillette, Metabolism, tissue distribution and covalent binding of tripelennamine and its N-nitroso derivative in the rat, *J. Pharmacol. Exp. Ther., 195,* 433-440 (1975).

Antihypertensives

101. M. Angelo, L. G. Dring, R. Lancaster, and R. L. Smith, The metabolism of debrisoquine in rat and man, *Biochem. Soc. Trans., 4,* 704-706 (1976).

102. C. R. Freed, R. J. Weinkam, K. L. Melmon, and N. Castagnoli, Chemical ionization mass spectrometric measurement of alpha-methyl dopa and s-dopa metabolites in rat brain regions, *Anal. Biochem, 78,* 319-332 (1977).

103. S. L. Malcolm, and T. R. Maten, Determination of debrisoquine and its 4-hydroxy metabolite in plasma by gas chromatography-mass spectrometry, *Anal. Chem., 48,* 807-809 (1976).

104. S. B. Zak, T. G. Gilleran, J. Karliner, and G. Lukas, Identification of two metabolites of hydralazine from human urine, *J. Med. Chem., 17,* 381-382 (1974).

105. H. Zimmer, R. Glaser, J. Kokosa, D. A. Garteiz, E. V. Hess, and A. Litwin, 3-Hydroxymethyl-*s*-triazolo[3,4-a] pthalazine, a novel urinary hydralazine metabolite in man, *J. Med. Chem., 18,* 1031-1033 (1975).

106. H. Zimmer, J. Kokosa, and D. A. Garteiz, Identification of 3-methyl-*s*-triazolo(3,4-a) phthalazine, a human hydrazaline metabolite, by gas chromatography-mass spectrometry, *Arzneim Forsch., 23,* 1028-1029 (1973).

Antiinflammatory (Nonsteroidal)

107. D. V. Awang, and A. Vincent, Mass spectral and pyrolytic behavior of the two main products of phenylbutazone degradation: simulation of unusual mass spectral fragmentation, *J. Pharm. Sci., 65,* 68-70 (1976).

108. D. V. Awang, A. Vincent, and F. Matsui, Pattern of phenylbutazone degradation, *J. Pharm. Sci., 62,* 1673-1676 (1973).

109. O. M. Bakke, G. H. Draffan, and D. S. Davies, The metabolism of phenylbutazone in the rat, *Xenobiotica, 4,* 237-254 (1974).

110. A. I. Cohen, I. Weliky, S. J. Lan, and S. D. Levine, Metabolism of (14C) niflumic acid. Isolation and identification of metabolites from human urine, *Biomed. Mass Spectrom., 1,* 1-4 (1974).

111. G. Forni, and G. Massarani, High-performance liquid chromatographic determination of colchicine and colchicoside in colchicum (Colchicum autumnale L.) seeds on a home-made stationary phase, *J. Chromatogr., 131,* 444-447 (1977).

112. G. G. Gallo, N. Rimorini, L. F. Zerilli, and P. Radaelli, Gas liquid chromatographic determination of the main urinary metabolite of diftalone, a new anti-inflammatory drug, *J. Chromatogr., 101,* 163-168 (1974).

113. P. M. Gros, H. J. Davi, L. F. Chasseaud, and D. R. Hawkins, Metabolic and pharmacokinetic study of bucloxic acid, I. Isolation and identification of the main excretion products in animals and in man, *Arzneim Forsch., 24,* 1385-1390 (1974).

114. D. G. Kaiser, S. R. Shaw, and G. J. Vangiessen, GLC determination of *dl*-2-(2-fluoro-4-biphenylyl) propionic acid (flurbiprofen) in plasma, *J. Pharm. Sci., 63,* 567-570 (1974).

115. Y. Manai, T. Kobayashi, and S. Tanayama, Metabolism of 6-chloro-6-cyclohexylindane-1-carboxylic acid (TAI-284), a new nonsteroidal anti-inflammatory agent. II. Isolation and identification of metabolites in the perfusate of the isolated liver perfusion of rats, *Xenobiotica, 3,* 657-670 (1973).

116. K. K. Midha, I. J. McGilveray, and C. Charette, GLC determination of plasma concentration of phenylbutazone and its metabolite oxphenbutazone, *J. Pharm. Sci., 63,* 1234-1239 (1974).

117. H. Noguchi, M. Okui, K. Noda, and R. Kato, Disposition and metabolic fate of tiaramide hydrochloride, a new anti-inflammatory agent in the rat, *Xenobiotica, 7,* 491-503 (1977).

118. P. S. Portoghese, K. Svanborg, and B. Samuelsson, Oxidation of oxyphenbutazone by sheep vesicular gland microsomes and lipoxygenase, *Biochem. Biophys. Res. Commun., 63,* 748-755 (1975).

Antineoplastic Agents

119. N. R. Bachur, R. C. Hildebrand, and R. S. Jaenke, Adriamycin and daunorubicin disposition in the rabbit, *J. Pharmacol. Exp. Ther., 191,* 331-340 (1974).

120. J. E. Bakke, C. E. Price, and R. G. Zaylskie, Mass spectral characterization of the glucuronide conjugates of crumfomate (4-5-butyl-2-chlorophenyl methyl methylphosphoramidate) metabolites, *Biomed. Mass Spectrom., 3,* 329-334 (1976).

121. A. Braufman, R. Bruni, V. Reinhold, D. Silveira, M. Chadwick, and D. Yesair, Characterization of the metabolites of ellipticine (NSC 71795) in rat bile, *Drug Metab. Dispos., 6,* 542-548 (1978).

122. T. A. Connors, P. J. Cox, P. B. Farmer, A. B. Foster, and M. Jarman, Some studies of the active intermediates formed in the microsomal metabolism of cyclophosphamide and isophosphamide, *Biochem. Pharmacol., 23,* 115-129 (1974).

123. T. A. Connors, P. J. Cox, P. B. Farmer, A. B. Foster, M. Jarman, and J. K. Macleod, Observation on the mechanism of hydroxylation of cyclophosphamide by rat liver microsomes: the metabolism of cyclophosphamide-4-d2, *Biomed. Mass Spectrom., 1,* 130-136 (1974).

124. T. A. Connors, J. A. Hickman, M. Jarman, D. H. Melzack, and W. C. Ross, The metabolism of the anitumor agent 1(1-aziridinyl)-2,4-dinitrobenzene (CB 1832), *Biochem., Pharmacol., 24,* 1665-1670 (1975).

125. P. J. Cox, P. B. Farmer, A. B. Foster, E. D. Gilby, and M. Jarman, The use of deuterated analogs in qualitative and quantitative investigation of the metabolism of cyclophosphamide (NSC-26271), *Cancer Treatment Rep., 60,* 483-491 (1976).

126. P. J. Cox, and L. Levin, Novel metabolic products of cyclophosphamide in human urine, *Biochem. Pharmacol., 24,* 1233-1235 (1975).

127. V. J. Feil, J. E. Bakke, and R. G. Zaylskie, Synthesis and mass spectrometry of crufomate metabolites and related compounds, *Biomed. Mass Spectrom., 3,* 316-328 (1976).

128. C. Fenselau, Review of the metabolism and mode of action of cyclophosphamide, *J. Assoc. Off. Anal. Chem., 59,* 1028-1036 (1976).

129. C. Fenselau, M. N. Kan, S. Billets, and M. Colvin, Identification of phosphorodiamidic acid mustard as a human metabolite of cyclophosphamide, *Cancer Res., 35,* 1453-1457 (1975).

130. C. Fenselau, M. N. Kan, S. S. Rao, A. Myles, O. M. Friedman, and M. Clovin, Identification of aldophosphamide as a metabolite of cyclophosphamide *in vitro* and *in vivo* in humans, *Cancer Res., 37,* 2538-2543 (1977).

131. L. J. Griggs, and M. Jarman, Synthesis of deuterium-labeled analogs of cyclophosphamide and its metabolites, *J. Med. Chem.*, *18*, 1102-1106 (1975).

132. I. Jardine, R. Brundrett, M. Colvin, and C. Fenselau, Approaches to the pharmacokinetics of cyclophosphamide (NSC-26271) quantitation of metabolites, *Cancer Treatment Rep.*, *60*, 403-408 (1976).

133. M. Jarman, D. H. Melzack, and W. C. Ross, The metabolism of the anti-tumor agent 5-(1-aziridinyl)-2, 4-dinitrobenzamide (CB 1954), *Biochem. Pharmacol.*, *25*, 2475-2478 (1976).

134. C. Mitoma, T. Onodera, T. Takegoshi, and D. W. Thomas, Metabolic disposition of chlorambucil in rats, *Xenobiotica*, *7*, 205-220 (1977).

135. M. N. Musa, G. M. Lower, Jr., P. G. Welling, B. M. Johnson, B. D. Kaiman, and G. T. Bryan, Absorption, tissue distribution, and excretion of the anti-neoplastic agent guanazole in the rat, *Res. Commun. Chem. Pathol. Pharmacol.*, *7*, 497-512 (1974).

136. H. T. Nagasawa, F. N. Shirota, and N. S. Mizuno, The mechanism of alkylation of DNA by 5-(3-methyl-1-tirazeno) imidazole-4-carboxamide (MIC), a metabolite of DIC (NSC-45388). Non-involvement of diazomethane, *Chem. Biol. Interact.*, *8*, 403-413 (1974).

137. N. Neuss, D. S. Fukuda, G. E. Mallett, D. R. Brannon, and L. L. Huckstep, Vinca alkaloids. XXXII. Microbiological conversions of vindoline, a major alkaloid from Vinca rosea L., *Helv. Chim. Acta*, *56*, 2418-2426 (1973).

138. V. N. Reinhold, and R. J. Bruni, Aromatic hydroxylation of ellipticine in rats: lack of an NIH shift, *Biomed. Mass Spectrom.*, *3*, 335-339 (1976).

139. T. S. Safonova, L. F. Linberg, Zh. F. Presnova, N. S. Kuriatov, and V. A. Chernov, Metabolism of the antineoplastic agent N-pyrimidly-2N1,N2-di-(ethylene)-triamide phosphate (fosfemid), *Dokl. Akad. Nauk SSSR*, *219*, 1271-1272 (1974).

140. R. F. Struck, M. C. Kirk, M. H. Witt, and W. R. Laster, Jr., Isolation and mass spectral identification of blood metabolites of cyclophosphamide: evidence of phosphoramide mustard as the biologically active metabolite, *Biomed. Mass Spectrom.*, *2*, 46-52 (1975).

141. S. Takanashi, and N. R. Bachur, Daunovibicin metabolites in human urine, *J. Pharmacol. Exp. Ther.*, *195*, 41-49 (1975).

Benzodiazepines

142. K. B. Alton, R. M. Grimes, C. Shaw, J. E. Patrick, and J. L. McGuire, Biotransformation of A 1,5-benzodiazepine, triflubazam, by man, *Drug Metab. Dispos.*, *3*, 352-360 (1975).

143. E. Bailey, M. Fenoughty, and L. Richardson, Automated high-resolution gas chromatographic analysis of psychotropic drugs in biological fluids using open-tubular glass capillary columns. I. Determination of nomifensine in human plasma, *J. Chromatogr.*, *131*, 347-355 (1977).

144. M. Hasegawa, H. Takai, and I. Matsubara, Metabolic fate of fluorazepam. II. A new metabolite obtained by *in vitro* liver drug metabolizing enzyme system, *J. Pharm. Sci.*, *64*, 1732-1732 (1975).

145. Y. Kanai, The biotransformation of 8-chloro-6-phenyl-4H-*s*-triazolo (4,3-a) (1,4) benzodiazepine (D-40TA), a new central depressant, in man, dog and rat, *Xenobiotica*, *4*, 441-456 (1974).

146. A. O. Niedermayer, and R. G. Koski, GLC determination of 7-chloro-5, 11-dihydrodibenz-[b,e]1,4-oxazepine-5-carboxamide in serum or plasma, *J. Pharm. Sci.*, *64*, 1531-1534 (1975).

147. S. Rendic, L. Kasinc, V. Sunjic, F. Kajfez, V. Kramer, and P. Mildner, Mass spectrometry of 1, 4 benzodiazepines, *Biomed. Mass Spec.*, *2*, 97-106 (1975).

148. M. C. Sânchez, J. Colomé, and E. Gelpi, Electron capture and multiple ion detection of benzodiazepine esters in pharmacokinetic studies, *J. Chromatog.*, *126*, 601-613 (1976).

149. G. Sedvall, G. Alfredsson, L. Bjerkenstedt, P. Eneroth, B. Fryö, C. Härnryd, and B. Wode-Helgodt, "Central Biochemical Correlates to Antipsychotic Drug Action in Man," in, *The Impact of Biology on Modern Psychiatry* (E. S. Gershon et al., Ed.), Plenum Press, New York, 1977, pp. 41-54.

150. M. Tateishi, and H. Shimizu, A new metabolic pathway of bromazepam involving attachment of a methylthio group, *Xenobiotica*, *6*, 431-439 (1976).

151. A. Trebbi, G. B. Gervasi, and V. Comi, Determination of pinazepam and its metabolites in serum, urine and brain by gas-liquid chromatography and mass spectrometry, *J. Chromatogr.*, *110*, 309-319 (1975).

152. L. W. Whitehouse, C. J. Paul, B. B. Coldwell, and B. H. Thomas, Effect of ethanol on diazepam distribution in rat, *Res. Commun. Chem. Pathol. Pharmacol.*, *12*, 221-225 (1975).

Diuretics

153. D. H. Chatfield, and W. H. Hunter, The metabolism of acetamidothiazoles in the rat. 2-acetamido-4-chloromethylthiazole, *Biochem. J.*, *134*, 879-884 (1973).

154. A. Karim, J. Hribar, W. Aksamet, M. Doherty, and L. J. Chinn, Spironolactone metabolism in man studied by gas chromatography-mass spectrometry, *Drug Metab. Dispos.*, *3*, 467-478 (1975).

155. K. Koch, M. Senn, and W. Steingross, Isolation and identification of a sulfur-containing metabolite of spironolactone from human urine, *Steroids, 28,* 467-480 (1976).

156. S. J. Kolis, T. H. Williams, and M. A. Schwartz, Identification of the urinary metabolites of ^{14}C-bumetamide in the rat and their excretion by rats and dogs, *Drug Metab. Dispos., 4,* 169-176 (1976).

157. W. Sadee, U. Abshagen, C. Finn, and N. Rietbrock, Conversion of spironolactone to canrenone and disposition kinetics of spironolactone and canrenoate-potassium in rats, *Naunyn Schmiedebergs Arch. Pharmacol., 283,* 303-318 (1974).

Field Desorption, Plasma Desorption

158. H. D. Beckey, "Determination of the Structures of Organic Molecules and Quantitative Analysis with the Field Ionization Mass Spectrometer," in *Biochemical Applications of Mass Spectrometry* (G. R. Waller, Ed.), Wiley-Interscience, New York, 1972, pp. 795-816.

159. D. A. Brent, P. de Miranda, and H. R. Schulten, Field desorption mass spectrometry of azathioprine and its metabolites, *J. Pharm. Sci., 63,* 1370-1375 (1974).

160. D. E. Games, M. P. Games, A. H. Jackson, A. H. Olavesen, M. Rossiter, and P. J. Winterburn, Field desorption mass spectrometry of sodium and potassium salts, *Tetrahedron Letters, 27,* 2377-2380 (1974).

161. J. F. Holland, B. Soltmann, and C. C. Sweeley, A model for ionization mechanisms in field desorption mass spectrometry, *Biomed. Mass Spectrom., 3,* 340-345 (1976).

162. W. D. Lehmann, H. D. Beckey, and H. R. Schulten, Qualitative and quantitative analysis of bansyl derivatives of dopamine and some of its metabolites in urine samples by electron impact and field desorption mass spectrometry, *Anal. Chem., 48,* 1572-1575 (1976).

163. W. D. Lehmann, and H. R. Schulten, Quantitative field desorption mass spectrometry, stable isotope dilution and cationization, *Angew. Chem. (Eng.), 16,* 184-185 (1977).

164. R. D. MacFarlane, and D. F. Torgerson, Califormium-252 plasma desorption mass spectroscopy, *Science, 191,* 920-925 (1976).

165. M. Przybylski, H. Ringsdorf, G. Voelcker, U. Draeger, G. Peter, and H. J. Hohorst, Characterization of cyclophosphamide (NSC-26271) metabolites and related derivatives by field desorption and electron impact mass spectrometry, *Cancer Treat. Rep., 60,* 509-516 (1976).

166. M. Przybylski, H. Ringsdorf, U. Lenssen, G. Peter, G. Voelcker, T. Wagner, and H. J. Horhorst, Mass spectrometric characterization of activated N-(2-chloroethyl) amido oxazaphosphorine derivatives, *Biomed. Mass Spectrom.*, *4*, 209-215 (1977).

167. H. R. Schulten, Field desorption mass spectrometry of metabolites of the anticancer drug cyclophosphamide, *Biomed. Mass Spectrom.*, *1*, 223-230 (1974).

168. H. R. Schulten, and D. E. Games, High resolution field desorption mass spectrometry. II Glycosides, *Biomed. Mass Spec.*, *1*, 120-123 (1974).

169. H. R. Schulten, Qualitative and quantitative investigations of cyclophosphamide (NSC-26271), cyclophosphamide metabolites, and related compounds by field desorption mass spectrometry, *Cancer Treatment Rep.*, *60*, 501-507 (1976).

170. H. R. Schulten, and W. D. Lehmann, High resolution field desorption mass spectrometry. Part VI. Organic sulphates and sulphate esters, *Anal. Chem. Acta*, *87*, 103-112 (1976).

171. D. F. Torgerson, R. P. Skowronski, and R. D. MacFarlane, New approach to the mass spectroscopy of non-volatile compounds, *Biochem. Biophys. Res. Commun.*, *60*, 616-621 (1974).

Hypoglycemics

172. V. Aggarwal, and I. Sunshine, Determination of sulfonylureas and metabolites by pyrolysis gas chromatography, *Clin. Chem.*, *20*, 200-204 (1974).

173. R. Blomstrand, and G. Ohman, Studies on the metabolism of LADH-inhibitor 4-methylpyrazole in the rat, *Life Sci.*, *13*, 107-112 (1973).

174. J. W. Kleber, J. A. Galloway, and B. E. Rodda, GLC determination of acetohexamide and hydroxyhexamide in biological fluids, *J. Pharm. Sci.*, *66*, 635-638 (1977).

175. S. B. Matin, J. H. Karam, P. H. Forsham, and J. B. Knight, Determination of phenformin in biological fluids using chemical ionization mass spectrometry, *Biomed. Mass Spectrom.*, *1*, 320-322 (1974).

176. S. B. Matin, and J. B. Knight, The use of solid probe chemical ionization mass spectrometry and gas chromatography-chemical ionization mass spectrometry in quantitation of drugs in biological fluids. Determination of tolbutamide and its metabolites in human plasma, *Biomed. Mass Spectrom.*, *1*, 323-328 (1974).

177. K. Sabih, High resolution mass spectroscopic analysis of methylated tolbutamide derivatives, *Biomed. Mass Spectrom*, *1*, 163-165 (1974).

178. D. J. Weber, High pressure liquid chromatographic analysis of tolbutamide in serum, *J. Pharm. Sci., 65,* 1502-1505 (1976).

Metabolic Pathways

179. H. C. Curtius, J. A. Völlmin, and K. Baerlocher, Study of metabolic pathways *in vivo* using stable isotopes, *Anal. Chem., 46,* 1107-1110 (1973).

180. D. E. Drayer, J. M. Strong, A. Sandler, and M. M. Reidenberg, *In vitro* acetylation of drugs by human blood cells, *Drug Metab. Dispos., 2,* 499-505 (1974).

181. D. E. Duggan, J. J. Baldwin, B. H. Arison, and R. E. Rhodes, N-glucoside formation as a detoxification mechanism in mammals, *J. Pharmacol. Exp. Ther., 190,* 563-569 (1974).

182. L. M. Hjelmeland, L. Aronow, and J. R. Trudell, Intramolecular determination of substituent effects in hydroxylations catalyzed by cytochrome P-450, *Mol. Pharmacol., 13,* 634-639 (1977).

183. D. M. Jerina, and J. W. Daly, Arene oxides: a new aspect of drug metabolism, *Science, 185,* 573-582 (1974).

184. W. J. Richter, K. O. Alf, W. Dieterle, J. W. Faigle, H. P. Kriemler, H. Mory, and T. Winkler, C-glucuronides, a novel type of drug metabolite, *Helv. Chem. Acta, 58,* 2572-2577 (1975).

185. J. E. Tomaszewski, D. M. Jerina, and J. W. Daly, Deuterium isotope effects during formation of phenols by hepatic monoxygenases. Evidence for an alternative to the arene oxide pathway, *Biochem., 14,* 2024-2031 (1975).

Miscellaneous Drugs

186. L. Almirante, B. Danieli, A. Frigerio, A. Mugnaini, and S. Picco, Metabolic fate of zolimidine. III. Chemical structure of the principal metabolites in human urine, *Farmaco [Sci.], 29,* 941-956 (1974).

187. C. C. Arnett, P. S. Callery, and N. Zenker, N-Methylation of benzimidazole by catechol-O-methyltransferase, *Biochem. Pharmacol., 26,* 377-380 (1977).

188. S. Baba, S. Morisita, and Y. Nabatsu, Studies on drug metabolism by use of isotopes. XVIII. Ion cluster technique for detection of urinary metabolites of 1-butyryl-4-cinnamyl-piperazine (author's trans.), *J. Pharm. Soc. Jpn., 96,* 1293-1297 (1976).

189. A. H. Beckett, and R. Achari, A note on the metabolism of respilene in man, *J. Pharm. Pharmacol., 29,* 253 (1977).

190. G. Belvedere, C. Pantarotto, V. Rovei, and A. Frigerio, Identification of 10, 11 epoxide and other cyclobenzaprine metabolites isolated from rat urine, *J. Pharm. Sci., 65,* 815-821 (1976).

191. R. Bouchér, and G. Lhoest, Isolation and identification by mass spectrometry of metabolites of bupivacaine, *Pharm. Acta Helv., 51,* 223-225 (1976).

192. I. Bornschein, H. H. Borchert, and S. Pfeifer, Biotransformation of iomeglamic acid (Falignost), *Pharmazie, 29,* 730-731 (1974).

193. J. E. Carter, R. V. Petersen, and D. B. Roll, Synthesis and PMR and mass spectra of potential metabolites and other derivatives of bis(2-ethylhexyl) phthalate, *J. Pharm. Sci., 66,* 546-549 (1977).

194. D. H. Chatfield, and W. H. Hunter, The metabolism of acetamidothiazoles in the rat. 2-acetamido-2-acetamido-4-methyl- and 2-acetamido-4-phenylthiazole, *Biochem. J., 134,* 869-878 (1973).

195. K. L. Clay, W. D. Watkins, and R. D. Murphy, Metabolism of pyrazole. Structure elucidation of urinary metabolites, *Drug Metab. Dispos., 5,* 149-156 (1977).

196. D. A. Cowan, G. Huizing, and A. H. Beckett, Identification of four new metabolic products of metoclopramide using mass spectrometry, *Xenobiotica, 6,* 605-616 (1976).

197. A. F. Delong, R. D. Smyth, A. Polk, R. K. Nayak, G. Martin, G. H. Douglas, and N. H. Reavey-Cantiwell, Comparative metabolism of fenclorac in rat, dog, monkey, and man, *Drug Metab. Dispos., 5,* 122-131 (1977).

198. P. DeMiranda, L. M. Beacham, III, T. H. Creagh, and G. B. Elion, The metabolic fate of the methylnitroimidazole moiety of azathioprine in the rat, *J. Pharmacol. Exp. Ther., 187,* 588-601 (1973).

199. R. M. DeSagher, A. P. DeLeenheer, A. A. Cruyl, and A. E. Claeys, Identification and urinary excretion of p-chlorophenoxyacetamide, a metabolite of iproclozide, in humans, *Drug Metab. Dispos., 3,* 423-429 (1975).

200. W. Dieterle, J. W. Faigle, H. Mory, W. J. Richter, and W. Theobald, Biotransformations and pharmacokinetics of sulfinpyrolzone (Anturan) in man, *J. Clin. Pharmacol., 9,* 135-145 (1975).

201. D. E. Duggan, J. J. Baldwin, B. W. Arison, and R. E. Rhodes, N-glucoside formation as a detoxification mechanism in mammals, *J. Pharmacol. Exp. Ther., 190,* 563-569 (1974).

202. G. C. Ford, N. J. Haskins, R. F. Palmer, M. J. Todd, and P. H. Buckley, The measurement of diphenoxylic acid in plasma following administration of <10 mg/ml diphenoxylate, *Biomed. M.S., 3,* 45-47 (1976).

203. A. A. Forist, and A. L. Pulliam, N-(2-(diethylamino)ethyl)-2-(p-hydroxyphenoxy)-acetamide, a metabolite of mefexamide in man, *Res. Commun. Chem. Pathol. Pharmacol., 8,* 385-388 (1974).

204. H. J. Förster, I. Kramer, K. H. Pook, and D. Wahl, Studies on the pharmacokinetics and biotransformation of ipratropium bromide in the rat and dog, *Arzneim Forsch.*, *26*, 992-1005 (1976).

205. G. G. Gallo, E. Beretta, and G. Pelizza, Isolation and structure determination of a metabolite of diftalone in humans, *Farmaco (Sci.)*, *29*, 534-540 (1974).

206. J. W. Gorrod, and D. J. Temple, The formation of an N-hydroxymethyl intermediate in the N-demethylation of N-methylcarbazole *in vivo* and *in vitro*, *Xenobiotica*, *6*, 265-274 (1976).

207. J. G. Leferink, I. Wagemaker-Engels, and R. A. Maes, Quantitative analysis of terbutaline in serum and urine at therapeutic levels using gas chromatography-mass spectrometry, *J. Chromatogr.*, *143*, 299-305 (1977).

208. F. J. Leinweber, F. C. Greenough, and J. D. Frederick, A new oxisuran metabolite, *Xenobiotica*, *6*, 617-624 (1976).

209. C. Lin, Y. Li, J. McGlotten, J. B. Morton, and S. Symchowicz, Isolation and identification of the major metabolite of albuterol in human urine, *Drug Metab. Dispos.*, *5*, 234-238 (1977).

210. K. K. Midha, and J. K. Cooper, GLC determination of plasma acenocoumarol levels, *J. Pharm. Sci.*, *66*, 799-802 (1977).

211. D. M. Morton, D. M. Fuller, and J. N. Green, Antiparasitic nitroimidazoles. 6. The metabolism and excretion of some 2-styryl-5-nitroimidazoles, *Xenobiotica*, *3*, 257-266 (1973).

212. R. Neidlein, and U. Stumpf, Biotransformation and pharmacokinetics of grandiflorenic acid (Kauradiene-9(11), 16-oic acid-18)/1st communication (author's trans.), *Arzneim Forsch*, *27*, 999-1004 (1977).

213. S. D. Nelson, G. D. Breck, and W. F. Trager, *In vivo* metabolite condensations. Formation of N1-ethyl-2-methyl-N3-(2,6-dimethylphenyl)-4-imidazolidinone from the reaction of a metabolite of alcohol with a metabolite of lidocaine, *J. Med. Chem.*, *16*, 1106-1112 (1973).

214. F. I. Onuska, M. E. Comba, and A. G. Harrison, Chemical ionization and high resolution electron impact mass spectra of 5,6-dihydro-2-methyl-1,4-oxathiin-3-carboxanilide and three of its metabolites, *Biomed. Mass Spectrom.*, *3*, 248-254 (1976).

215. S. Picco, A. Bonaldi, A. Frigerio, B. Danieli, and L. Almirante, Metabolic fate of zolimidine. II. Analytical profile, tissue levels and urinary excretion in animals and in man, *Farmaco [Sci.]*, *29*, 848-860 (1974).

216. L. R. Pohl, R. Haddock, W. A. Garland, and W. F. Trager, Synthesis and thin-layer chromatographic ultraviolet, and mass spectral properties of the anticoagulant phenprocoumon and its monohydroxylated derivatives, *J. Med. Chem.*, *18*, 513-519 (1975).

217. L. R. Pohl, S. D. Nelson, W. A. Garland, and W. F. Trager, The rapid identification of a new metabolite of warfarin via a chemical ionization mass spectrometry ion doublet technique, *Biomed. Mass Spectrom.*, *2*, 23-30 (1975).

218. M. W. Sauerhoff, J. P. Heeschen, R. A. Nyquist, and W. H. Braun, Pharmacokinetic profile of diamidfos in rats, *Food Cosmet. Toxicol.*, *14*, 401-408 (1976).

219. H. Schröder, and B. M. Schröder, Isolation and determination of a hydroxylated metabolite of sulphapyridine from human urine, *Acta Pharm. Suec.*, *10*, 263-268 (1973).

220. K. N. Scott, M. W. Couch, B. J. Wilder, and C. M. Williams, The use of mass spectroscopy and 13C and 1H nuclear magnetic resonance in the characterization of 1-methyl-(2,6-xylyloxy)ethylamine metabolites, *Drug Metab. Dispos.*, *1*, 506-515 (1973).

221. G. G. Skellern, B. I. Knight, F. M. Luman, J. B. Stenlake, D. G. McLarty, and M. J. Hooper, Identification of 3-methyl-2-thiohydantoin, a metabolite of carbimazole, in man, *Xenobiotica*, *7*, 247-253 (1977).

222. B. Sjoquist, B. Lindstrom, and E. Anggard, Mass fragmentographic determination of 4-hydroxy-3-methoxyphenylglycol (HMPG) in urine, cerebrospinal fluid, plasma and tissues using a deuterium-labelled internal standard, *J. Chromatogr.*, *105*, 309-316 (1975).

223. J. M. Strong, D. E. Mayfield, A. J. Atkinson, Jr., B. C. Burris, F. Raymon, and L. T. Webster, Jr., Pharmacological activity, metabolism, and pharmacokinetics of glycinexylidide, *Clin. Pharmacol. Ther.*, *17*, 184-194 (1975).

224. H. R. Sullivan, J. G. Page, and S. L. Due, Species differences in the metabolism of alpha-l-trans-4-dimethyl-aminotetrahydro-3-furyl cyclohexanephenyl-glycolate, an experimental anticholinergic agent, *Biomed. Mass Spectrom.*, *2*, 53-58 (1975).

225. F. F. Sun, and J. E. Stafford, Metabolic fate of N,N-dimethyl-N-(p-phenoxyphenyl) sulfamide in rat, dog, and human, *J. Pharm. Sci.*, *63*, 539-544 (1974).

226. N. R. Trenner, J. R. Carlin, B. H. Arison, and W. J. VandenHeuvel, Urinary metabolites of D,L-1,2-bis(5-(6)-methoxy-2-benzimidazolyl)-1,2-ethanediol in mice, *J. Pharm. Sci.*, *63*, 1312-1313 (1974).

227. D. Trülzsch, J. Roboz, H. Grein, P. Czygan, J. Rudick, F. Hutterer, F. Schaffner, and H. Popper, Hydroxylation of taurolithocholate by isolated human liver microsomes. I. Identification of metabolic product, *Biochem. Med.*, *9*, 158-166 (1974).

228. W. J. VandenHeuvel, J. R. Carlin, R. L. Ellsworth, F. J. Wolf, and R. W. Walker, Source of the carboxamide nitrogen atom in 2-carboxamido-5-isopropoxy-carbonylaminobenz-imidazole, a metabolite of cambendazole, *Biomed. Mass Spectrom.*, *1*, 190-191 (1974).

229. K. O. Vollmer, and H. Achenbach, Metabolism of ethyl DL-trans-2-dimethylamino-1-phenylcyclohex-3-ene-trans-1-carboxylate hydrochloride. 1. Mass spectroscopic structural elucidation of metabolites I and II and their determination in various biological fluids, *Arzneim Forsch., 24,* 1237-1242 (1974).

230. K. O. Vollmer, A. V. Hodenberg, A. Poisson, R. Bode, and E. Schauerte, Metabolism and pharmacokinetics of piprozoline in the rat, dog and man. 1. Studies with the radioactively-labeled substance, *Arzneim Forsch., 27,* 502-508 (1977).

231. L. Witte, H. Nau, J. H. Fuhrhop, A. Doenicke, and B. Grote, Quantitative analysis of trifluoroacetic acid in body fluids of patients treated with halothane, *J. Chromatogr., 143,* 329-334 (1977).

232. B. A. Wood, D. Rycroft, and A. M. Monro, The metabolism of tinidazole in the rat and dog, *Xenobiotica, 3,* 801-812 (1973).

Narcotics, Narcotic Antagonists

233. R. C. Baselt, and M. H. Bickel, Biliary excretion of methadone by the rat: identification of a para-hydroxylated major metabolite, *Biochem. Pharmacol., 22,* 3117-3120 (1973).

234. G. D. Bellward, P. M. Warren, W. Howland, J. E. Axelson, and F. S. Abbott, Methadone maintenance: effect of urinary pH on renal clearance in chronic high and low doses, *Clin. Pharmacol. Ther., 22,* 92-99 (1977).

235. U. Boerner, R. L. Roe, and C. E. Becker, Detection, isolation and characterization of normorphine and norcodeine as morphine metabolites in man, *J. Pharm. Pharmacol., 26,* 393-398 (1974).

236. N. Chatterjie, J. M. Fujimoto, C. E. Inturrisi, S. Roerig, R. I. Wang, D. V. Bowen, F. H. Field, and D. D. Clarke, Isolation and stereochemical identification of a metabolite of naltrexone from human urine, *Drug Metab. Dispos., 2,* 401-405 (1974).

237. P. A. Clarke, and R. L. Foltz, Quantitative analysis of morphine in urine by gas chromatography-chemical ionization-mass spectrometry, with [N-C^2H_3] morphine as an internal standard, *Clin. Chem., 20,* 465-469 (1974).

238. W. J. Cole, J. Parkhouse, and Y. Y. Yousef, Application of the extractive alkylation technique to the pentafluorobenzylation of morphine (a heroin metabolite) and surrogates, with special reference to the quantitative determination of plasma morphine levels using mass fragmentography, *J. Chromatogr., 136,* 409-416 (1977).

239. E. J. Cone, C. W. Gorodetzky, and S. Y. Yeh, Biosynthesis, isolation and identification of 6-Beta-hydroxynaltrexone, a major human metabolite of naltrexone, *J. Pharm. Sci., 64,* 618-621 (1975).

240. W. O. Ebbighasnsen, J. H. Mowat, H. Stearns, and P. V. Vestergaard, Mass fragmentography of morphine and 6-monoacetylmorphine in blood with a stable isotope internal standard, *Biomed. Mass Spectrom.*, *1*, 305-311 (1974).

241. L. Gyarmati, A. David, Z. Meszaros, I. Racz, and P. Szentmiklosi, Data on the metabolism of benzylisoquinoline derivatives, The metabolism of drotaverin, *Pharmazie*, *30*, 174-177 (1975).

242. R. L. Hawks, I. J. Kopin, R. W. Colburn, and N. B. Thoa, Norcocaine: a pharmacologically active metabolite of cocaine found in the brain, *Life Sci.*, *15*, 2189-2195 (1974).

243. A. Klutch, A chromatographic investigation of morphine metabolism in rats. Confirmation of N-demethylation of morphine and isolation of a new metabolite, *Drug Metab. Dispos.*, *2*, 23-30 (1974).

244. E. G. Leightly, and A. F. Fentiman, Jr., Metabolism of cocaine to norcocaine and benzoyl ecgonine by an *in vitro* microsomal enzyme system, *Res. Commun. Chem. Pathol. Pharmacol.*, *8*, 65-74 (1974).

245. C. Lindberg, and C. Bogentoft, Mass spectrometric and gas chromatographic properties of pethidine N-oxide, *Acta Pharm. Suec.*, *12*, 507-510 (1975).

246. C. Lindberg, C. Bogentoft, U. Bondesson, and B. Danielsson, Mass spectrometric identification of a new metabolite of pethidine, *J. Pharm. Pharmacol.*, *27*, 975-976 (1975).

247. C. Lindberg, C. Bogentoft, and B. Danielsson, Mass spectrometry of pethidine and pethidine metabolites. A study utilizing specifically deuterated pethidine, *Acta Pharm. Suec.*, *11*, 201-210 (1974).

248. R. K. Lynn, G. D. Olsen, R. M. Leger, W. P. Gordon, R. G. Smith, and N. Gerber, The secretion of methadone and its major metabolite in the gastric juice of humans: comparison with blood and salivary concentrations, *Drug Metab. Dispos.*, *4*, 504-509 (1976).

249. G. Norhein, and E. Bye, X-ray crystallographic and gas chromatographic-mass spectrometric studies on norpropoxyphene, *Arch. Toxicol. (Berl.)*, *36*, 97-102 (1976).

250. W. G. Stillwell, C. S. Myron, and J. T. Stewart, Meperidine metabolites: identification of N-hydroxynormeperidine and a hydroxy-methoxy derivative of meperidine in biological fluids, *Res. Commun. Chem. Pathol. Pharmacol.*, *14*, 605-619 (1976).

251. H. R. Sullivan, S. L. Due, and R. E. McMahon, Methadone N-oxide in the urine of methadone maintenance subjects--an artifact?, *J. Pharm. Pharmacol.*, *25*, 1009-1010 (1973).

252. H. R. Sullivan, F. J. Marshall, R. E. McMahon, L.-M., Gunne, J. H. Holmstrand, and E. Anggard, Mass fragmentographic determination of unlabelled and deuterium labelled methadone in human plasma, *Biomed. Mass Spectrom.*, *2*, 197-200 (1975).

253. K. Verebely, M. A. Chedekel, S. J. Mulc, and D. Rosenthal, Isolation and identification of a new metabolite of naltrexone in human blood and urine, *Res. Commun. Chem. Pathol. Pharmacol.*, *12*, 67-84 (1975).

254. J. E. Zarembo, B. Douglas, J. Valenta, and J. A. Weisbach, Metabolites of metragynine, *J. Pharm. Sci.*, *63*, 1407-1415 (1974).

Psychedelics

255. Z. Ben-Zvi, J. R. Bergen, and S. Burstein, Cannabinol-7-oic acid: a metabolite of delta-1-tetrahydrocannabinol in the rhesus monkey, *Res. Commun. Pathol. Pharmacol.*, *9*, 201-204 (Sept. 1974).

256. Z. Ben-Zvi, and S. Burstein, 7-oxo-delta-1-tetrahydrocannabinol: a novel metabolite of delta-1-tetrahydrocannabinol, *Res. Commun. Chem. Pathol. Pharmacol.*, *8*, 223-229 (June 1974).

257. Z. Ben-Zvi, S. Burstein, and J. Zikopoulos, Metabolism of delta-1-tetrahydrocannabinol by mouse hepatic microsomes: identification of 6-alpha-hydroxy-tetrahydrocannabinol (letter), *J. Pharm. Sci.*, *63*, 1173-1174 (July 1974).

258. M. Binder, "Identification of Hydroxylated Cannabinoids by PMR and Mass Spectroscopy," in *Marihuana: Chemistry, Biochemistry, and Cellular Effects* (G. G. Nahas et al., Eds.), Springer, New York, 1976, pp. 159-168.

259. J. C. Craig, and L. D. Gruenke, "A Contribution to Psychoactive Drug Measurement Techniques," in *Pharmacokinetics of Psychoactive Drugs* (L. A. Gottschalk, and S. Merlis, Eds.), Spectrum Publications, New York, 1976, pp. 15-21.

260. K. Fonseka, and M. Widman, Dihydroxylated metabolites of cannabinol formed by rat liver *in vitro*, *J. Pharm. Pharmacol.*, *29*, 12-14 (Jan. 1977).

261. D. J. Harvey, B. R. Martin, and W. D. Paton, Identification of the glucuronides of cannabidiol and hydroxycannabidiols in mouse liver, *Biochem. Pharmacol.*, *25*, 2217-2219 (Oct. 1976).

262. D. J. Harvey, and W. D. Paton, Characterization of three monohydroxyacid and two dihydroxyacid metabolites of Δ^1-THC in mouse liver, *Res. Commun. Chem. Pathol. Pharmacol.*, *13*, 585-599 (1970).

263. D. J. Harvey, B. R. Martin, and W. D. Paton, Identification of glucuronides as *in vivo* conjugates of seven cannabinoids and some of their hydroxy and acid metabolites, *Res. Commun. Pathol. Pharmacol.*, *16*, 265-279 (Feb. 1977).

264. D. J. Harvey, Cyclic alkylboronates as derivatives for the characterization of cannabinolic acids by combined gas chromatography and mass spectrometry, *Biomed. Mass. Spectrom.*, *4*, 88-93 (Apr. 1977).

265. D. J. Harvey, and W. D. Paton, "Examination of the Metabolites of Δ^1-Tetrahydrocannabinol in Mouse Liver, Heart, and Lung by Combined Gas Chromatography, and Mass Spectrometry," *Marihuana: Chemistry, Biochemistry, and Cellular Effects* (G. G. Nahas et al., Eds.), Springer, New York, pp. 93-109.

266. L. E. Hollister, S. L. Kanter, R. D. Board, and D. E. Green, Marihuana metabolites in urine of man. III. Unchanged delta-9-tetrahydrocannabinol, *Res. Commun. Chem. Pathol. Pharmacol., 8,* 579-584 (Aug. 1974).

267. G. Jones, M. Widman, S. Agurell, and J. E. Lindgren, Monohydroxylated metabolites of delta-1-tetrahydrocannabinol in mouse brain. Comparison with *in vitro* liver metabolites, *Acta Pharm. Suec., 11,* 283-294 (June 1974).

268. W. W. Just, and M. Wiechmann, Proceedings: Detection of tetrahydrocannabinol in saliva of man and studies on its metabolism in the monkey, *Naunyn Schmiedebergs Arch. Pharmacol., 282,* R43 (Mar. 22, 1974).

269. S. L. Kanter, L. E. Hollister, F. Moore, and D. E. Greene, Marihuana metabolites in urine of man. II. Undescribed metabolite following oral ingestion of delta-9-tetrahydrocannabinol, *Res. Commun. Chem. Pathol. Pharmacol., 7,* 79-84 (Jan. 1974).

270. J. A. Kelley, and K. P. Arnold, Detection of urinary cannabis metabolites: a preliminary investigation, *J. Forensic Sci., 21,* 252-262 (1976).

271. J. J. Kettenes-van den Bosch, and C. A. Salemink, Cannabis. XVI. Constituents of marihuana smoke condensate, *J. Chromatogr., 131,* 422-424 (1977).

272. E. G. Leighty, A. F. Fentiman, Jr., and R. L. Foltz, Long retained metabolites of Δ^9 and Δ^8 tetrahydrocannabinols as novel fatty acid conjugates, *Res. Commun. Chem. Pathol. Pharmacol., 14,* 13-28 (1976).

273. M. A. Lyle, S. Pallante, K. Head, and C. Fenselau, Synthesis and characterization of glucuronides of cannabinol, cannabidiol, Δ^9-tetrahydrocannabinol and Δ^8-tetrahydrocannabinol, *Biomed. Mass Spec., 4,* 190-196 (1977).

274. B. R. Martin, D. J. Harvey, and W. D. Paton, Biotransformation of cannabidiol in mice. Identification of new acid metabolites, *Drug Metab. Dispos., 5,* 259-267 (1977).

275. T. Niwaguchi, T. Inoue, and T. Sakai, Studies on the *in vitro* metabolism of compounds related to lysergic acid diethylamide (LSD), *Biochem. Pharmacol., 23,* 3063-3066 (1974).

276. M. Nordquist, J. E. Lindgren, and S. Agurell, A method for the identification of acid metabolites of tetrahydrocannabinol (THC) by mass fragmentography, *Natl. Inst. Drug Abuse Res. Monogr. Sci., 7,* 64-69 (1976).

277. M. Nordqvist, S. Agurell, M. Binder, and I. M. Nilsson, Structure of an acidic metabolite of delta-1-tetrahydrocannabinol isolated from rabbit urine, *J. Pharm. Pharmacol.*, *26*, 471-473 (1974).

278. J. Rosenfeld, "Mass Fragmentographic Assays for the Cannabinoids and Their Metabolites," in *Marihuana: Chemistry, Biochemistry, and Cellular Effects* (G. G. Nahas et al., Eds.), Springer, New York, 1976, pp. 87-92.

279. N. Seiler, and L. Demisch, Oxidative metabolism of mescaline in the central nervous system. 3. Side chain degradation of mescaline and formation of 3,4,5-trimethoxy-benzoic acid *in vivo*, *Biochem. Pharmacol.*, *23*, 259-271 (1974).

280. M. Widman, M. Nordqvist, S. Agurell, J. E. Lindgren, and F. Sandberg, Biliary excretion of delta-1-tetrahydrocannabinol and its metabolites in the rat, *Biochem. Pharmacol.*, *23*, 1163-1172 (1974).

281. L. K. Wong, and K. Biemann, Metabolites of phencyclidine in humans, *Biomed. Mass Spectrom.*, *2*, 204-205 (1975).

Reviews, General Methods

282. V. Aggarwal, R. Bath, and I. Sunshine, Technique for rapidly separating drugs from biological samples, *Clin. Chem.*, *20*, 307-309 (1974).

283. M. L. Bastos, G. L. Kannanen, R. M. Young, J. Montforte, and I. Sunshine, Detection of organic bases and their metabolites in urine, *Clin. Chem.*, *16*, 931 (1970).

284. G Belvedere, A. Frigerio, C. Pantarotto, and M. Zanol, Chemical ionization mass spectrometry and its use in the pharmacological field, *Boll. Chim. Farm.*, *115*, 419-431 (1976).

285. R. Binks, R. J. Goodfellow, J. MacMillan, and R. J. Poyce, Acetyl tri-n-butyl citrate. A common laboratory contaminant, *Chem. and Ind.*, 565-566 (1970).

286. M. Claeys, S. P. Markey, and W. Maenhaut, Variance analysis of error in selected ion monitoring assays using various internal standards. A practical study case, *Biomed. Mass Spectrom.*, *4*, 122-128 (1977).

287. E. J. Cone, and C. W. Gorodetzky, Analytical controls in drug metabolic studies. A commentary, *Drug Metab. Dispos.*, *3*, 417-418 (1975).

288. C. E. Costello, H. S. Hertz, T. Sakai, and K. Biemann, Routine use of a flexible gas chromatograph-mass spectrometer-computer system to identify drugs and their metabolites in body fluids of overdose victims, *Clin. Chem.*, *20*, 255-265 (1974).

289. C. E. Costello, T. Sakai, and K. Biemann, "Identification of Drugs in Body Fluids, Particularly in Emergency Cases of Acute Poisoning," in *Techniques of Combined Gas Chromatography/Mass Spectrometry* (W. H. McFadden, Ed.), Wiley, New York, 1973, pp. 377-384.

290. A. P. DeLeenheer, and R. R. Roncucci, Eds., *Quantitative Mass Spectrometry in Life Sciences,* Elsevier, Amsterdam, 1977.

291. B. J. Gudzinowicz, and M. J. Gudzinowicz, *Analysis of Drugs and Metabolites by Gas Chromatography-Mass Spectrometry, Vols. 1-3,* Marcel Dekker, New York, 1977.

292. P. E. Haywood, and P. Chalmers, Less common "doping" agents and substances encountered during routine screening for drugs, *Br. J. Sports Med., 10,* 171-176 (1976).

293. M. G. Horning, C. M. Butler, J. Nowlin, and R. M. Hill, Drug metabolism in the human neonate, *Life Sci., 16,* 651-671 (1975).

294. E. C. Horning, D. I. Carroll, I. Dzidic, K. D. Haegele, S. Lin, C. U. Oertli, and R. N. Stillwell, Development and use of analytical systems based on mass spectrometry, *Clin. Chem., 23,* 13-21 (1977).

295. M. G. Horning, J. Nowlin, M. Stafford, K. Lertratanangkoon, K. R. Sommer, R. M. Hill, and R. N. Stillwell, The use of gas chromatographic-mass spectrometric-computer in pharmacokinetic studies, *J. Chromatogr., 112,* 605-615 (1975).

296. M. G. Horning, P. Gregory, J. Nowlin, M. Stafford, K. Lertratanangkoon, C. Butlis, W. G. Stillwell, and R. M. Hill, Isolation of drugs and drug metabolites from biological fluids by use of salt-solvent pairs, *Clin. Chem., 20,* 282-287 (1974).

297. J. V. Jackson, "Extraction Methods in Toxicology," in *Isolation and Identification of Drugs* (E. G. C. Clarke, Ed.), The Pharmaceutical Press, London, Vol. 2, 1975, pp. 914-920, Vol. 1, 1969, pp. 16-30.

298. S. T. Kozlev, and E. A. Rudzit, Use of chromato-mass-spectrometry in pharmacokinetic research (review of the literature), *Farmakol Toksikol, 39,* 750-755 (1976).

299. A. M. Lawson, The scope of mass spectrometry in clinical chemistry, *Clin. Chem., 21,* 803-824 (1975).

300. A. M. Lawson, and G. F. Draffan, Gas liquid chromatography-mass spectrometry in biochemistry, pharmacology, and toxicology, *Prog. Med. Chem., 12,* 1-103 (1975).

301. M. G. Lee, and B. J. Millard, Qualitative aspects of mass spectrometry and drug metabolism, *Biochem. Soc. Trans., 3,* 462-465 (1975).

302. G. W. A. Milne, "The Application of Mass Spectrometry to Problems in Medicine and Biochemistry," in *Mass Spectrometry Techniques and Applications* (G. W. A. Milne, Ed.), Wiley Interscience, New York, 1971, pp. 327-372.

303. H. Nau, "Determination of Drugs and Metabolites by a GC-MS Computer System," in *New Approaches to the Evaluation of Abnormal Embryonic Development* (D. Neubert, and H. J. Merker, Eds.), Thieme, Stuttgart, 1975, pp. 280-297.

304. A. Prox, Some applications of mass spectrometry in drug metabolism studies, *Kenobiotica, 3,* 473-492 (1973).

305. G. Spiteller, and G. Remberg, Drug metabolism studies using mass spectrometry, *Naturwissenschaften, 61,* 491-499 (1974).

306. R. N. Stillwell, and W. G. Stillwell, Gas chromatographic-mass spectrometric-computer methods in the study of drug metabolism, *J. Chromatogr., 126,* 547-557 (1976).

307. R. E. Stoner, and C. Parker, Single pH extraction procedure for detecting drugs of abuse, *Clin. Chem., 20,* 309-311 (1974).

308. C. C. Sweeley, N. D. Young, J. F. Holland, and S. C. Gates, Rapid computerized identification of compounds in complex biological mixtures by gas chromatography-mass spectrometry, *J. Chromatogr., 99,* 507-517 (1974).

309. J. E. Wallace, K. Blum, and J. M. Singh, Determination of drugs in biologic specimens--a review, *Clin. Toxicol., 7,* 477-495 (1974).

310. J. T. Watson, *Introduction to Mass Spectrometry: Biomedical, Environmental and Forensic Applications,* Raven Press, New York, 1976.

311. L. K. Wong, C. E. Costello, and K. Biemann, Methylation artifacts in the gas chromatography of serum extracts. Competitive suppression with trimethyl-d_9 anilinium hydroxide, *J. Chromatogr., 116,* 321-331 (1976).

Sedative--Hypnotics

312. J. J. Ambre, and L. J. Fischer, Identification and activity of the hydroxy metabolite that accumulates in the plasma of humans intoxicated with glutethimide, *Drug Metab. Dispos., 2,* 151-158 (1974).

313. B. D. Andersen, J. L. Templeton, R. H. Hammer, and H. L. Panzik, Synthesis and characterization of a catechol metabolite of glutethimide (Doriden) in human urine, *Res. Commun. Chem. Pathol. Pharmacol., 14,* 259-268 (1976).

314. B. D. Andersen, J. L. Templeton, R. H. Hammer, H. L. Panzik, and M. J. Moldowan, Definitive characterization of the *p*-hydroxyphenyl metabolite of glutethimide in human urine, *Res. Commun. Chem. Pathol. Pharmacol., 12,* 627-634 (1975).

315. W. Baldeo, J. N. Gilbert, J. W. Powell, L. V. Jones, and M. J. Whitehouse, A new metabolite of amylobarbitone: 5-(3'-carboxybutyl)-5-ethylbarbituric acid, *J. Pharm. Pharmacol.*, *29*, 254 (1977).

316. C. Bogentoft, O. Ericsson, and B. Danielsson, Methaqualone metabolites. Synthesis of two hydroxymethyl analogues of methaqualone- a quinazoline-benzodiazepine rearrangement, *Acta Pharm. Suec.*, *11*, 59-66 (1974).

317. C. Bogentoft, D. Ericsson, and B. Danielsson, Methaqualone metabolites. A mass spectrometric investigation of the monohydroxy derivatives of methaqualone, *Acta Pharm. Suec.*, *11*, 513-522 (1975).

318. R. Bonnichsen, R. Hjälm, Y. Moarde, M. Möller, and R. Ryhage, Metabolism studies of chlormethiazole by gas chromatography-mass spectrometry, *Z. Rechtsmed.*, *73*, 225-233 (1973).

319. R. Bonnichsen, R. Dimberg, Y. Moarde, and R. Ryhage, Variations in human metabolism of methaqualone given in therapeutic doses and in overdose cases studied by gas chromatography-mass spectrometry, *Clin. Chim. Acta*, *60*, 67-75 (1975).

320. J. A. Cella, Preparation of 2-methyl-3-(2-hydroxymethyl)phenyl-4(3H)-quinazalone: a metabolite of methaqualone, *J. Pharm. Sci.*, *63*, 1627 (1974).

321. W. J. Cole, R. G. Mitchell, and R. F. Salamonsen, Isolation, characterization, and quantitation of chloral hydrate as a transient metabolite of trichloroethylene in man using electron capture gas chromatography and mass fragmentography, *J. Pharm. Pharmacol.*, *27*, 167-171 (1975).

322. P. Daenens, and M. van Boven, Biotransformation of mecloqualone in man. Synthesis and identification of some major metabolites, *Arzneim Forsch.*, *24*, 195-202 (1974).

323. S. J. Dickson, The determination of methyprylon and its metabolites in biological fluids by gas chromatography, *Forensic Sci.*, *4*, 177-182 (1974).

324. S. Ebel, and H. Schütz, Studies on the detection of clonazepam and its main metabolites considering in particular thin-layer chromatography discrimination of nitrazepam and its major metabolic products (author's transl.), *Arzneim Forsch.*, *27*, 325-337 (1977).

325. H. Ehrsson, Gas chromatographic determination of barbituates after extractive methylation in carbon disulfide, *Anal. Chem.*, *46*, 922-924 (1974).

326. L. J. Fischer, and J. J. Ambre, Possible interference by a metabolite in gas chromatographic assays for glutethimide which employ certain nonselective liquid phases, *J. Chromatogr.*, *87*, 379-386 (1973).

327. J. N. Gilbert, W. L. Hetherington, J. W. Powell, and W. B. Whalley, Metabolism of quinalbarbitone, *J. Pharm. Pharmacol.*, *27*, 343-347 (1975).

328. J. N. Gilbert, B. J. Millard, J. W. Powell, W. B. Whalley, and B. J. Wilkins, Metabolism of nealbarbitone, *J. Pharm. Pharmacol.*, *26*, 119-122 (1974).

329. J. N. Gilbert, and J. W. Powell, Mass spectrometric determination of butobarbitone and its metabolites in man, *Biomed. Mass Spectrom.*, *1*, 142-144 (1974).

330. J. N. Gilbert, B. J. Millard, J. W. Powell, and W. B. Whalley, Metabolism of heptabarbitone, *J. Pharm. Pharmacol.*, *26*, 123-125 (1974).

331. J. N. Gilbert, J. W. Powell, and J. Templeton, A study of the human metabolism of *sec*-butobarbitone, *J. Pharm. Pharmacol.*, *27*, 923-927 (1975).

332. M. A. Goldberg, J. Gal, B. J. Hodshon, A. K. Cho, and D. Jenden, A clinical pharmacology of eterobarb: a gas chromatographic-mass spectrometric study, *Trans. Am. Neurol. Assoc.*, *101*, 204-207 (1976).

333. J. Grove, P. A. Toseland, G. H. Draffan, R. A. Clare, and F. M. Williams, Butobarbitone metabolism in man: identification of 3'-ketobutobarbitone, *J. Pharm. Pharmacol.*, *26*, 175-178 (1974).

334. A. R. Hansen, and L. J. Fischer, Gas-chromatographic simultaneous analysis for glutethimide and an active hydroxylated metabolite in tissues, plasma, and urine, *Clin. Chem.*, *20*, 236-242 (1974).

335. J. W. Kleber, L. D. Bechtol, M. K. Brunson, and S. M. Chernish, GLC determination of ethinamate and its hydroxy derivative in biological fluids, *J. Pharm. Sci.*, *66*, 992-994 (1977).

336. J. H. McReynolds, H. d'A. Heck, and M. Anbar, Determination of picomole quantities of methaqualone and 6-hydroxymethaqualone in urine, *Biomed. Mass Spectrom.*, *2*, 299-303 (1975).

337. J. F. Menez, F. Berthou, D. Picart, L. Bardou, and H. H. Floch, Gas chromatographic analysis of dialkylbarbituric acids, *J. Chromatogr.*, *129*, 155-169 (1976).

338. R. G. Moore, A. V. Robertson, M. P. Smyth, J. Thomas, and J. Vine, Metabolism and urinary excretion of chlormethiazole on humans, *Xenobiotics*, *5*, 687-696 (1975).

339. S. Pfeifer, H. Pohlmann, P. Bornschein, and R. Kraft, Analysis and biochemistry of aza analogs of methaqualone. 4. Biotransformation of 2-methyl-3-(4'-chlorophenyl)-4-oxo-3,4-dihydropyrido-2,3-Dpyrimidine, *Pharmazie*, *30*, 290-296 (1975).

340. W. G. Stillwell, Metabolism of glutethimide in the human, *Res. Commun. Chem. Pathol. Pharmacol.*, *12*, 25-41 (1975).

341. W. G. Stillwell, M. Stafford, and M. G. Horning, Metabolism of glutethimide (Doriden) by the epoxide-diol pathway in the rat and guinea pig, *Res. Commun. Chem. Pathol. Pharmacol., 6,* 579-590 (1973).

342. S. R. Sun, and A. H. Chun, Determination of pentobarbital in serum by electron-capture GLC, *J. Pharm. Sci., 66,* 477-480 (1977).

343. B. K. Tang, T. Inaba, and W. Kalow, Amobarbital metabolism in man. Determination of N-hydroxyamobarbital and 3'-hydroxyamobarbital in urine by gas chromatography chemical ionization mass spectrometry, *Biomed. Mass Spectrom., 4,* 73-76 (1977).

344. L. B. Turnbull, C. P. Johnson, III, Y. H. Chen, L. F. Sancilio, and R. B. Bruce, Disposition and metabolism of 4-methyl-2(4-phenylbenzyl)-2-oxazoline-4-methanol in the rat and dog, *J. Med. Chem., 17,* 45-48 (1974).

345. M. Van Boven, G. Janssen, and P. Daenens, Mass spectrometry-gas chromatographic determination of mecloqualone metabolites from urine extracts, *Mikrochim. Acta, 4,* 603-610 (1974).

346. M. Vore, B. J. Sweetman, and J. T. Bush, The metabolism of 1-*n*-butyl-5,5-diethylbarbituric acid (N-*n*-butylbarbital) in the rat, *Pharmacol. Exp. Ther., 190,* 384-394 (1974).

Stable Isotope Methods, Isotope Effects, Isotope Problems

347. E. Anggard, Proceedings: Use of stable isotopes and GC-MS to measure kinetics in the steady-state, *Acta Pharm. Suec., 11,* 638-639 (1975).

348. M. I. Blake, H. L. Crespi, and J. J. Katz, Studies with deuterated drugs, *J. Pharm. Sci., 64,* 367-391 (1975).

349. T. Cronholm, Isotope effects and hydrogen transfer during simultaneous metabolism of ethanol and cyclohexanone in rats, *Biochem., 43,* 189-196 (1974).

350. E. Dagne, L. Gruenke, and N. Castagnoli, Jr., Deuterium isotope effects in the *in vivo* metabolism of cotinine, *J. Med. Chem., 17,* 1330-1333 (1974).

351. G. A. Digenis, Stable and short-lived isotopes in the study of tissue distribution, *J. Toxicol. Environ. Health, 2,* 757-785 (1977).

352. G. C. Ford, S. J. Grigson, and N. J. Haskins, The replacement of trideuteromethyl by methyl groups when using methane as a reagent gas in chemical ionization mass spectrometry, *Biomed. Mass Spectrom., 3,* 230-231 (1976).

353. A. Frigerio, G. Lambelon, G. Jacques, M. J. Simon, and R. Roncucci, Stable isotopes in the biopharmaceutical research, *Bull. Chim. Farm., 114,* 684-699 (1976).

354. W. A. Garland, S. D. Nelson, and H. A. Sasami, Primary and β secondary deuterium isotope effects in the O-de-ethylation of phenacetin, *Biochem. Biophys. Res. Commun.*, *72*, 539-545 (1976).

355. C. T. Gregg, Some application of stable isotopes in clinical pharmacology, *Eur. J. Clin. Pharmacol.*, *7*, 315-319 (1974).

356. M. G. Horning, W. G. Stillwell, J. Nowlin, K. Lertratanangkoon, D. Carroll, I. Dzidic, R. N. Stillwell, and E. C. Horning, The use of stable isotopes in gas chromatography-mass spectrometric studies of drug metabolism, *J. Chromatogr.*, *91*, 413-423 (1974).

357. M. Iio, Clinical application of stable isotope and its problems (author's transl.), *Jpn. J. Clin. Pathol.*, *25*, 296-300 (1977).

358. P. D. Klein, J. R. Haumann, and D. L. Hachey, Stable isotope and ratio-meter-multiple ion detector unit for quantitative and qualitative stable isotope studies by gas chromatography-mass spectrometry, *Clin. Chem.*, *21*, 1253-1257 (1975).

359. M. G. Lee, and B. J. Millard, A comparison of unlabelled and labelled internal standards for quantification by single and multiple ion monitoring, *Biomed. Mass Spec.*, *2*, 78-81 (1975).

360. S. D. Nelson, L. R. Pohl, and W. F. Trager, Primary and β secondary deuterium isotope effects in N-deethylation reactions, *J. Med. Chem.*, *18*, 1062-1065 (1975).

361. J. F. Pickup, and K. McPherson, Theoretical considerations in stable isotope dilution mass spectrometry for organic analysis, *Anal. Chem.*, *48*, 1885-1890 (1976).

362. D. A. Schoeller, A review of the statistical considerations involved in the treatment of isotope dilution calibration data, *Biomed. Mass Spectrom.*, *3*, 265-271 (1976).

363. G. E. Von Unruh, D. J. Hauber, D. A. Schoeller, and J. M. Hayes, Limits of detection of carbon-13 labelled drugs and their metabolites in human urine, *Biomed. Mass Spectrom.*, *1*, 345-349 (1974).

Steroids

364. U. Abshagen, and H. Rennekamp, Isolation and identification of a sulfur-containing metabolite of spironolactone from human urine, *Steroids*, *28*, 467-480 (1976).

365. S. Baba, K. Kawai, and Y. Shida, Studies on drug metabolism by use of isotopes. XII. Quantitative analysis of 22H-labelled compounds on benzene ring by mass spectrometry (author's transl.), *J. Pharm. Soc. Jpn.*, *94*, 826-832 (1974).

366. H. L. Bradlow, B. Zumoff, C. Monder, H. J. Lee, and L. Hellman, Isolation and identification of four new carboxylic acid metabolites of cortisol in man, *J. Clin. Endocrinol. Metab.*, *37*, 811-818 (1973).

367. W. E. Braselton, T. J. Lin, T. M. Mills, J. O. Ellegood, and V. B. Mahesh, Identification and measurement by gas chromatography-mass spectrometry of norethindrone and metabolites in human urine and blood, *J. Steroid Biochem.*, *8*, 9-18 (1977).

368. H. Breuer, D. E. Kime, and R. Knuppen, Metabolism of 6-chloro-9β, 10α-pregna-1,4,6-triene-3,20-dione in rat, rabbit, monkey, and man, *Acta Endocrinol* (Kbh), *74*, 127-143 (1973).

369. P. Bournot, B. F. Maume, and P. Padieu, Sex-linked specificity of the hepatic metabolism of steroids in rats. Mass fragmentography as a method for the assay of hydrogenated metabolites of corticosterone in the liver, *Biomed. Mass Spectrom.*, *1*, 29-39 (1974).

370. E. M. Chambaz, G. Defaye, and C. Madani, Trimethylsilyl ether-enol-trimethyl silyl ether- a new type of derivative for the gas phase study of hormonal steroids, *Anal. Chem.*, *46*, 1090-1098 (1973).

371. J. R. Chapman, and E. Bailey, Use of chloromethylsilyl ether derivatives for the determination of hydroxylated steroids by gas chromatography-mass spectrometry, *Anal. Chem.*, *45*, 1636-1641 (1973).

372. K. Einarsson, J. A. Gustafsson, and K. Hellström, Effects of clofibrate on the metabolism of progesterone and oestradiol in rat liver microsomal fraction. *Biochem. J.* *136*, 622-628 (1973).

373. K. Einarsson, J. A. Gustafsson, J. Sjövall, and E. Zietz, Dose-dependent effects of ethinyloestradiol, diethylstilbesterol and oestradiol on the metabolism of 4-androstene-3, 17-dione in rat liver microsomes, *Acta Endocrinol. (Kbh)*, *78*, 54-64 (1975).

374. J. M. Fromson, S. Pearson, and S. Bramah, The metabolism of tamoxifen (I.C.I. 46, 474). I. In laboratory animals, *Xenobiotica*, *3*, 693-709 (1973).

375. H. Gleispach, The use of different silylating agents for structure analysis of steroids, *J. Chromatogr.*, *91*, 407-412 (1974).

376. R. W. Handy, K. H. Plamer, M. E. Wall, and C. Piantadosi, The metabolism of antifertility steroids. The *in vitro* metabolism of chlormadinone acetate, *Drug Metab. Dispos.*, *2*, 214-220 (1974).

377. S. Honma, and T. Nambara, Isolation and characterization of biliary metabolites of estrone in the rat, *Chem. Pharm. Bull. (Tokyo)*, *22*, 387-695 (1974).

378. J. W. Honour, and C. H. Shackleton, Mass spectrometric analysis for tetrahydroaldosterone, *J. Steroid Biochem.*, *8*, 299-305 (1977).

379. I. Huhtaniemi, Metabolism of pregnenolone and pregnenolone sulphate in human foetal liver tissue *in vitro*, *Acta Endocrinol. (Kbh)*, *76*, 525-538 (1974).

380. H. J. Lee, Acidic metabolite of prednisolone, *Experientia*, *33*, 253-254 (1977).

381. B. F. Maume, P. Bournot, J. C. Lhugnenot, C. Baron, F. Barbier, G. Maume, M. Prost, and P. Padieu, Mass fragmentographic analysis of steroids, catecholamines, and amino acids in biological materials, *Anal. Chem., 45,* 1073-1082 (1973).

382. T. Nambara, J. Ishiguro, Y. Kawarada, and H. Tajima, Isolation and characterization of biliary metabolites of estriol in the rat, *Chem. Pharm. Bull. (Tokyo), 22,* 889-893 (1974).

383. J. C. Plasse, and B. P. Lisboa, Studies on the metabolism of steroids in the foetus. Metabolism of testosterone in the human foetal adrenals, *Eur. J. Biochem., 39,* 449-454 (1973).

384. M. Prost, P. Bournot, and B. F. Maume, Adrenal and liver metabolites of 18-hydroxy-11-deoxycorticosterone in the rat. Identification of reduced compounds (18-OH-TH-DOC), *Steroids, 25,* 177-188 (1975).

385. A. Roukonen, and R. Vihko, Steroid metabolism in human and boar testis tissue. Steroid concentrations and the position of the sulfate group in steroid sulfates, *Steroids, 23,* 1-16 (1974).

386. C. H. Shackleton, Progesterone and oestrogen metabolism in the pregnant marmoset (Callithrix jacchus), *J. Steroid Biochem., 5,* 597-600 (1974).

387. C. H. Shackleton, and J. W. Honour, Identification and measurement of 18-hydroxycorticosterone metabolites by gas chromatography-mass spectrometry, *J. Steroid Biochem., 8,* 199-203 (1977).

388. L. Siekmann, S. Martin, and H. Breuer, Gas-liquid chromatography and mass spectrometry of derivatives of aldosterone and its metabolites, *Z. Klin. Chem. Klin. Biochem., 12,* 260 (1974).

389. M. C. Williams, E. D. Helton, and J. W. Goldzieher, The urinary metabolites of 17α-ethynylestradiol-9α-, 11xi-3H in women. Chromatographic profiling and identification of ethynyl and non-ethynyl compounds, *Steroids, 25,* 229-246 (1975).

Sympathomimetic Amines, Amphetamines, Catecholamines

390. A. H. Beckett, R. T. Coutts, and F. A. Ogunbona, Metabolism of amphetamines. Identification of N-oxygenated products by gas chromatography and mass spectrometry, *J. Pharm. Pharmacol., 25,* 708-717 (1973).

391. A. H. Beckett, R. T. Coutts, and F. A. Ogunbona, The structure of nitrones derived from amphetamines, *J. Pharm. Pharmacol., 26,* 312-316 (1974).

392. A. H. Beckett, and K. K. Midha, The identification of four metabolic products after incubation of p-methoxyamphetamine with liver preparations of various species, *Xenobiotica, 4,* 297-311 (1974).

393. L. Bertilsson, and L. Palmer, Determination of isomeric acid dopamine metabolites in human cerebrospinal fluid by mass fragmentography, *Life Sci., 13,* 859-866 (1973).

394. R. L. Bronaugh, S. E. Hattox, M. M. Hoehn, R. L. Murphy, and C. O. Rutledge, The separation and identification of dopamine 3-O-sulfate and dopamine-4-O-sulfate in urine of Parkinsonian patients, *J. Pharmacol. Exp. Ther., 195,* 441-452 (1975).

395. F. Cattabeni, G. Racagni, and R. Paoletti, A mass fragmentographic study of amphetamine and metabolites in rat brain, *J. Psychiatr. Res., 11,* 45-51 (1974).

396. R. T. Coutts, G. W. Dawson, C. W. Kazakoff, and J. Y. Wong, *In vivo* phenolic metabolites of N-alkyl-amphetamines in the rat. Evidence in favor of catechol formation, *Drug Metab. Dispos., 4,* 256-261 (1976).

397. R. T. Coutts, and S. H. Kovach, Metabolism *in vitro* of N-methylamphetamine with rat liver homogenates, *Biochem. Pharmacol., 26,* 1043-1049 (1977).

398. J. K. Faulkner, and K. J. Smith, Characteristics of the mass spectra of some aliphatic amine N-oxides, *J. Pharm. Pharmacol., 26,* 473-474 (1974).

399. J. Gal, B. J. Hodshon, C. Pintauro, B. L. Flamm, and A. K. Cho, Pharmacokinetics of methylphenidate in the rat using single-ion monitoring GLC-mass spectrometry, *J. Pharm. Sci., 66,* 866-869 (1977).

400. M. T. Gilbert, and C. J. W. Brooks, Characterization of diastereomeric and enantiomeric ephedrines by gas chromatography combined with electron impact mass spectrometry and isobutane chemical ionization mass spectrometry, *Biomed. Mass Spec., 4,* 226-231 (1977).

401. J. A. Jonsson, Hydroxylation of amphetamine to parahydroxyamphetamine by rat liver microsomes, *Biochem. Pharmacol., 23,* 3191-3197 (1974).

402. B. Lindeke, A. K. Cho, T. L. Thomas, and L. Michelson, Microsomal N-hydroxylation of phenylalkylamines. Identification of N-hydroxylated phenylalkylamines as their trimethylsilyl derivatives by GC-MS, *Acta Pharm. Suec., 10,* 493-506 (1973).

403. S. B. Martin, S. H. Wan, and J. B. Knight, Quantitative determination of enantiomeric compounds. I--Simultaneous measurement of the optical isomers of amphetamine in human plasma and saliva using chemical ionization mass spectrometry, *Biomed. Mass Spectrom., 4,* 118-121 (1977).

404. K. K. Midha, Identification of two *in vitro* metabolites of 3,4-methylenedioxyamphetamine by gas-liquid chromatography-mass spectrometry, *J. Chromatogr., 101,* 210-214 (1974).

405. R. M. Milberg, K. L. Rinehart, Jr., R. L. Sprague, and E. K. Sleator, A reproductive gas chromatographic mass spectrometric assay for low levels of methylphenidate and ritalinic acid in blood and urine, *Biomed. Mass Spectrom.*, *2*, 15-22 (1975).

406. H. Miyazaki, Y. Hashimoto, M. Iwanaga, and T. Kubodera, Analysis of biogenic amines and their metabolites by gas chromatography-chemical ionization mass spectrometry, *J. Chromatogr.*, *99*, 575-586 (1974).

407. F. A. Muskiet, D. C. Fremouw-Ottevangers, B. G. Wolthers, and J. A. Vries, Gas-chromatographic profiling of urinary acidic and alcoholic catecholamine metabolites, *Clin. Chem.*, *23*, 863-867 (1977).

408. C. J. Parli, and B. Schmidt, Metabolism of 4-chloroamphetamine to 3-chloro-4-hydroxyamphetamine in rat: Evidence for an *in vivo* NIH shift of chlorine, *Res. Commun. Chem. Pathol. Pharmacol.*, *10*, 601-604 (1975).

409. K. H. Poole, K. L. Rominger, and D. Arndts, Mass spectral analysis of glucuronides from sympathomimetic hydroxyphenylalkylaminoethanols, *J. Pharm. Sci.*, *65*, 1513-1518 (1976).

410. A. D. Sherman, and E. M. Gál, Mass-spectrographic evidence of the conversion of p-chloroamphetamine to 3,4-dimethoxyamphetamine, *Psychopharmacol. Commun.*, *2*, 421-427 (1976).

411. B. Sjoquist, and E. Anggard, Analysis of catecholamine metabolites by gas chromatography-mass spectrometry, *Lakartioningen.*, *73*, 630-637 (1976).

412. J. Wilner, H. F. LeFever, and E. Costa, Assay by multiple ion detection of phenylethylamine and phenylethanolamine in rat brain, *J. Neurochem.*, *23*, 857-859 (1974).

Techniques--Chemical, Instrumental

413. R. P. Barron, W. R. Benson, and J. A. Roach, Mass spectral coding technique for identification of drugs and related compounds in reference compendia, *J. Assoc. Off. Anal. Chem.*, *60*, 635-641 (1977).

414. J. E. Biller, and K. Biemann, Reconstructed mass spectra, a novel approach for the utilization of gas chromatograph-mass spectrometer data, *Anal. Letters*, *7*, 515 (1974).

415. K. Blau, and G. S. King (Eds.), *Handbook of Deviations for Chromatography*, Heydon & Sons, London, 1977.

416. E. Costa, Multiple ion detection: a new tool for neurochemistry and biological psychiatry, *J. Psychiatr. Res.*, *11*, 295-300 (1974).

417. H. Ehrsson, T. Walle, and S. Wikström, Gas chromatography of ether glucuronides as methyl-trifluoroacetyl derivatives, *J. Chromatogr.*, *101*, 206-209 (1974).

418. A. E. Gordon, and A. Frigerio, Mass fragmentography as an application of gas-liquid chromatography-mass spectrometry in biological research, *J. Chromatogr.*, *73*, 401 (1972).

419. L. D. Gruenke, J. C. Craig, and D. M. Bier, Multiple ion detection by accelerating voltage alternation in conjunction with voltage sweeping, *Biomed. Mass Spectrom.*, *1*, 418-422 (1974).

420. C. G. Hammar, B. Holmstedt, and R. Rjhage, Mass fragmentography, *Anal. Biochem.*, *25*, 532 (1968).

421. E. C. Horning, M. G. Horning, D. I. Carroll, I. Dzidic, and R. N. Stillwell, A new program detection system based on a mass spectrometer with an external ionization source at atmospheric pressure, *Anal. Chem.*, *45*, 936-943 (1973).

422. E. C. Horning, and M. G. Horning, Metabolic profiles: gas-phase methods for analysis of metabolites, *Clin. Chem.*, *17*, 802-809 (1971).

423. M. G. Horning, P. Gregory, J. Nowlin, M. Stafford, K. Lertratanangkoon, C. Butler, W. G. Stillwell, and R. M. Hill, Isolation of drugs and drug metabolites from biological fluids by use of salt-solvent pairs, *Clin. Chem.*, *20*, 282-287 (1974).

424. D. F. Hunt, G. C. Stafford, Jr., F. W. Crow, and J. W. Russell, Pulsed positive negative ion chemical ionization mass spectrometry, *Anal. Chem.*, *48*, 2098-2104 (1976).

425. M. I. Kelsey, A. Keskingen, and E. A. Moscatelli, Gas-liquid chromatographic analysis of fluphenazine and fluphenazine sulfoxide in the urine of chronic schizophrenic patients, *J. Chromatogr.*, *75*, 294-297 (1973).

426. B. A. Knights, M. Legendre, J. L. Laseter, and J. S. Storer, Use of high resolution open tubular glass capillary columns to separate acidic metabolites in urine, *Clin. Chem.*, *21*, 888-891 (1975).

427. J. Meola, and M. Vanko, A simple system for drug analysis, *Clin. Chem.*, *17*, 556 (1971).

428. N. Narasimhachari, The use of isomeric compounds as internal standards for quantitative mass fragmentography of biological samples. Determination of homovanillic acid in cerebrospinal fluid and plasma, *Biochem. Biophys. Res. Commun.*, *56*, 36-46 (1974).

429. H. Nau, and K. Biemann, Computer-assisted assignment of retention indices in gas chromatography-mass spectrometry and its application to mixtures of biological origin, *Anal. Chem.*, *46*, 426-434 (1974).

430. J. Nowlin, K. Lertratanangkoon, P. Kellaway, and T. E. Zion, Use of saliva in therapeutic drug monitoring, *Clin. Chem.*, *23*, 157-164 (1977).

431. K. Ohya, and M. Sano, Analysis of drugs by pyrolysis, *Biomed. Mass Spectrom.*, *4*, 241-247 (1972).

432. R. Reimendal, and J. Sjövall, Computer evaluation of gas chromatographic-mass spectrometric analyses of steroids from biological materials, *Anal. Chem.*, *45*, 1083-1089 (1973).

433. D. Sohn, J. Simon, M. Hanna, and G. Ghali, Preparatory procedures in screening for drugs of abuse: "clean-up" methods in current use, *J. Chromatogr. Sci.*, *10*, 294-296 (1972).

434. R. M. Thompson, N. Gerber, R. A. Seibert, and D. M. Desiderio, A rapid method for the mass spectrometric identification of glucuronides and other polar drug metabolites in permethylated rat bile, *Drug Metab. Dispos.*, *1*, 489-505 (1973).

435. N. Weissman, M. L. Lowe, J. M. Beattie, and J. A. Demetriou, Screening method for detection of drugs of abuse in human urine, *Clin. Chem.*, *17*, 875-881 (1971).

Tranquilizers

436. G. Alfredsson, B. Wode-Helgodt, and G. Sedvall, A mass fragmentographic method for the determination of chlorpromazine and two of its active metabolites in human plasma and CSF, *Psychopharmacology*, *48*, 123-131 (1976).

437. L. Baiocchi, A. Frigerio, M. Giannangeli, and G. Palazzo, Basic metabolites of trazodone in humans, *Arzneim Forsch*, *24*, 1699-1706 (1974).

438. U. Breyer, A. Prox, R. Bertele, and H. J. Gaertner, Tissue metabolites of trifluoperazine, fluphenazine, prochlorperazine, and perphenazine in the rat: identification and synthesis, *J. Pharm. Sci.*, *63*, 1842-1848 (1974).

439. U. Breyer, and H. J. Gaertner, Tissue accumulation of metabolites during chronic administration of piperazine-substituted phenothiazine drugs, *Adv. Biochem. Psychopharmacol.*, *9*, 167-173 (1974).

440. U. Breyer, H. J. Gaertner, and A. Prox, Formation of identical metabolites from piperazine- and dimethylamino-substituted phenothiazine drugs in man, rat and dog, *Biochem. Pharmacol.*, *23*, 313-322 (1974).

441. J. C. Craig, W. A. Garland, L. D. Gruenke, L. R. Kray, and K. A. Walker, The use of combined gas chromatography-mass spectrometry techniques for the identification of hydroxylated and dihydroxylated metabolites of phenothiazine drugs, *Adv. Biochem. Psychopharmacol.*, *9*, 405-411 (1974).

442. J. Dreyfuss, J. J. Ross, Jr., J. M. Shekosky, and E. C. Schreiber, Metabolism in dogs of the chloro- and trifluoromethyl-analogues of a piperazine-substituted dihydrobenzoxazepine, *Xenobiotica*, *1*, 29-41 (1971).

443. A. M. Duffield, The use of mass spectrometry for the identification of metabolites of phenothiazines, *Adv. Biochem. Psychopharmacol., 9,* 111-117 (1974).

444. G. Eckhardt, S. Goenechea, and W. Gielsdorf, Isolation and identification of some metabolites of mephenoxalone (Control-OM) from human urine (author's transl.), *J. Clin. Chem. Clin. Biochem., 15,* 115-120 (1977).

445. E. E. Essien, D. A. Cowan, and A. H. Beckett, Metabolism of phenothiazines: Identification of N-oxygenated products by gas chromatography and mass spectrometry, *J. Pharm. Pharmacol., 27,* 334-342 (1975).

446. G. A. Fusaro, D. A. Garteiz, K. Y. Chan, and G. A. Leeson, Identification of basic metabolites of 4-[4-(p-chlorobenzoyl) piperidino]-4'-fluorobutyrophenone, an experimental neuroleptic agent, *Drug Metab. Dispos., 4,* 562-567 (1976).

447. L. D. Gruenke, and J. C. Craig, Identification of a metabolite of thioridazine and mesoridazine from human plasma, *Res. Commun. Chem. Pathol. Pharmacol., 10,* 221-225 (1975).

448. D. C. Hobbs, W. M. Welch, M. J. Short, W. A. Moody, and C. D. Vander Velde, Pharmacokinetics of thiothiazene in man, *Clin. Pharmacol. Ther., 16,* 473-478 (1974).

449. D. Karashima, A. Shigematsu, and T. Furukawa, Esterification of trifluoroacetic acid with phenyldiazomethane for quantitative gas chromatographic analysis. Methods involving separation from biological materials, *J. Chromatogr., 130,* 77-86 (1977).

450. N. Narasimhachari, and R. L. Lin, A possible mechanism for the antischizophrenic action of chlorpromazine: inhibition of the formation of dimethyltryptamine by chlorpromazine metabolites, *Res. Commun. Chem. Pathol. Pharmacol., 8,* 341-351 (1974).

451. M. A. Sprites, Two types of metabolically produced triflupperazine N-oxides, *Adv. Biochem. Psychopharmacol., 9,* 399-404 (1974).

452. C. Yamato, T. Takahashi, and T. Fujita, Studies on metabolism of trazodone. I. Metabolic fate of (14C) trazodone hydrochloride in rats, *Xenobiotica, 4,* 313-326 (1974).

Vasodilators

453. S. Baba, and K. Kawai, Studies on drug metabolism by use of isotopes. XI. Studies on synthesis and physiochemical features of 2H-labeled l-ephedrine on benzene ring (author's transl.), *J. Pharm. Soc. Jpn., 94,* 783-787 (1974).

454. F. M. Belpaire, M. G. Bogaert, M. T. Rosseel, and M. Anteunis, Metabolism of papaverine. 1. Identification of metabolites in rat bile, *Xenobiotica, 5,* 413-420 (1975).

455. B. D. Cameron, L. F. Chasseaud, D. R. Hawkins, and T. Taylor, The metabolic fate of the coronary vasodilator 4-(3,4,5-trimethoxycinnamoyl)-1-(N-pyrrolidino carbonylmethyl) piperazine (cinepazide) in the rat, dog, and man, *Xenobiotica, 6,* 441-455 (1976).

456. S. Higuchi, H. Sasaki, Y. Shiobara, and T. Sado, Absorption, excretion and metabolism of a new dihydropyridine diester cerebral vasodilator in rats and dogs, *Xenobiotica, 7,* 469-479 (1977).

457. G. A. Meeson, D. A. Garteiz, W. C. Knapp, and G. J. Wright, N-methylation, a newly identified pathway in the dog for the metabolism of oxprenolol, a beta-receptor blocking agent, *Drug Metab. Dispos., 1,* 565-568 (1973).

458. W. Neuldermans, W. Lauwers, A. Knaeps, and J. Heykants, *In vitro* metabolism of lidoflazine by rat and dog liver fractions, *Arzneim Forsch, 27,* 828-832 (1977).

459. D. Pitre, Isolation and characterization of metabolites of 7-[2-(3-pyridylmethylamino)ethyl]theophylline nicotinate (Teonicon) in the rat and human urine, *Farmaco (Prat), 29,* 46-60 (1974).

460. W. Reiss, H. Huerzeler, and F. Raschdorf, The metabolites of oxprenolol (Trasicor) in man, *Xenobiotica, 4,* 365-373 (1974).

461. M. T. Rosseel, and F. M. Belpaire, "Identification of Papaverine Metabolites in Human Urine by Mass Spectrometry," in *Quantitative Mass Spectrometry in Life Sciences* (A. P. deLeenheer and R. R. Roncucci, Eds.), Elsevier, Amsterdam, 1977, pp. 133-144.

462. M. Sheehan, R. H. Hertel, and C. T. Kelly, Gas-chromatographic/mass-spectrometric determination of theophylline in whole blood, *Clin. Chem., 23,* 64-68 (1977).

463. B. Spiegelhalder, and M. Eichelbaum, Determination of verapamil in human plasma by mass fragmentography using stable isotope-labelled verapamil as internal standard, *Arzneim Forsch, 27,* 94-97 (1977).

464. F. M. Williams, R. H. Briant, C. T. Dollery, and D. S. Davies, The influence of the route of administration on urinary metabolites of isoetharine, *Xenobiotica, 4,* 345-353 (1974).

Chapter 2

CHEMICAL DERIVATIZATION IN GAS CHROMATOGRAPHY-MASS SPECTROMETRY

PAUL VOUROS

Department of Chemistry and
Institute of Chemical Analysis,
Applications and Forensic Science
Northeastern University
Boston, Massachusetts

I. INTRODUCTION

One of the severest limitations of mass spectrometry (MS) and combined gas chromatography-mass spectrometry (GC-MS) is the requirement for sample vaporization. It is important that the compound under investigation be vaporized into the ion source of the MS or the GC-MS system in its "intact" form without thermal degradation or molecular rearrangement. Alteration of the structural characteristics of the compound will invariably result in erroneous interpretation of both the chromatographic and mass spectrometric data, particularly if little prior information is available as to their origin and consistency. Chemical derivatization to improve volatility thus was recognized early as an essential consideration for much of the analytical work involving mass spectrometry and GC-MS. With the increased use of GC-MS in the biomedical, forensic, and other areas, numerous new and progressively more difficult problems have evolved, and have required the development of new and more sophisticated methodology. While improved instrumentation has often provided the means for solving complex problems, it is more often a chemical background and personal ingenuity that have permitted the analyst to circumvent many of the difficulties encountered in an analytical problem. Chemical derivatization is one of many such examples in the field of analytical chemistry and certainly it has been put to extensive use in MS and GC-MS, as attested by the numerous publications over the past two decades. Some representative examples are discussed in this chapter.

II. GENERAL CONSIDERATIONS

A. Chemical Derivatization in Gas Chromatography

In gas chromatography chemical derivatization is used to provide the following.

1. It affords improved volatility by decreasing the polarity of functional groups. For example, the sodium salt of an amino acid (e.g., glycine, $H_2N\text{-}CH_2\text{-}COONa$) cannot be subjected to gas chromatographic analysis, but its methyl ester can be analyzed quite readily by GC.
2. It provides thermal stability. For example, as seen in Scheme 1, tetrahydrocortisone, 1, decomposes during GC and is eluted as its corresponding 17-ketosteroid, 2, presumably with the indicated concommitant elimination of a hydroxy-aldehyde.

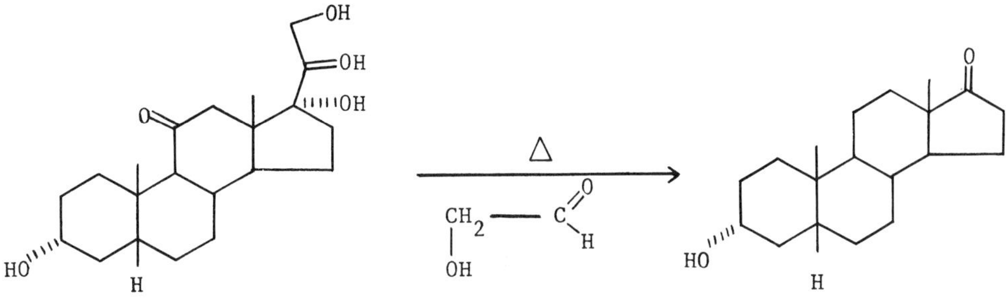

1 2

Scheme 1

On the other hand, its methoxy-trimethylsilyl (MO-TMS) derivative, 3, even though of much higher molecular weight, is eluted without thermal degradation under similar GC conditions [1].

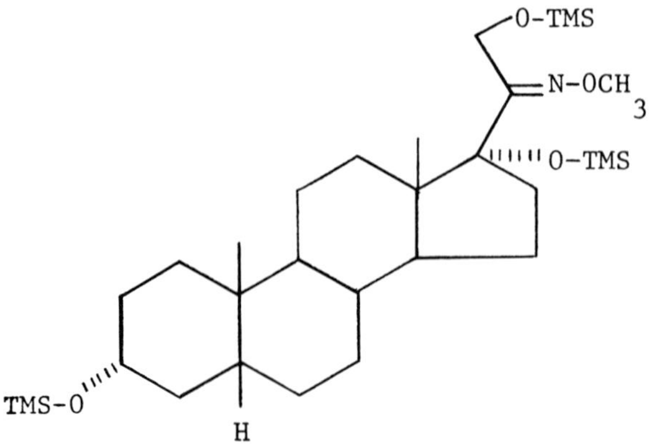

3

3. It enables improved resolution of otherwise unresolved gas chromatographic peaks by selection of the proper type of derivative. For example, the trifluoroacetyl (TFA) derivatives of the 3- and 4-O-methyl ethers of dopamine, 4 and 5, are not separable by GC, whereas their isothiocyanate derivatives are sufficiently resolved to permit their identification by gas chromatography (Table 1) [2]. In effect this is accomplished by altering the properties of the substance rather than resorting to different operating conditions of the gas chromatographic system.

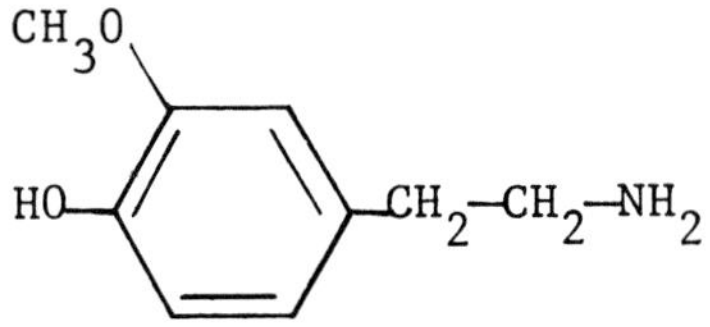

4

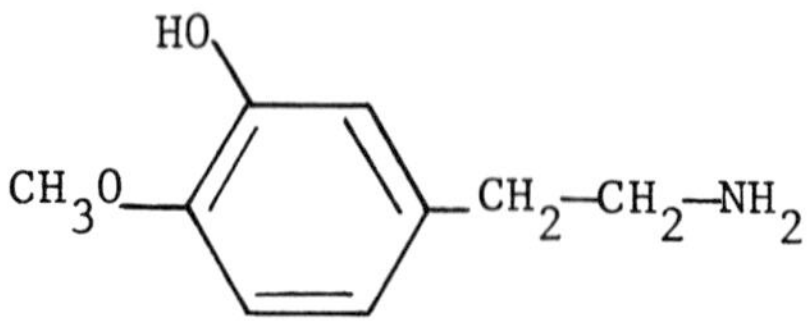

5

In addition to the separation of structural isomers, derivative formation in gas chromatography can be also employed to shift the retention time of an entire group of compounds having a common reactive function. Thus ketosteroids with a sterically unhindered carbonyl group can be converted to their benzyloxime (BO) rather than the methyloxine (MO) derivatives, and the increase in molecular weight leads to a significant increase in retention time; compare Figs. 1 and 2 [3].

TABLE 1

Retention data* for TFA and NCS Derivatives of 4-O-Methyldopamine and 3-Methoxytyramine

	1% OV-101	2.5% OV-225
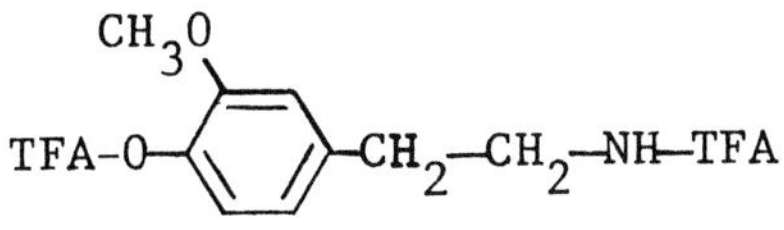	17.73	24.19
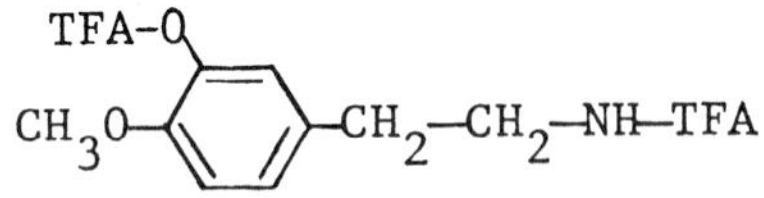	17.73	24.19
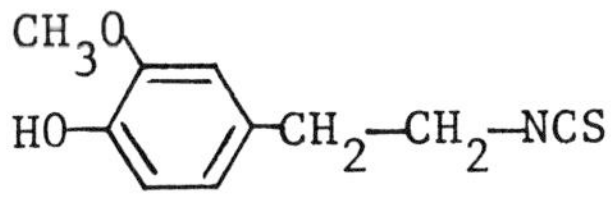	18.27	26.87
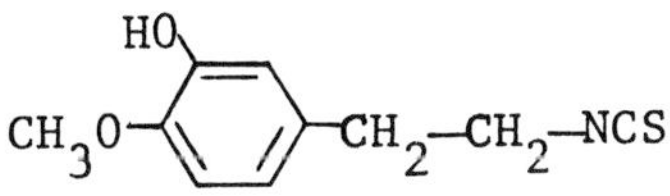	18.44	27.11

*Retention given in methylene units.

B. Chemical Derivatization in Mass Spectrometry

In view of the growing use of combined GC-MS, chemical derivatization should be considered in terms of its influence on both the gas chromatographic and the mass spectrometric properties of the compound. It is fortunate that usually the type of derivative used for gas chromatography is also suitable for mass spectrometric analysis. Nevertheless, specific functionalities can direct selectively the mass spectral fragmentation of an organic molecule and, as a result, the mass spectra of different derivatives of the same compound can vary significantly. This variation is reflected not only in the shift of the molecular ion peaks and a change in spectral patterns, but also in the mode and extent of breakdown of the basic skeleton of the original molecule. Consequently, selection of the derivative for a specific functional group can be made on the basis of the structural information desired, and it is thus well worth knowing in advance its effect

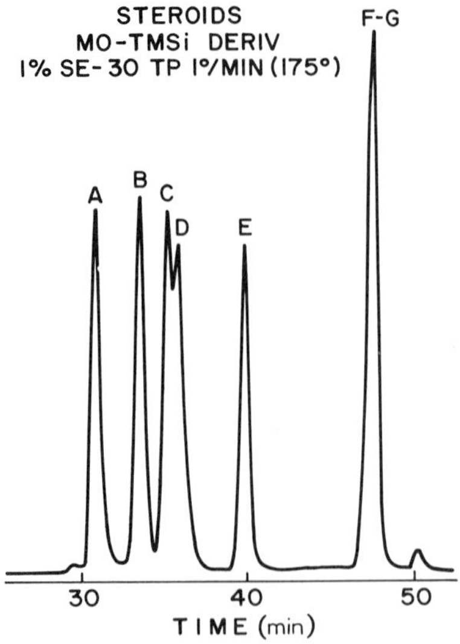

FIG. 1. Separation of steroids as TMS and MO-TMS derivatives on a 12-ft 1% SE 30 column with temperature programming at 1°/min from 175°C: (A) 5β-androstan-3α-ol-17-one; (B) 5-androsten-3β,16α-diol; (C) 5-androsten-3β,17β-diol; (D) 5β-androstan-3α-ol-11, 17-dione; (E) 5β-pregnan-3β-ol-20-one; (F) 5-pregnen-3β,20α-diol; (G) 5-androsten-3β,17β-diol-16-one. (Reprinted with permission from *Analytical Biochemistry*.)

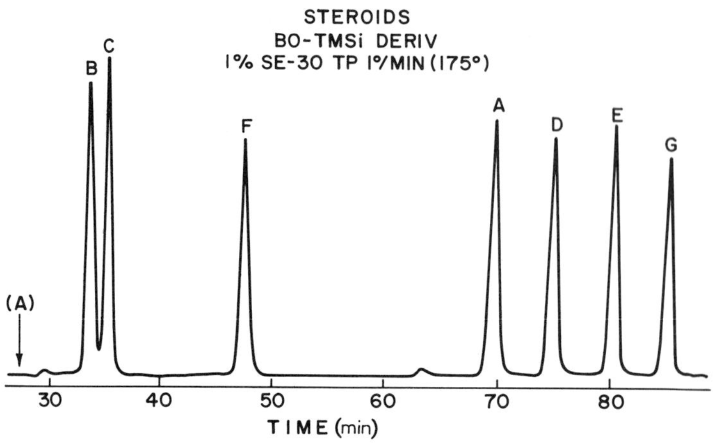

FIG. 2. Separation of same mixture of steroids as in Fig. 1, under same conditions and with same column, but with TMS and BO-TMS derivatives in place of MO-TMS derivatives. Arrow indicates retention time for TMS derivative of 5β-androstan-3α-ol-17-one; this would be present if conversion to BO derivatives were less than quantitative. (Reprinted with permission from *Analytical Biochemistry*.)

on the fragmentation of a compound. It is in this latter context that the subject of chemical derivatization in mass spectrometry and gas chromatography-mass spectrometry will be presented in this chapter. The reader interested in the subject of chemical derivatization strictly from the point of view of gas chromatography is referred to the recent comprehensive and well-documented review by Drozd [4].

From the standpoint of mass spectrometry, derivative formation can provide additional features to supplement the advantages offered by the increased volatility and improved thermal stability. Some of these features may be summarized as follows:

1. Introduction of a group of low ionization potential can result in increased abundance of the molecular ion, thus providing for more definitive determination of the molecular weight of a compound [5]. This has been attributed to an increase in energy spread between the internal energy of the molecular ions formed upon electron impact and the lowest energy of activation for decomposition [6]. The net result is a reduction in the rate of unimolecular decomposition so that more molecular ions will reach the collector of the mass spectrometer.
2. A new functionality may direct the fragmentation of a compound and thus produce intense fragment ions characteristic of the position of substitution. For example, the mass spectrum of 5α-androstan-3-one, 6, contains a multitude of fragment ions of comparable relative abundance [7], while that of its ethylene ketal derivative, 7, is dominated by four major peaks (Table 2).

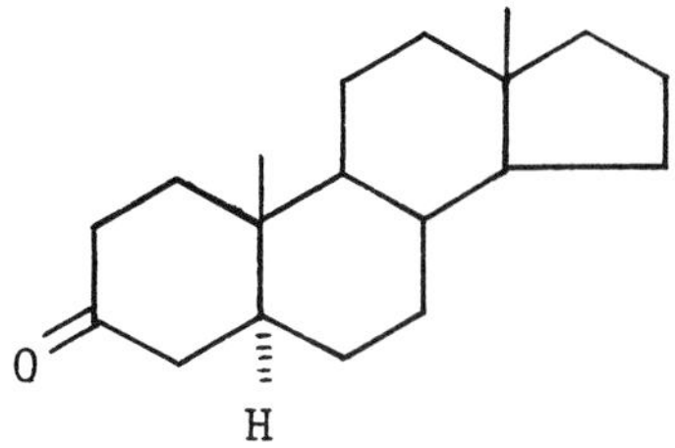

6

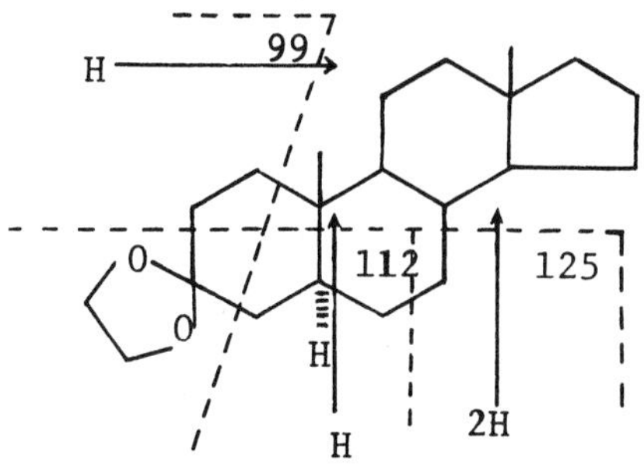

7

The simplicity of the spectrum will obviously draw the investigator's attention to the interpretation of the origin of those ions. Reasonable mechanisms, based on initial charge localization on the ethylene ketal moiety, can be written for the formation of these ions [8], thus identifying the site of the original keto substitution (7).

TABLE 2

Partial Mass Spectrum of 5-Androstan-3-One Ethylene Ketal

m/e	% RI*
318 $[M]^{+\cdot}$	10
125	45
112	13
99	100

*RI = relative intensity.

3. Methyl, ethyl, halogen, methoxy, or other groups can be introduced in a molecule to induce a mass shift of selected fragment ions in a mass spectrum and thus help define the portion(s) of the molecule involved in their formation. It is essential that for this purpose no major change occur in either relative ion abundances or the mode of fragmentation following group substitution. An example of the application of this shift rule is illustrated in the comparison of the mass spectra of the heptafluorobutyrylmethyl ester derivatives of 3,3',5,5'-tetraiodothyroxine (T_4) and 3',5'-dibromo-3,5-diiodothyroxine (Br_2T_2) (Figs. 3 and 4). The spectrum of the T_4 derivative exhibits intense peaks corresponding to $[M-213]^{+\cdot}$ (*m/e* 970) and $[M-284]^{+\cdot}$ (*m/e* 899). To ascertain whether the I_2 loss involved iodines from the same or opposite phenyl rings, the dibromothyroxine derivative (Br_2T_2) was examined (Fig. 4), and the observed losses of IBr (groups of peaks at *m/e* 597 and 668) confirmed the involvement of iodine atoms from different rings [9]. In this particular example, the bromine atoms served the dual role of providing both an easy tracer due to isotopic labeling (^{81}Br and ^{79}Br isotopic ratios) as well as the desired mass shift (*m/e* 645 → *m/e* 597, *m/e* 599; *m/e* 716 → *m/e* 688, *m/e* 670).

4. The distinction of structural or positional isomers may often be possible via the use of a specific derivative or a combination of derivatives. For example, the mass spectra of the 3- and 4-O methyl ethers of dopamine, 4 and 5, are identical and so are those of their isothiocyanate derivatives. On the other hand, trimethylsilylation of the phenolic OH group following formation of the isothiocyanate derivative of the amino function altered the molecular properties sufficiently to permit their differentiation. As shown in Fig. 5, despite many similarities, the mass spectra of the combined NCS-TMS derivatives of the two isomers are sufficiently unique to be distinguished. This is notably reflected in the ratios of *m/e* 193: *m/e* 179 and the distinct presence of $[M-15]^{+\cdot}$ and $[M-30]^{+\cdot}$ peaks in the spectrum of the 4-O-methyl derivative [2].

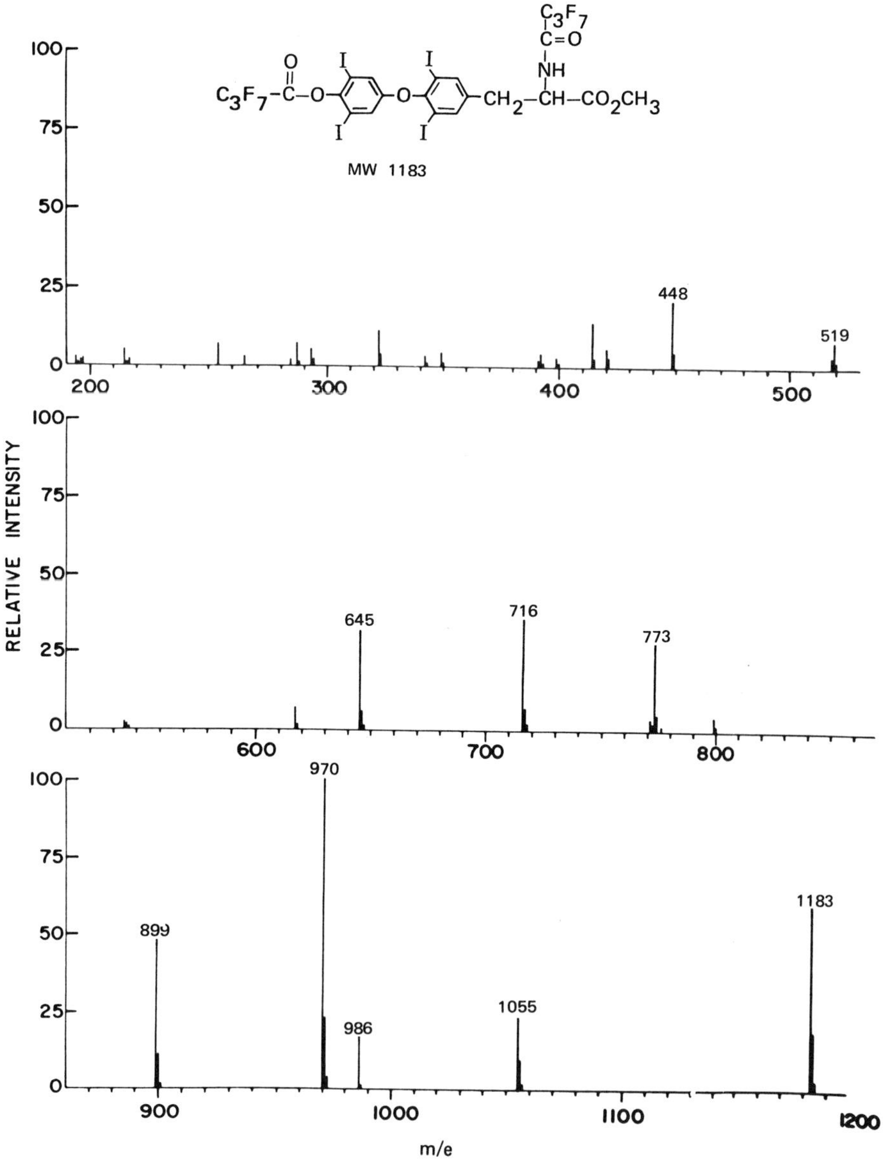

FIG. 3. Mass spectrum of heptafluorobutyryl-methyl ester derivative of 3,3',5,5'-tetraiodothyronine. (Reprinted from *Analytical Chemistry* with permission from the publisher.)

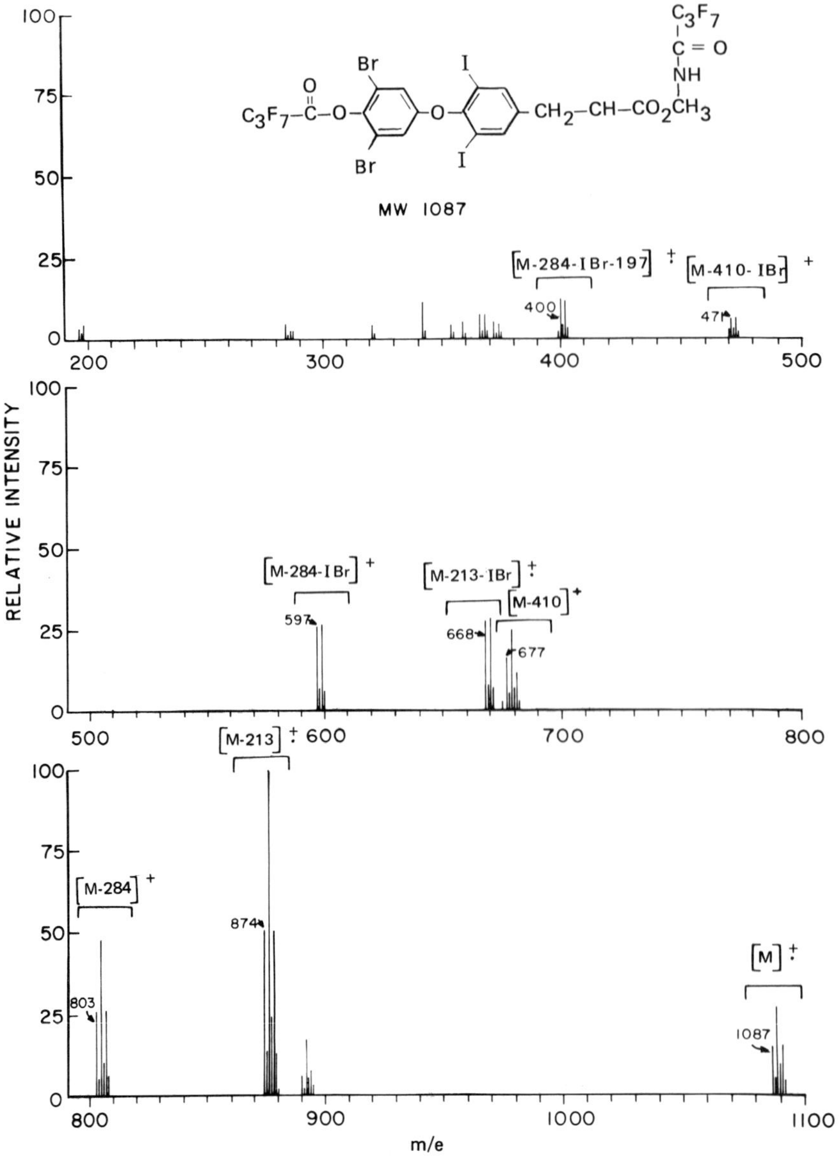

FIG. 4. Mass spectrum of heptafluorobutyryl-methyl ester derivative of 3',5'-dibromo-3,5-diiodothyrozine. (Reprinted from *Analytical Chemistry* with permission from the publisher.)

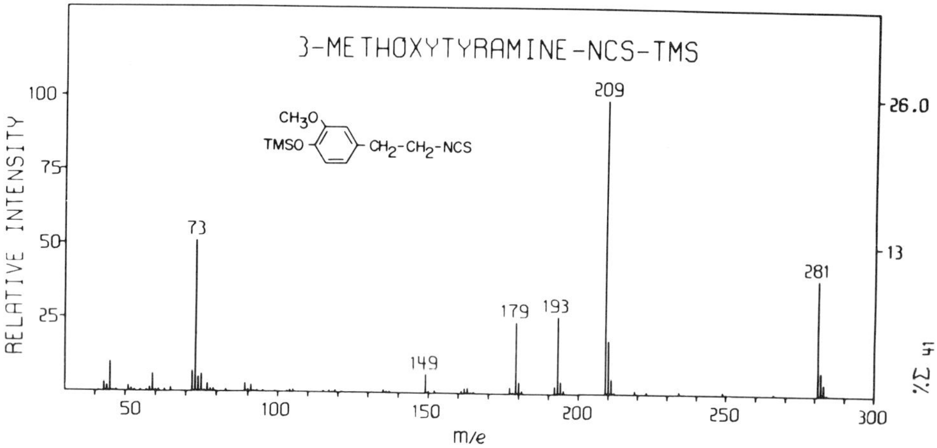

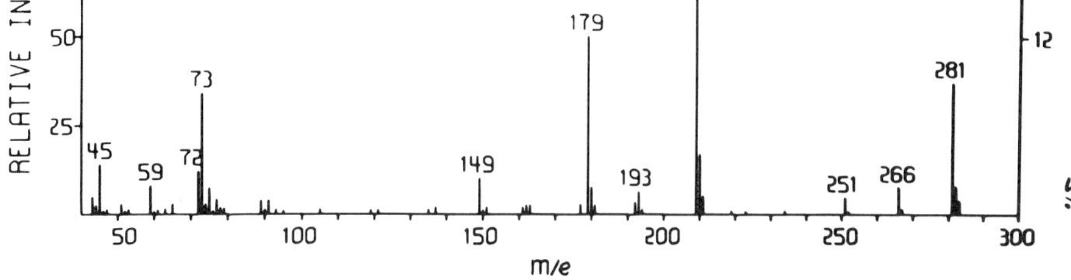

FIG. 5. Mass spectra of isothiocyanate-trimethylsilyl derivatives of 3-methoxytyramine (top) and 3-hydroxy-4-methoxy phenylethylamine (bottom). (Reprinted from the *Journal of Chromatography* with permission from the publisher.)

5. Location of double-bond position in a long-chain hydrocarbon or fatty acid by mass spectrometry is best accomplished following chemical derivatization.

   Whereas olefins usually present no major problem as far as their GC analysis is concerned, identification of the position of a double bond in a long-chain system by mass spectrometry is made extremely difficult by the tendency of the double bond to "migrate" upon ionization. It has been rationalized that this migration is induced by consecutive hydride and hydrogen radical shifts following formation of the radical ion site in the molecular ion [10]. In the absence of authentic samples, it is thus virtually impossible to determine unequivocally the position of a double bond in such a system.

   Examples of different types of derivatives employed for the mass spectrometric determination of double-bond positions in long-chain fatty acid or other systems are discussed below.

6. Isotopic labeling is a form of derivatization that has been most effectively utilized in mass spectrometry. It can be used to help elucidate fragmentation mechanisms, to provide internal standards for GC-MS, or to offer structural information on an unknown compound by, for example, identifying the number and type of exchangeable hydrogens. Typical stable isotopes used in mass spectrometry are ^{2}H, ^{13}C, ^{15}N, and ^{18}O. A large and increasing variety of reagents containing these isotopes is now readily available and, in view of the small sample quantities required for mass spectrometric analysis, the cost of conducting simple chemical derivatization reactions with many of these isotopes may be minimal. As discussed below, isotopic exchange reactions can often be conducted on-column, thus minimizing further the sample requirements.

III. THE PREPARATION OF DERIVATIVES

A. Requirements for Derivatization Reaction

The requirements for a derivatization reaction may vary depending on the conditions or the expectations from a specific analysis. However, it is reasonable to assume that the following criteria will play a major role in the selection of the particular method of derivative formation:

1. It is important that only one product be derived from the derivatization reaction.
2. Completeness of reaction, although not essential when an internal standard is employed, is highly desirable. When an internal standard is used, it is assumed that reaction conditions--e.g., temperature, catalyst, etc.--influence both the unknown and the

standard in an identical fashion. While this may not always be so with GC where structurally different standards are often used, it is certainly the case in GC-MS where isotopically labeled analogs are usually employed as internal standards. As the isotopes are preferably incorporated at positions removed from or not involved in the derivatization reaction, isotopic effects are of little concern.

3. When dealing with trace quantities--e.g., subnanogram--it is important that the quantitativeness of the reaction be established at the physiological or natural level of occurrence of the compound to be determined. Unnecessary losses at that level can influence significantly the accuracy of an analysis.
4. Ideally, the derivatization reaction should be easy to perform, preferably by simple mixing of reagents and solvent in a vial. It is indeed fortuitous that some of the most versatile derivatives (e.g., silyl, perfluoroacyl) are in fact of that variety. This feature certainly enhances their utility for routine analysis.
5. Time may be of the essence depending on the particular circumstances. To that effect the rate of a reaction needs to be taken into account whether it is to attain a quantitative yield or a yield sufficient for reproducible analysis.
6. Stability of the derivative may be a point of concern, particularly if final analysis needs to be delayed. In that respect it should be noted that selection of pure solvents and reagents should always be a consideration.
7. The atomic mass unit (amu) increment introduced by a functional derivative may often have to be considered. While the requirements summarized previously pertain to both gas chromatography and mass spectrometry, this last one is largely a mass spectrometric problem. For a compound of, for example, molecular weight 600, which contains several replaceable hydrogens, it is preferable, if possible, to introduce a functionality providing a 14-amu (methyl) rather than a 96-amu (trifluoroacetyl) increment per reactive site. The latter derivative can easily increase the molecular weight of the compound well beyond the scanning range of many mass spectrometers. It should be noted, however, that while generally a small amu increment is desired, the presence of intense ion peaks at the high mass (e.g., m/e > 700 amu) region of the spectrum can provide significant advantages for trace level detection by SIM, because of the scarcity of background peaks due to column bleed or pump oil at that mass range [9].

B. Typical Reactive Groups and Functional Derivatives

A large variety of derivatization techniques have been developed for the study of organic compounds by GC-MS. Many of these procedures rely on standard classical reactions with various modifications introduced over the years to deal with specific problems. A partial summary of functional groups and reagents is given in Table 3. Additional examples of derivatives as related to more specific needs will be included with the discussion of the various classes of compounds. For details of the most common derivatization reactions, the reader is referred to the monograph by Drozd [4] and references cited therein. It is worth noting here that in recent years the field of chemical derivatization for GC and GC-MS has advanced to such an extent that numerous reagents, even many of the more specialized variety, are now commercially available. In fact, commercial catalogs (such as Pierce or Supelco) contain comprehensive details and bibliography pertaining to the subject. Some representative examples will be presented here.

1. Carboxylic acids

Methyl ester formation is the most common form of derivatization of carboxylic acids.

Reaction with diazomethane--usually released from the decomposition of N-nitroso-N-methyl-p-toluenesulfonamide by reaction with alkali--is perhaps the simplest and easiest of methods, Eq. (1)

$$\text{R-COOH} + CH_2N_2 \xrightarrow{Et_2O} \text{R-COOCH}_3 + N_2 \qquad (1)$$

although care should be exercised to avoid the presence of moisture since diazomethane reacts with water, and also because of its toxicity and explosive nature. In a variation of this reaction, benzyl esters of long-chain fatty acids have been prepared with phenyldiazomethane [11]. The latter derivative has proved particularly useful in the analysis of low-molecular-weight acids. For example, the GC-MS analysis of trifluoroacetic acid in body tissues was performed best by conversion into a derivative of relatively lower volatility by reaction with phenyldiazomethane [11a].

Reaction with methanol in the presence of either BF_3 or acid (HCl or H_2SO_4) catalyst is a common alternative to diazomethane and does not require excessive precautions.

TABLE 3

Summary of Representative Functional Groups and Derivatives

Reactive group(s)	Reagent	Reaction product
R-COOH	CH_2N_2	$R\text{-}COOCH_3$
	CH_3OH/BF_3	$R\text{-}COOCH_3$
	CH_3OH/HCl	$R\text{-}COOCH_3$
	$CH_3OH/SOCl_2$	$R\text{-}COOCH_3$
	$(CH_3)_4N^+OH^-$	$R\text{-}COOCH_3$
	Silylation	$R\text{-}COOSiR_3$
	$(R'O)_2\text{-}CH\text{-}N(CH_3)_2$	R-COOR'
RR'-C=O	$H_2N\text{-}OCH_3 \cdot HCl$	$RR'\text{-}C{=}N\text{-}OCH_3$
	$H_2N\text{-}OCH_2\phi \cdot HCl$	$RR'\text{-}C{=}N\text{-}OCH_2\phi$
	$H_2N\text{-}NR^1R^2$	$RR'\text{-}C{=}N\text{-}NR^1R^2$
R-OH	Silylation	$R\text{-}O\text{-}SiR_3$
	$CH_3CO\text{-}O\text{-}COCH_3$	$R\text{-}O\text{-}COCH_3$
	$F(CF_2)_nCO\text{-}O\text{-}CO(CF_2)_nF$	$R\text{-}O\text{-}CO(CF_2)_nF$
	CH_3I/AgO	$R\text{-}O\text{-}CH_3$
$R\text{-}NH_2$	Silylation	$R\text{-}N\text{-}(SiR_3)_2$
	$F(CF_2)_nCO\text{-}O\text{-}CO(CF_2)_nF$	$R\text{-}NH\text{-}CO(CF_2)_nF$
	CS_2	R-N=C=S
	Dansyl chloride	$R\text{-}NH\text{-}SO_2$– 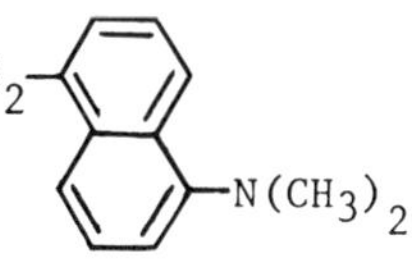
RR'NH	Silylation	$RR'N\text{-}SiR_3$
	$F(CF_2)_nCO\text{-}O\text{-}CO(CF_2)_nF$	$RR'N\text{-}CO\text{-}(CF_2)_nF$

(continued)

TABLE 3 (continued)

Reactive group(s)	Reagent	Reaction product
$R-\overset{O}{\overset{\|}{P}}-(OH)_2$ (Sugar phosphates or phosponates)	Silylation	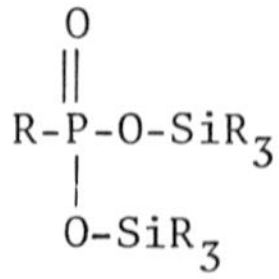
$H_2N-(R)-COOH$ (Amino acids)	$R'OH/H^+$; $F(CF_2)_nCO-O-CO(CF_2)_nF$	$F(CF_2)_n-CO-NH-(R)-COOR'$
	Silylation	$(R_3'Si)_2-N-(R)-COOSiR'_3$ and/or $(R'_3Si-NH-(R)-COOSiR'_3$
	$(R'O)_2-CH-N(CH_3)_2$	$(CH_3)_2N-CH=N-(R)-COOR'$
Hydroxy ketosteroids	$H_2N-OCH_3 \cdot HCl$/silylation	e.g. 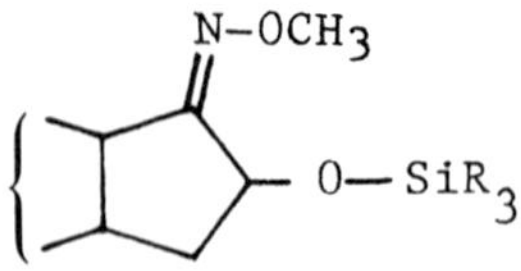
$HX-(R)-OH$ where: X = NH or X = O	$R'-B(OH)_2$	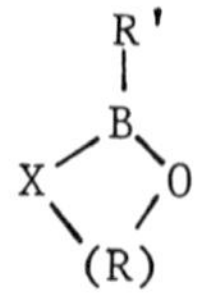
$R-CH=CH-R'$	OsO_4; Trimethylsilylation	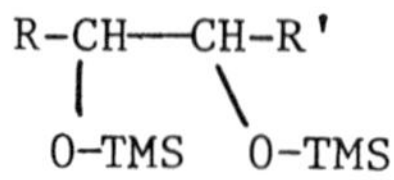
	Epoxidation with a peracid	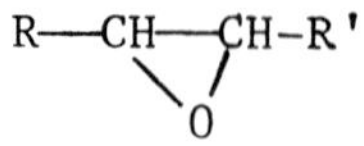
	OsO_4; NaH, DMSO, CH_3I	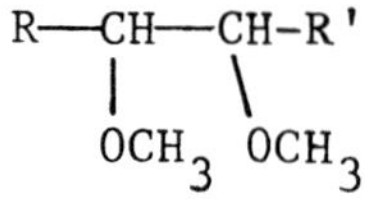
	OsO_4; CH_3COCH_3	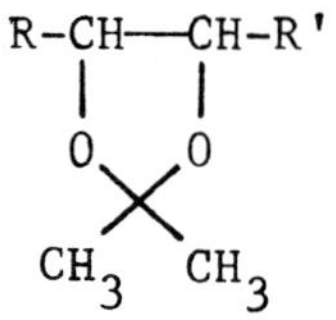

Good results have been reported by Thenot and Horning for the esterification of carboxylic groups using dimethylformamide dialkyl acetals; Eq. (2). The reaction is characterized by its simplicity and rapidity (10 min at 60°), but has not as yet been employed as extensively as the former esterification methods to evaluate its general applicability. The alkyl group R' can be varied as desired to prepare various homologs.

$$\text{R-COOH} + \begin{matrix}\text{R'O}\\ \text{R'O}\end{matrix}\!\!>\!\text{CH-N(CH}_3)_2 \longrightarrow \text{R-COOR'} + \text{HCON(CH}_3)_2 \qquad (2)$$

A convenient on-column esterification reaction involves the decomposition of the tetramethylammonium or trimethylanilinium salts of acids at the injection port of the gas chromatograph; Eq. (3) [12,13].

$$\text{R-COOH} \xrightarrow{(\text{CH}_3)_4\text{N}^+\text{OH}^-} \text{R-COO}^-(\text{CH}_3)_4\text{N}^+ \xrightarrow{\Delta} \text{R-COOCH}_3 \qquad (3)$$

2. *Carbonyl Groups*

The carbonyl group usually presents no serious problem for GC-MS, at least as far as its volatility is concerned. It is often the case, however, that the carbonyl function can introduce chemical or thermal instability in a molecule thus inhibiting its effective derivatization. Such is the case with the E-type prostaglandins, which contain a β-keto ring structure, 8, and readily dehydrate to form the conjugated ketone, 9 [14]. Conversion to a methyl oxime derivative on the other hand stabilizes the system and permits further reaction with other derivatizing reagents [15]. Enolization of a keto group is another problem frequently encountered in the silylation of hydroxy-

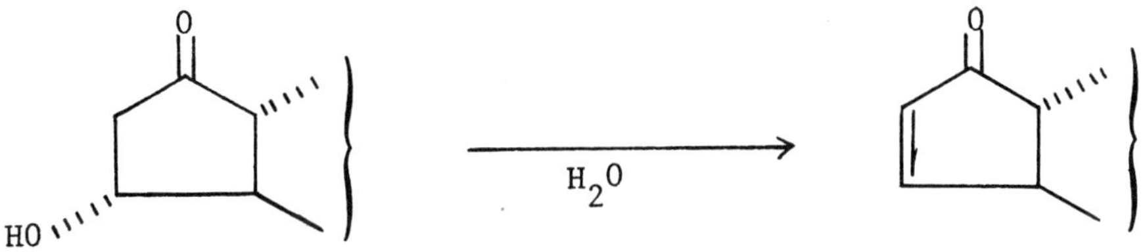

8 9

Scheme 2

keto compounds (e.g., corticosteroids, 11-ketosteroids), and results in the introduction of one or more silyl functionalities, in addition to those replacing the original hydroxylic hydrogens. On occasion this has been taken advantage of, as in corticosteroids where the system was stabilized for GC-MS by silylation under controlled base-catalyzed conditions [16]. Similarly, 3-keto-Δ^4-steroids have been converted to their enolheptafluorobutyryl (HFB) derivatives, which showed good GC properties and satisfactory mass spectral patterns [17,18]. The difficulties that may be encountered in working with 3-keto-Δ^4-steroids are further demonstrated from the work of Devaux and Horning [19]. Reaction of testosterone, 10, with heptafluorobutyric anhydride (HFBA), in an attempt to convert to a derivative suitable for electron-capture detection, yielded two products identified as isomers 11 and 12. It was significant that the ratio of the two isomeric enol forms varied with the choice of solvent and reaction conditions, a further indication of the problems involved with the derivatization of enolizable carbonyl groups.

In view of the above, it is generally preferable to convert the keto group to a stable derivative prior to any further sample treatment. In addition, as pointed out earlier, derivatization of the keto group is often carried out to separate carbonyl compounds from the interfering peaks during GC as demonstrated from the work of Devaux and Horning [3], Vogh [20], and others.

Reaction of carbonyl compounds with N,N-dialkyl hydrazines affords derivatives with good gas chromatographic properties, Eq. (4).

$$\begin{matrix}R \\ R'\end{matrix}\!\!>\!N\!-\!NH_2 + O\!=\!C\!\!<\!\!\begin{matrix}R^1 \\ R^2\end{matrix} \xrightarrow[H_2O]{} \begin{matrix}R \\ R'\end{matrix}\!\!>\!N\!-\!N\!=\!C\!\!<\!\!\begin{matrix}R^1 \\ R^2\end{matrix} \qquad (4)$$

Hydrazones, however, are susceptible to decomposition in the presence of impurities, which somewhat limits their utility, particularly for trace level studies [21].

It is generally accepted that methyl oximes (MO) provide the simplest and most versatile derivatives of carbonyl functions for GC-MS. Their volatility is low, thus not impairing their analysis by GC; furthermore, their mass increment (29 amu) and fragmentation

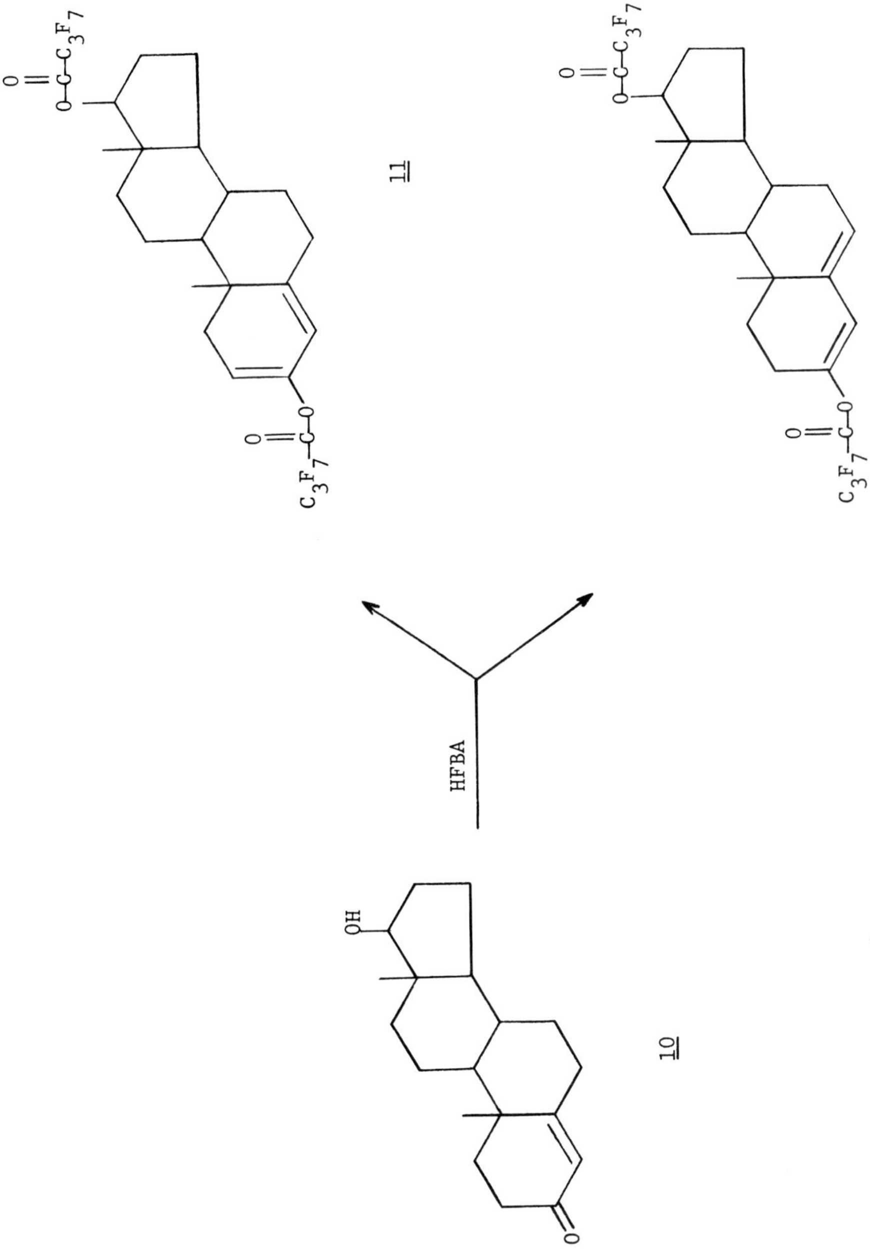

Scheme 3

pattern (typically loss of a methoxy group from $M^{+\cdot}$) do not complicate the mass spectrometric analysis. The derivatization, Eq. (5),

$$R\text{-}O\text{-}NH_2 + O{=}C\begin{smallmatrix}\diagup R'' \\ \diagdown R'\end{smallmatrix} \xrightarrow{\text{Pyridine}} R\text{-}O\text{-}N{=}C\begin{smallmatrix}\diagup R'' \\ \diagdown R'\end{smallmatrix} + H_2O \tag{5}$$

is easy to perform and involves an overnight reaction at room temperature between excess methoxylamine and the carbonyl compound in pyridine solvent [22]. Ordinary oximes (R=H) have also been prepared, but are in principle impractical as they require further derivatization of the hydroxylic function. On the other hand, benzyloximes (BO) have been effectively utilized for GC separations as noted above [3], while O-butyloximes and O-pentyloximes have GC and MS properties intermediate between those of the MO and BO derivatives and have been used in the analysis of keto steroids [23].

In working with oxime derivatives of carbonyl compounds, the analyst should beware of the formation of syn and anti isomers. These are in fact frequently separated chromatographically and the resulting GC doublets may lead to occasional confusion.

3. *Silylation Reactions*

Silyl derivatives are unquestionably the most versatile types employed in GC-MS. Because of their applicability to a large variety of functional groups (e.g., OH, COOH, amino, thiol, etc.) and the numerous silylating reagents that have become commercially available in recent years, they are considered separately in this section. Details on the silylation of organic compounds can be found in a book on this subject [24]. With a few exceptions, silyl derivatives have been primarily employed in the form of their trimethylsilyl (TMS) counterparts [25]. The virtues and versatility of TMS derivatives for GC-MS have been discussed by Brooks and Middleditch [26], and are in part summarized as follows:

1. They are easy to prepare.
2. They can be used to protect many different types of functionalities against thermal decomposition during GC.

3. They have very low polarity, which renders them ideal for GC, thus providing excellent GC peaks.
4. They are sterically sensitive, which may provide different chromatographic properties for stereochemical isomers or diastereomers, and thus afford their resolution during GC.
5. They are not easily adsorbed on surfaces and their high volatility permits quick removal from the ion source of the mass spectrometer.
6. Their mass spectra often exhibit abundant molecular ions, or at least indirect evidence about the molecular weight of a compound can be obtained due to the favorable loss of a TMS methyl group and the resulting intense peaks corresponding to $[M-15]^{+}$.
7. The trimethylsilyl group tends to direct the fragmentation of a compound and structurally diagnostic ions can thus be obtained by examining the spectrum of a TMS derivative rather than that of the parent compound.
8. The availability of deuterium-labeled trimethylsilylating reagents, or other types of silyl derivatives such as dimethylsilyl or chloromethyl(dimethyl)silyl, permits the preparation and parallel examination of other silyl analogs, which assist greatly in the interpretation of the mass spectra of TMS derivatives.
9. In most cases, TMS derivatives can be converted back to their parent compound to conduct a different derivatization or other reaction, if needed.

On the negative side, the following drawbacks of trimethylsilyl derivative formation should be noted.

1. They provide a high mass increment (72 amu per reactive group) to the parent compound.
2. They are easily susceptible to hydrolysis.
3. In the mass spectra of compounds containing more than one TMS group, abundant rearrangement ions are often found, which can complicate the mass spectral interpretation. As noted below, however, this feature can be used to an advantage, as such rearrangements have been shown to be structurally and even stereochemically dependent.

Several reagents for the preparation of TMS derivatives are now available commercially and their structures are given in Table 4. Standard abbreviations for these reagents have become routine in chemical jargon and are indicated in the same table. Additional silylation reagents, some of more specialized use, are also included. Many of these have been used for purposes of functional group labeling

TABLE 4

List of Common Silylating Reagents

Silylating reagent	Structure	Abbreviation
N,O-Bis(trimethylsilyl)-acetamide	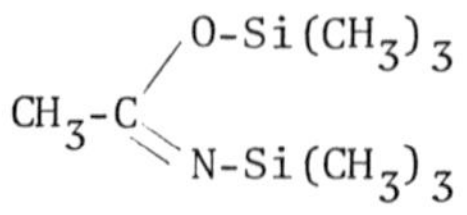	BSA
Hexamethyldisilazane	$(CH_3)_3Si-NH-Si(CH_3)_3$	HMDS
Trimethylchlorosilane	$(CH_3)_3SiCl$	TMCS
Bis(trimethylsilyl)trifluoro-acetamide	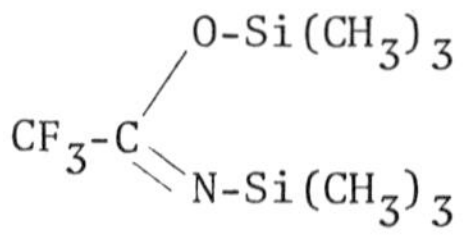	BSTFA
Trimethylsilylimidazole	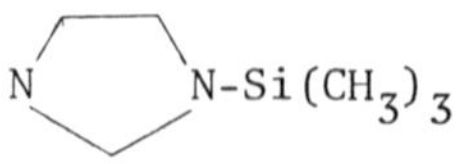	TSIM
N-Trimethylsilyldiethylamine	$(CH_3)_3SiN(C_2H_5)_2$	TMSDEA
N-Trimethylsilyldimethylamine	$(CH_3)_3SiN(CH_3)_2$	TMSDMA
N-Methyl-N-trimethylsilyl-trifluoroacetamide	$CF_3CON(CH_3)[Si(CH_3)_3]$	MSTFA
Dimethylchlorosilane	$H(CH_3)_2SiCl$	DMCS
Tetramethyldisalazane	$H(CH_3)_2Si(CH_3)_2H$	TMDS
Chloromethyldimethylchlorosilane	$(CH_2Cl)(CH_3)_2SiCl$	CMDMCS

to clarify the origin of fragment ions in the mass spectra of TMS derivatives or improve the quality of gas chromatographic data. More recently, silyl derivatives containing combinations of alkyl substituents larger than methyl have also been investigated. Silyl ethers of cannabinols [27], steroids [28], nucleosides [29], and fatty acids [30] have been examined in this fashion and they are noted for their higher resistance to hydrolysis as compared with ordinary TMS ethers. Special mass spectral features of these derivatives are discussed below.

One of the most important features of the various trimethylsilylating reagents is their wide variation in reactivity. Selective silylation reactions can thus be conducted according to one's chromatographic and/or mass spectrometric interests. In addition, volatility differences are quite marked for the different reagents and this can be of significance in trace analysis of materials that elute near the tailing end of the elution peak of the solvent and reagents. The volatility of the reaction by-products also needs to be considered in the selection of a specific silylation process. For example, BSA and BSTFA are of comparable reactivity, but the fluorinated by-products of BSTFA, i.e., monotrimethylsilyltrifluoroacetamide and trifluoroacetamide, are considerably more volatile than their protonated counterparts from BSA and are thus less likely to interfere with GC analysis. On the other hand, TMSDMA apparently reacts faster than TMSDEA because its by-product, dimethylamine $[HN(CH_3)_2]$, is a gas and its rapid elimination drives the reaction to completion faster.

Silylation reactions can be carried out neat or in the presence of solvent. Pyridine, acetonitrile, dimethylformamide, tetrahydrofuran, and dimethylsulfoxide are among the most commonly employed solvents. Pyridine is the least desirable as far as its GC properties and volatility are concerned, but it also acts as a good catalyst in most silylation reactions. In cases where methyloxime formation precedes or accompanies silylation, pyridine is the most reasonable solvent of choice in view of its applicability to both reactions.

4. *Acylation Reactions*

Acylation is a common form of derivatization of hydroxylic, amino (secondary or primary), and thiol groups, usually conducted by reaction with the corresponding acid anhydride. While preparation of acetyl derivatives--particularly of sugars--is common, the higher volatility and good electron-capture properties of perfluoroacyl derivatives generally favor the use of the latter compounds. Perfluoroacyl anhydrides of high purity are now commercially available for the preparation of trifluoroacetyl (TFA), pentafluoropropionyl (PFP), and heptafluorobutyryl (HFB) derivatives. The derivatization reactions are generally marked by the same overall simplicity that characterizes most silylation processes and are carried out in pyridine solvent. PFP derivatives of biogenic amines have been prepared directly from the stable salt of the amine by reaction with perfluoropropionic anhydride in ethyl acetate [31].

Perfluoroacyl derivatives are frequently used for the derivatization of amino acids. Acylation of the amino group is normally preceded by esterification of the carboxylic function to form their n-butyl [32] or other homologous alkyl esters. The same basic approach has also been used for the derivatization of thyroid hormones. Following esterification of the carboxylic function, N,O-dipivalyl [$(CH_3)_3C-C=O$] derivatives were prepared by acylation of both the hydroxylic and the amino function [33]. More recently, it was shown that the HFB-methyl ester derivatives of thyroid hormones possessed excellent gas chromatographic [34] and mass spectrometric [4] properties. Detection limits of 0.2 pg and 0.5 pg were attained using GC-electron-capture detection and GC-MS-single-ion monitoring respectively.

In addition to using the appropriate acid anhydride for acylation, derivatization reactions have also been carried out using the corresponding N-perfluoroacylimidazole. The following N-acylimidazoles are commercially available: acetyl ($CH_3C{=}O$), trifluoroacetyl ($CF_3C{=}O$), pentafluoropropionyl ($CF_3-CF_2-C{=}O$), heptafluorobutyryl ($CF_3CF_2CF_2-C{=}O$), and pentafluorobenzoyl ($C_6F_5-C{=}O$). The N-acylimidazoles react with hydroxylic groups as well as primary and secondary amines. Mixed derivatives have also been prepared using combinations of silylating reagents and N-acylimidazoles. For example, in catecholamines the hydroxylic groups were first reacted with trimethylsilylimidazole (TSIM), followed by derivatization of the amine functions with heptafluorobutyrylimidazole [35]. More recently, selective acylation and silylation reactions were used by Donike for the preparation of

N-TFA/O-TMS derivatives of biogenic amines from their salt forms using N-methyl-N-TMS-trifluoroacetamide (MSTFA) and N-methyl-bis-trifluoroacetamide (MBTFA) [36]. The reaction proceeds as shown in Scheme 4. It was previously shown that if silylation of the amine

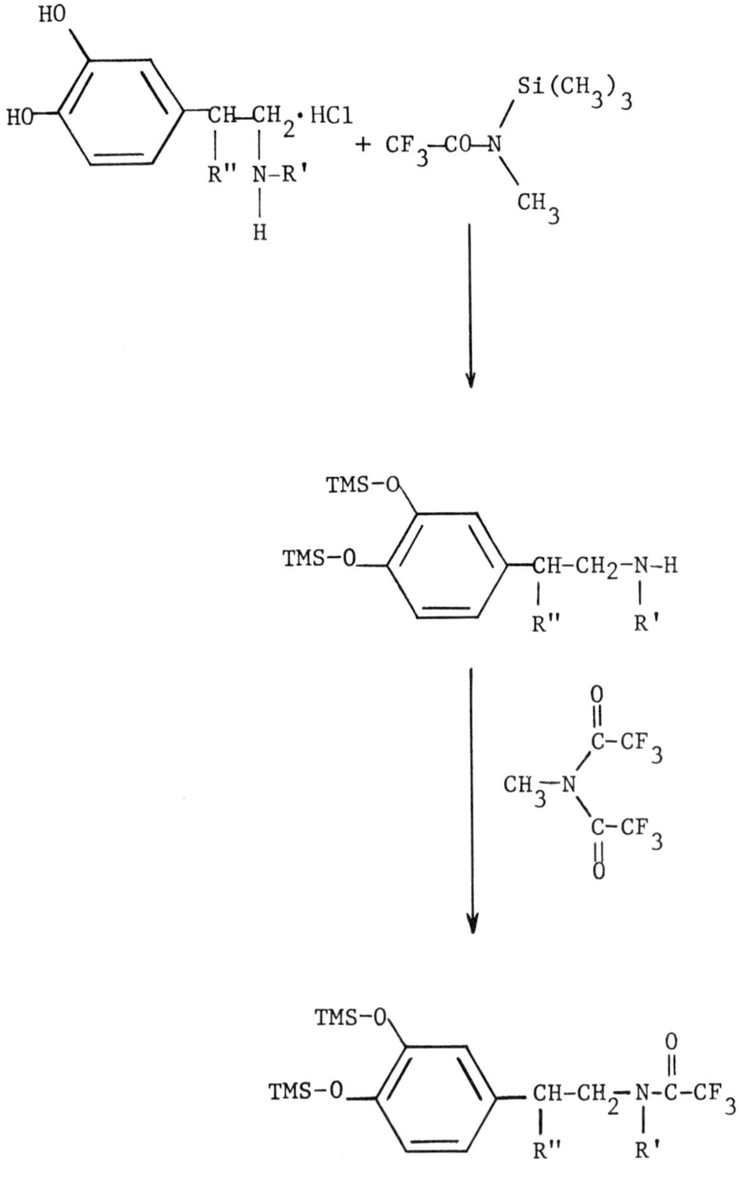

Scheme 4

function should occur during the first step of the reaction, the N-TMS group can be selectively exchanged for a TFA group during acylation [37].

5. *Cyclic Boronates*

Alkyl boronates are normally used with multifunctional compounds to derivatize two vicinal functionalities with a common boronate substituent. In that respect boronate formation is highly selective and can be used to distinguish between cis and trans or stereoisomers as in sugars [38,39] or other cyclic diols [40]. n-Butyl- or phenylboronic acids are typical reagents used for this reaction, which has also been applied to the GC-MS analysis of α- and β-hydroxy acids [41], catecholamines [42], and prostaglandins [43]. In corticosteroids and related compounds, n-butylboronates have been used to stabilize the dihydroxyacetone side chain by formation of 17,21-cyclic n-butylboronates [41]. The boronates are sufficiently stable to carry out further derivatization of free hydroxyl and/or keto groups in the rest of the steroid structure. For example, the 17,21-cyclic n-butylboronate derivative, 14, of 5β-pregnane-3α, 17α, 21-triol-20-one, 13, can be further converted to its 20-MO-3α-OTMS counterpart, 16, as shown in Scheme 5.

6. *On-Column Derivatization Reactions*

On-column reactions have been developed in an attempt to simplify the problems associated with sample handling, particularly when small quantities are involved. Reviews describing various precolumn reactions have been published [44]. In addition to the esterification of fatty acids cited earlier [12,13], trimethylanilinium hydroxide in methanolic solution has also been used for the methylation of cannabis metabolites [45], diphenylhydantoin [46], phenolic alkaloids [47], xanthine bases [48], and aromatic carboxylic acids such as phenylacetic acid and 5-hydroxyindoleacetic acid [49]. For the last compound, methylation of both the carboxylic group and the phenolic group was complete.

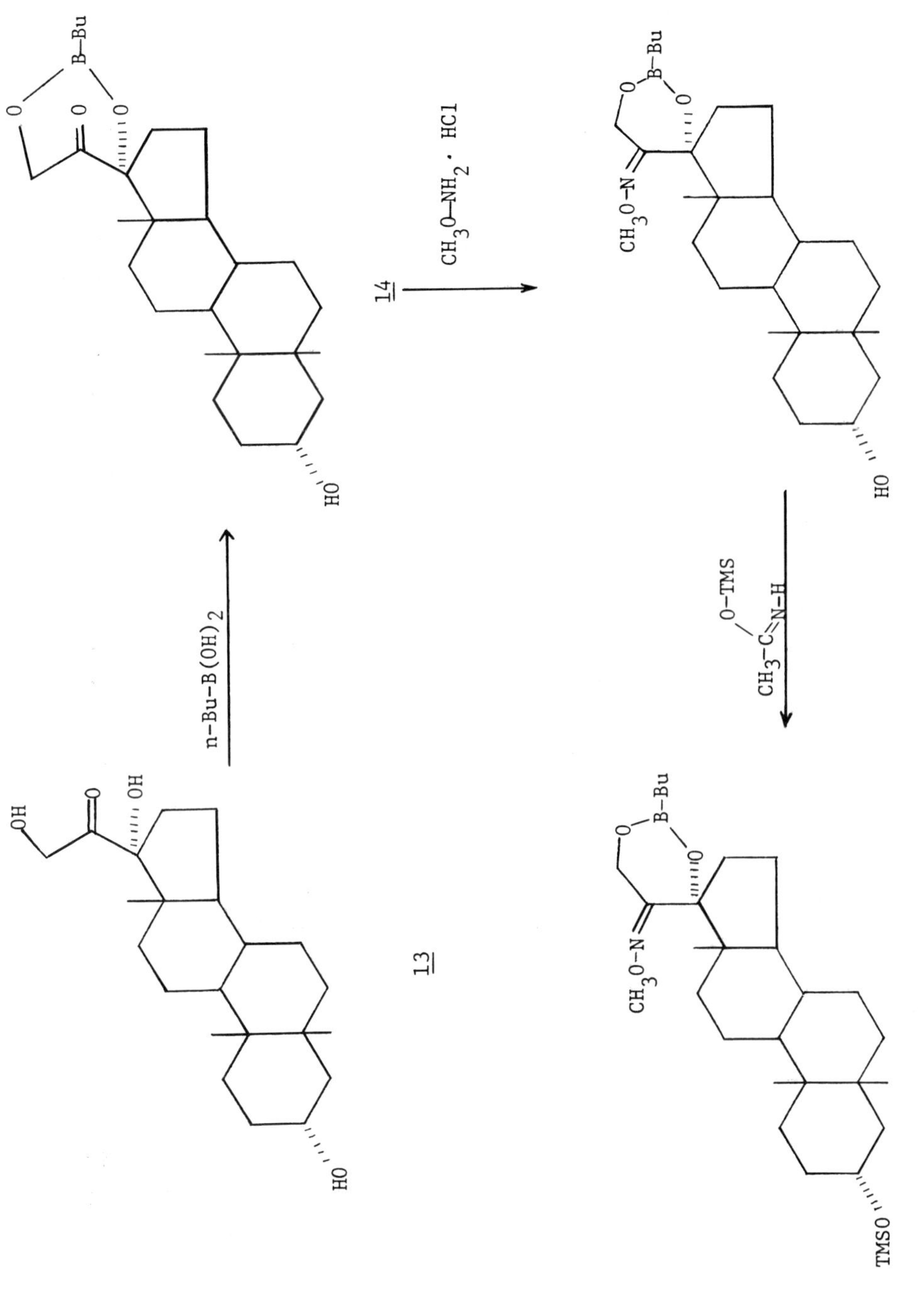

13 14 15 16

Scheme 5

Methylation using trimethylanilinium hydroxide is the method of choice for the GC-MS analysis of barbiturates, as it is routinely used in this field [50,51]. Caution, however, should be exercised in interpreting data of metabolic methylation of barbiturates and fatty acids when using the on-column derivatization method [53]. Artifacts of this type were detected through the use of trimethyl-d_9-anilinium hydroxide, which was able to distinguish between endogenous methylated material vs that produced by methylation upon GC injection.

A selective on-column N-methylation of purines and pyrimidines without attack of the carbonyl oxygen has been reported. The procedure may have bearing in the field of GC-MS of nucleosides in general and utilizes a solution of methyl iodide and sodium methoxide in DMSO [53]. An example of the methylation of thymine, 17, is shown in Scheme 6.

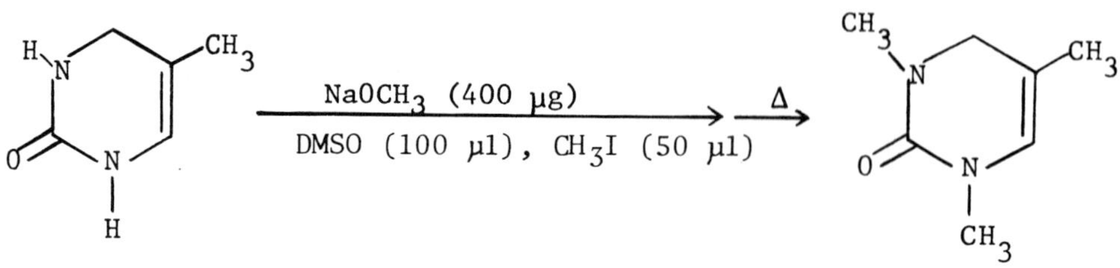

17 (50 μg)

Scheme 6

In an attempt to achieve on-column derivatization of sterols, VandenHeuvel et al. [54] converted them to their methanesulfonate derivatives by reacting with methanesulfonyl chloride in ethyl acetate using pyridine as catalyst. Upon injection into the GC, methanesulfonic acid was quantitatively eliminated and the steroids eluted as their olefinic analogs. In several instances, however, two or more olefinic steroids or steroids containing a 3,5-cyclo system were observed. While limiting its usefulness for steroid analysis by GC-MS, the method provided an approach to the study of elimination reactions involving nonclassical carbonium ion intermediates at the microgram scale [54-56].

A method for the on-column derivatization of hydroxamic acids has also been reported. It involves reaction with acetic anhydride and thermal conversion to an isocyanate at the GC injection port; Eqs. (6) and (7) [57].

$$\text{R-CO-NHOH} + \text{CH}_3\text{CO-O-COCH}_3 \xrightarrow{\text{Pyridine}} \text{CH}_3\text{COOH} + \text{R-CO-NH-O-COCH}_3 \tag{6}$$

$$\text{R-CO-NH-O-COCH}_3 \xrightarrow{\Delta} \text{R-N=C=O} + \text{CH}_3\text{COOH} \tag{7}$$

It should be noted that deuterium isotope exchange reactions are also frequently carried out on-column during GC-MS, whether it is for structural elucidation purposes or for the clarification of mass spectral patterns. The advantage of this on-line approach is reflected in the ability to work with minute amounts of sample without prior need for sample isolation. Specific examples are given below in the section under isotopic labeling.

C. Artifacts or Structural Transformations in the GC-MS Analysis of Derivatives

An obvious requirement for the derivatization reaction is to cause no structural transformation to the compound under investigation. For this reason, the reactions chosen are usually straightforward and the reaction products easily predictable. Nevertheless, unforeseen reaction products and derivatives can be formed even under mild derivatization conditions. In other words, the analyst must always be on the alert for the unusual whether during the derivatization reaction or during the GC-MS analysis. This may include artifacts encountered in the mass spectrum and introduced from pump oil or background [58], plasticizers [59], and gas chromatographic column or septum bleed [60]. However, even more serious can be the problems associated with sample handling and derivative preparation. In this text, derivatization artifacts are classified under two general categories--those occurring during the actual derivative preparation and those occurring in the gas chromatographic injection port. The following examples from the recent literature provide a representative, albeit limited, picture of the types of structural transformations that a molecule can undergo upon derivatization and GC-MS analysis, and demonstrate the complications and even usefulness that may result from this occurrence.

1. Structural Transformations in Solution

a. Aromatization of the A-ring of norethynodrel. The product normally expected from persilylation of norethynodrel, 18, with TSIM is the 3,5(10)-dien-3,17-bis-trimethylsilyl derivative, 19. However, when the reaction was carried out under more vigorous conditions (e.g., 100°C), it was discovered that 19 was merely an intermediate, and that after 4 hr norethynodrel was quantitatively converted to the bis-trimethylsilyl derivative of the ethynylestradiol, 20 (Scheme 7).

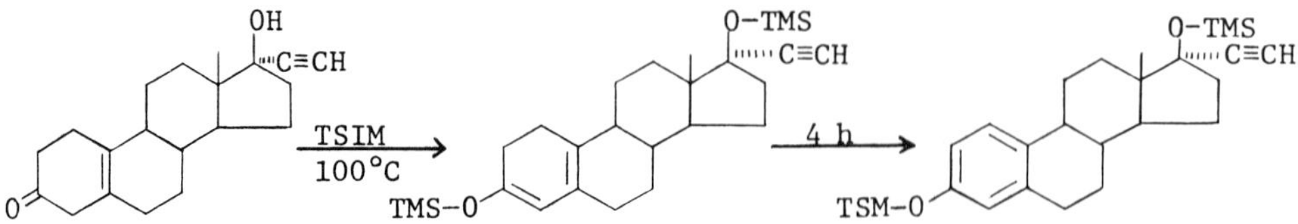

18 19 20

Scheme 7

Analysis of the reaction mixture even immediately after mixing TSIM with 18 at room temperature revealed a 20% yield of 20, whereas milder silylating reagents such as BSA or BSTFA failed to derivatize 18. No such aromatization was observed for norethindrone, the Δ^4-isomer of 18 upon reaction with TSIM, indicating a structural selectivity for the reaction [61].

b. Functional group transformations in the derivatization of carbamates and urethanes. VandenHeuvel et al. [62] recently provided several examples of structural transformation of the parent compound as a result of a simple derivatization reaction.

Derivatization of p-ureido-p'-aminodiphenylsulfone, 21, a therapeutic agent effective against Marck's disease, was considered necessary for the GC-MS analysis of this compound in chicken tissue.

Reaction of 21 with heptafluorobutyric anhydride resulted in a quantitative yield of the bis-HFB derivative of the diamine, 22. This transformation was indeed fortuitous as it was in fact considered and deemed desirable to hydrolyze 21 to its diamine and then convert to 22 for GC-MS analysis.

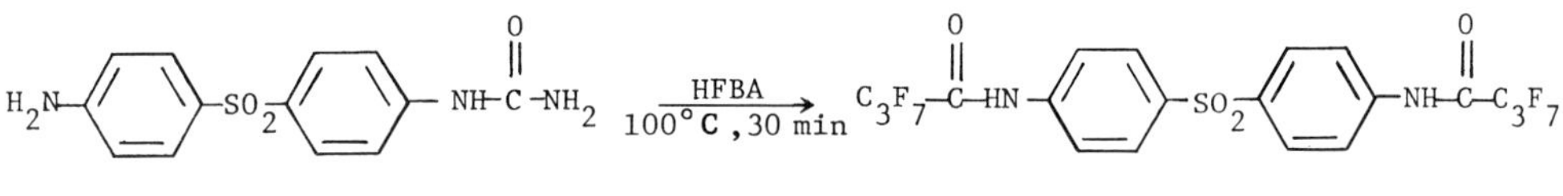

21 22

Scheme 8

Reaction of ronidazole, 23, an antiparasitic compound, with BSTFA in pyridine and GC-MS analysis of the product identified the latter as the TMS derivative of the alcohol 24. Additional examination of the reaction product with infrared spectroscopy, thin-layer chromatography, and MS analysis via the direct inlet probe confirmed that conversion of 23 to the alcohol occurred during silylation (Scheme 9) rather than as a result of an on-column reaction. It is

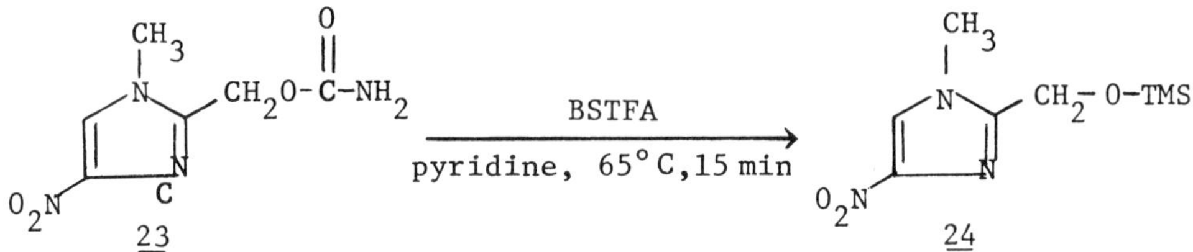

23 24

Scheme 9

interesting, however, that transformation of 23 to the free alcohol of 24 can also be thermally effected on the GC injection port via elimination of isocyanic acid.

c. Oxygen incorporation during silylation. The trimethylsilylation of 7-methylpurine nucleosides revealed an unexpected phenomenon of oxygen incorporation to produce an 8-oxo compound (Scheme 10). Isotopic labeling data using $^{18}O_2$ confirmed the participation of dissolved oxygen in the reaction [63]. The mass

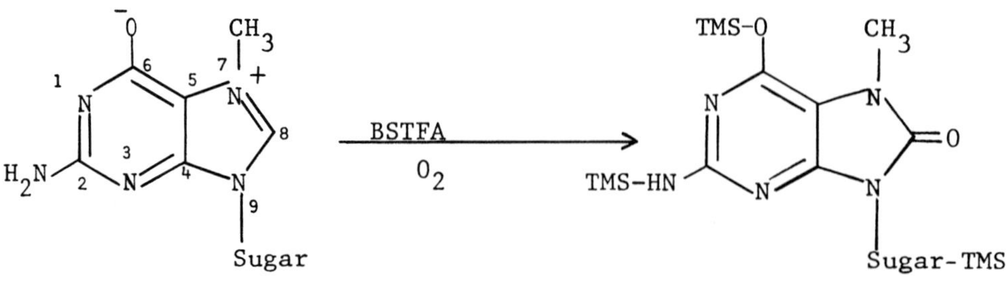

25 26

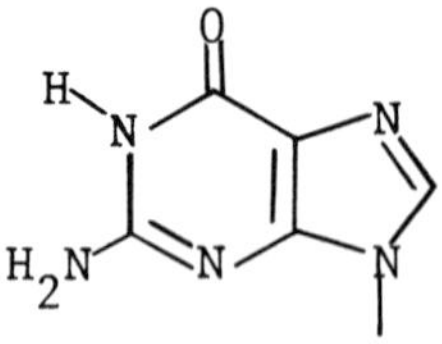

27

Scheme 10

spectrum of the TMS derivative 26 was compared with that of the TMS derivative of guanosine, 27. The molecular ion in the spectrum of 26, as well as the fragment ions corresponding to the base, were found to occur 30 amu (i.e., CH_2 + 0) higher than in guanosine-TMS, thus supporting the structural assignment of 26. It appears that this transformation to an 8-oxo derivative is highly characteristic of alkylation in the 7-position of purine nucleosides. In that respect, the phenomenon can be structurally diagnostic and was in fact utilized to identify the presence of N^2,N^2,7-trimethylguanosine at the 5'-terminal nucleotide of U-2 ribonucleic acid [64].

d. Group interchange in the derivatization of some sulfur-containing amino acids. The derivatization of the disulfide amino acids (cystine and homocystine, 27a and 27b) using a direct esterification-acylation procedure resulted in the formation of a new species in addition to the TFA-butyl ester derivatives of 27a and 27b, namely, the TFA-butyl ester derivative of the cysteine-homocysteine disuffide, 27c (Scheme 10a). The new compound, as expected, eluted between the derivatives of the two parent compounds in the

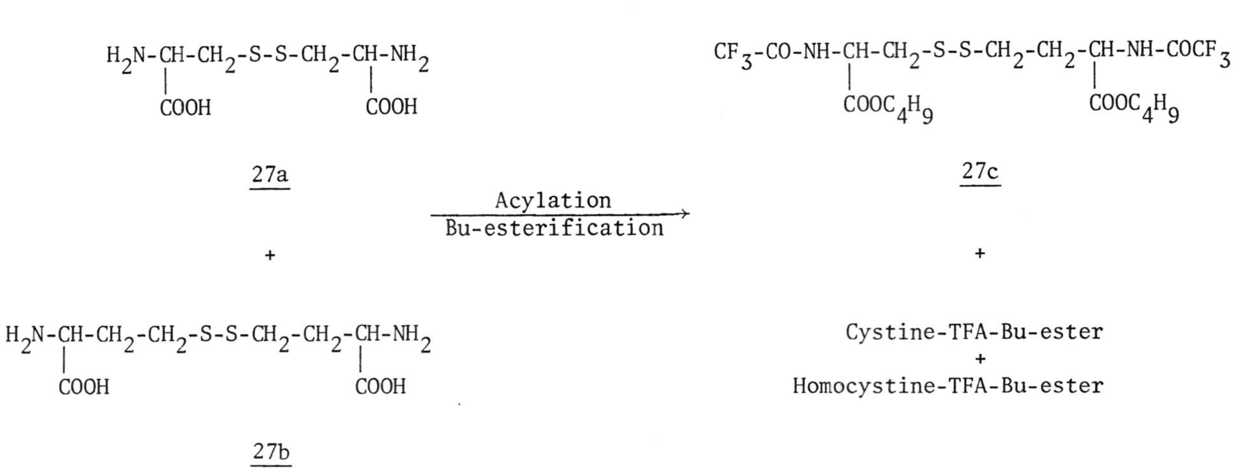

Scheme 10a

$$\text{RS-SR} + \text{R'S-SR'} \xrightarrow[\text{Bu-esterification}]{\text{Acylation}} \begin{array}{c}\text{TFA-Bu-ester-RS-SR}\\ + \\ \text{TFA-Bu-ester-RS-SR'}\\ + \\ \text{TFA-Bu-ester-R'S-SR'}\end{array}$$

Scheme 10b

gas chromatograph. To distinguish between an "on-column" vs a "during derivatization" product several experiments were carried out. In one of them cystine and homocystine were derivatized independently to obtain their respective TFA-butyl ester derivatives. Coinjection of a mixture of the latter into the GC/MS revealed no appreciable formation of the mixed disulfide 27c. This evidence is strongly supportive of a conversion during the actual derivatization process. The general reaction for this phenomenon can be described by the reaction in Scheme 10b. It is apparent that quantitative analysis of cystine and homocystine in the presence of each other, other disulfides, or free sulfhydryl groups can lead to erroneous results arising from such intermolecular group transfers [65].

2. *On-Column Transformations*

a. *Degradation of phenobarbital.* The on-column methylation method using trimethylanilinium hydroxide has been widely adopted in clinical laboratories for the analysis of barbiturates. The chromatographic analysis of phenobarbital, 28, was marked by the presence of an early extraneous peak, which has been recently identified as N-methyl-2-phenylbutyramide, 29. It is presumably formed by degradation of the phenobarbital (Scheme 11) [66].

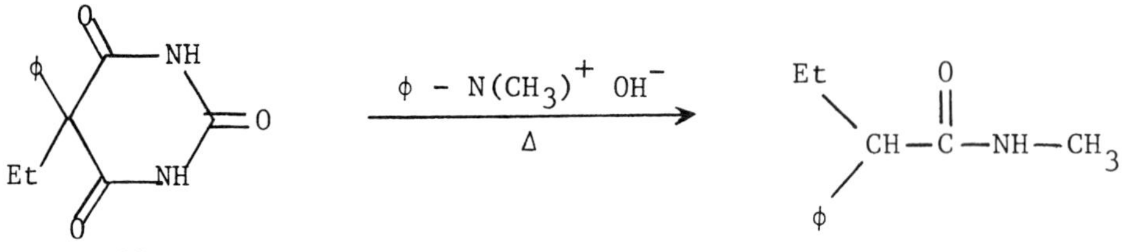

28 29

Scheme 11

b. *Methylation of barbiturates extracted from serum.* N-Methyl derivatives of phenobarbital and diphenylhydantoins in serum extracts were observed during GC-MS studies. These N-methyl compounds were found not to be endogenous metabolites, but rather artifacts generated upon injection of the serum extract into the GC. The presence of such an endogenous methylating reagent was confirmed in a series of experiments using phenobarbital-ethyl-d_5, 30a, and trimethyl-d_9-anilinium hydroxide as outlined in the following sequence.

1. When a serum extract was mixed with phenobarbital-ethyl-d_5, 30a, N-methyl-phenobarbital-ethyl-d_5, 30b, was detected along with its unlabeled analog, thus supporting the presence of a methylating agent in the blood.

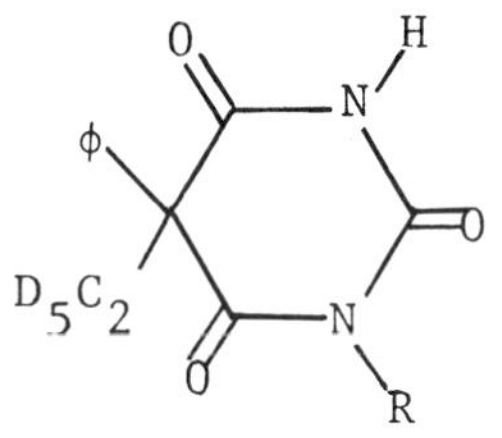

30a, R = H

30b, R = CH_3

30

2. GC-MS analysis of a serum sample devoid of any drug components, but spiked with phenobarbital, indicated the presence of small amounts of N-methyl phenobarbital.
3. Steps 1 and 2 proved that *at least part* of the N-methyl derivative was due to an endogenous methylation. To prove that *all* of it was indeed due to an endogenous agent, a third experiment was carried out to examine whether endogenous N-methylation could be totally suppressed by addition of an alternative N-alkylating reagent. The absence of any N-methyl compound would prove that no in vivo methylation had occurred and that the N-methylphenobarbital detected was merely an artifact due to GC. Trimethyl-d_9-anilinium hydroxide was added to the serum extract, and indeed no N-methyl or N-methyl-N'-methyl-d_3-phenobarbital was observed. Instead, complete conversion to N,N'-dimethyl-d_6-phenobarbital was indicated. It was ascertained that lecithin, a choline derivative present in serum and which contains a quaternary amine moiety just like trimethylanilinium hydroxide, was responsible for the N-methylation of phenobarbital [52].

c. *Thermally induced degradation of a carboxamide.* The carboxamide, 31, was derivatized with BSTFA for analysis by GC-MS. The following observations were made.

1. After a 3-min reaction at 65° two GC peaks--no. 1 and no. 2--were observed with a ratio of ~1:20. The mass spectrum of peak no. 1 (MW 346) exhibited no ion at [M-15-42], that is, no loss of C_3H_6 was observed to indicate the presence of an isoproproxy group. The spectrum of peak no. 2 (MW 406) was found to contain two TMS groups corresponding to structure 32.

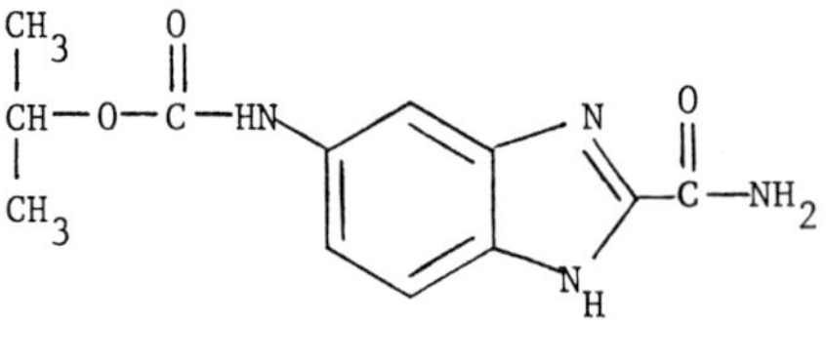

31

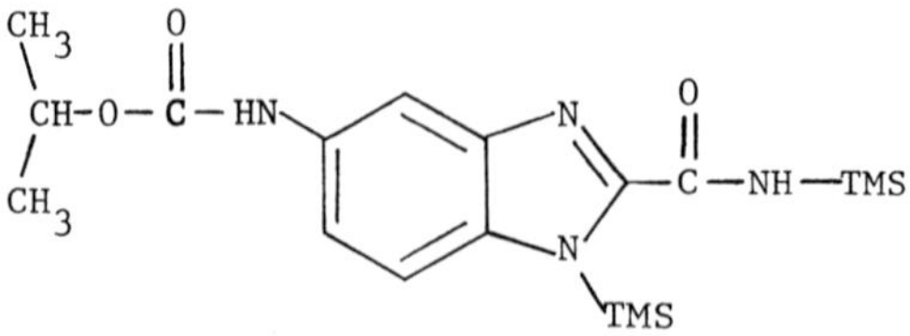

32

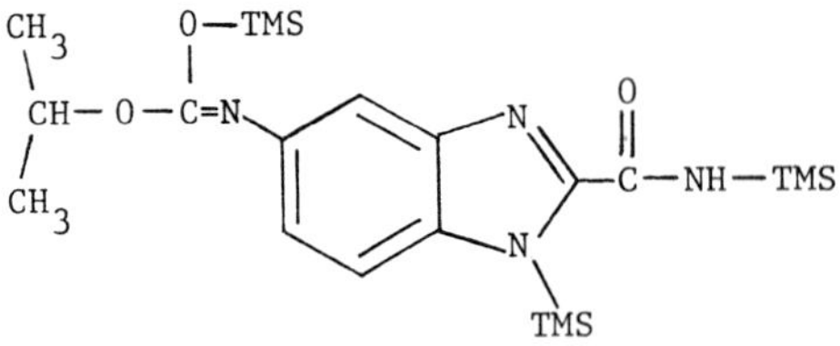

33

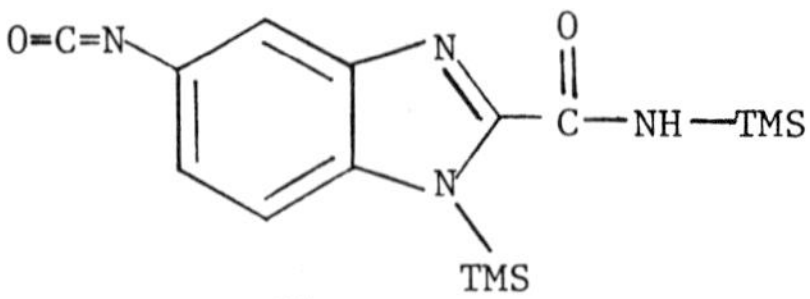

34

2. After a 15-min reaction the chromatogram showed more than 90% conversion to peak no. 1.
3. Mass spectra of the mixture obtained via the direct inlet probe indicated that after a 3-min reaction a compound of MW 406 was the principal component (i.e., 32) while after 15 min the principal component was a tri-TMS compound of MW 478, presumably with structure 33, as its mass spectrum contained an intense [M-15-42] ion.

On the basis of the above observations and data, it was speculated that the isoproproxy group in 33 was thermally eliminated at the GC injection port as $(CH_3)_2$CH-OTMS to yield the isocyanate derivative 34. This was further confirmed by conducting the GC-MS analysis at reduced injector and column temperatures, in which case a single GC peak of MW 478--i.e., compound 33--was observed [62].

D. Isotopic Labeling in GC-MS

The use of stable isotopes in mass spectrometry has been reviewed by Caprioli [67]. The article examines, among other topics, common methods of preparation of isotopically labeled compounds (e.g., isotopic exchange in D_2O or $H_2{}^{18}O$, reactions with deuterium-labeled reagents, etc.), the use of stable isotopes in biochemistry, determination of isotopic abundances, and isotope effects in reaction processes. In this section we shall confine the discussion of isotopic labeling to qualitative applications associated with on-line GC-MS techniques and some examples of the utility and technique for preparation of isotopically labeled derivatives.

1. On-Line Exchange of Labile Protons in GC-MS

The exchange of labile hydrogens (e.g., OH, NH_2) with deuterium in D_2O or CH_3OD medium, using "wet" chemical methods, is a well-known technique and has been employed extensively in organic mass spectrometric studies [68]. A similar exchange reaction can be conducted on-column during GC-MS, using a D_2O or CH_3OD conditioned column [69]. In addition to minimizing the sample requirements, the latter approach has the added advantage of permitting work with mixtures, as the sample can be deuteriated and separated from impurities in a single step. Mass spectrometric fragmentation mechanisms can be readily studied in this fashion. For example, participation of the 17α-hydroxylic hydrogen in the elimination of the elements of methane from the molecular ion of the 3β,20α-bis-TMS derivative of pregn-5-ene-3β,17α,20α-triol (Scheme 12) was confirmed by on-column exchange during GC-MS. A 6-ft, 1% SE-30 column was used and the steroid was introduced into the system immediately following injection of 5 μl of CH_3OD (99.9% isotopic purity). A 60% isotope incorporation in the final product was sufficient to resolve this mechanistic problem [70].

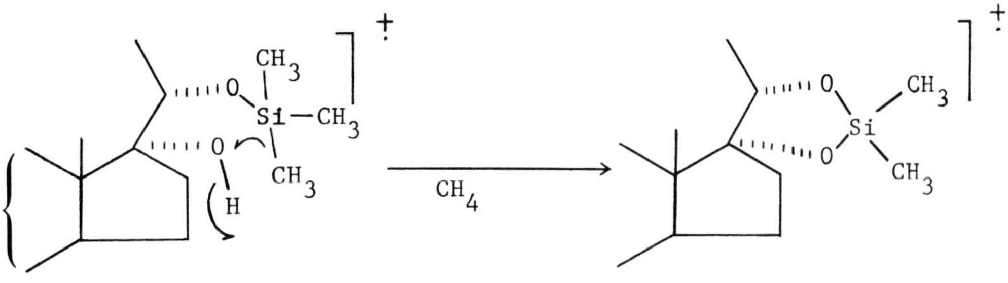

Scheme 12

The technique of on-column isotope exchange during GC has been applied on preparative scale [71] for almost quantitative (>90%) deuteriation of not only hydroxylic and amino hydrogens, but also

$\rangle\overset{O}{\overset{\|}{P}}\text{-H}$ to $\rangle\overset{O}{\overset{\|}{P}}\text{-D}$ in O,O-dialkyl hydrogen phosphonates. The "vapor phase" exchange method is of particular significance in the latter case because the use of a "wet" labeling method may result in hydrolysis of the phosphonate.

Enolizable hydrogens (α- to a carbonyl group) can be exchanged either in a basic or acidic medium. Anthony and Brooks [72] have used base (KOH or $Ba(OH)_2$) impregnated columns to effect exchange of enolizable hydrogens in ketosteroids during GC-MS. Apiezon-L was selected as the stationary phase because of its known compatibility with KOH in the GC analysis of free amines. Typical phases consisted of:

1. Apiezon-L on Gas Chrom Q, incorporating 0.5% KOH
2. OV-1 (1%) + Apiezon-L (1%) + $Ba(OH)_2$ (1%)

Trimethylsilyl derivatives could be analyzed in the latter system, which was used to recognize the environment of carbonyl groups on a steroid nucleus during GC-MS.

2. *Isotopically Labeled Functional Derivatives*

Deuterium-labeled derivatizing reagents [e.g., CD_3I, $(CD_3O)_2CO$] are routinely used to assist in the interpretation of the mass spectra of the corresponding methyl or acetyl derivatives of organic compounds. Of particular significance are the perdeuteriotrimethylsilyl (TMS-d_9) derivatives, the examination of which is virtually essential for the interpretation of the mass spectra of their TMS counterparts. The use of TMS-d_9 derivatives in mass spectrometry was first introduced by McCloskey et al. [73]. A number of perdeuteriotrimethylsilylating reagents are now available commercially, including HMDS-d_{18}, BSA-d_{18}, TSIM-d_9, and TMCS-d_9. In view of the wide variation in reactivity of the various TMS-d_o and TMS-d_9 reagents, it has been possible to prepare mixed TMS-d_o-TMS-d_9 derivatives in which the labeled and unlabeled groups occupy specific positions with virtually 100% selectivity [74]. The usefulness of these mixed derivatives for the interpretation of the mass spectra of TMS derivatives is illustrated in Fig. 6, which shows the mass spectra of the 3β,17β-bis-TMS,

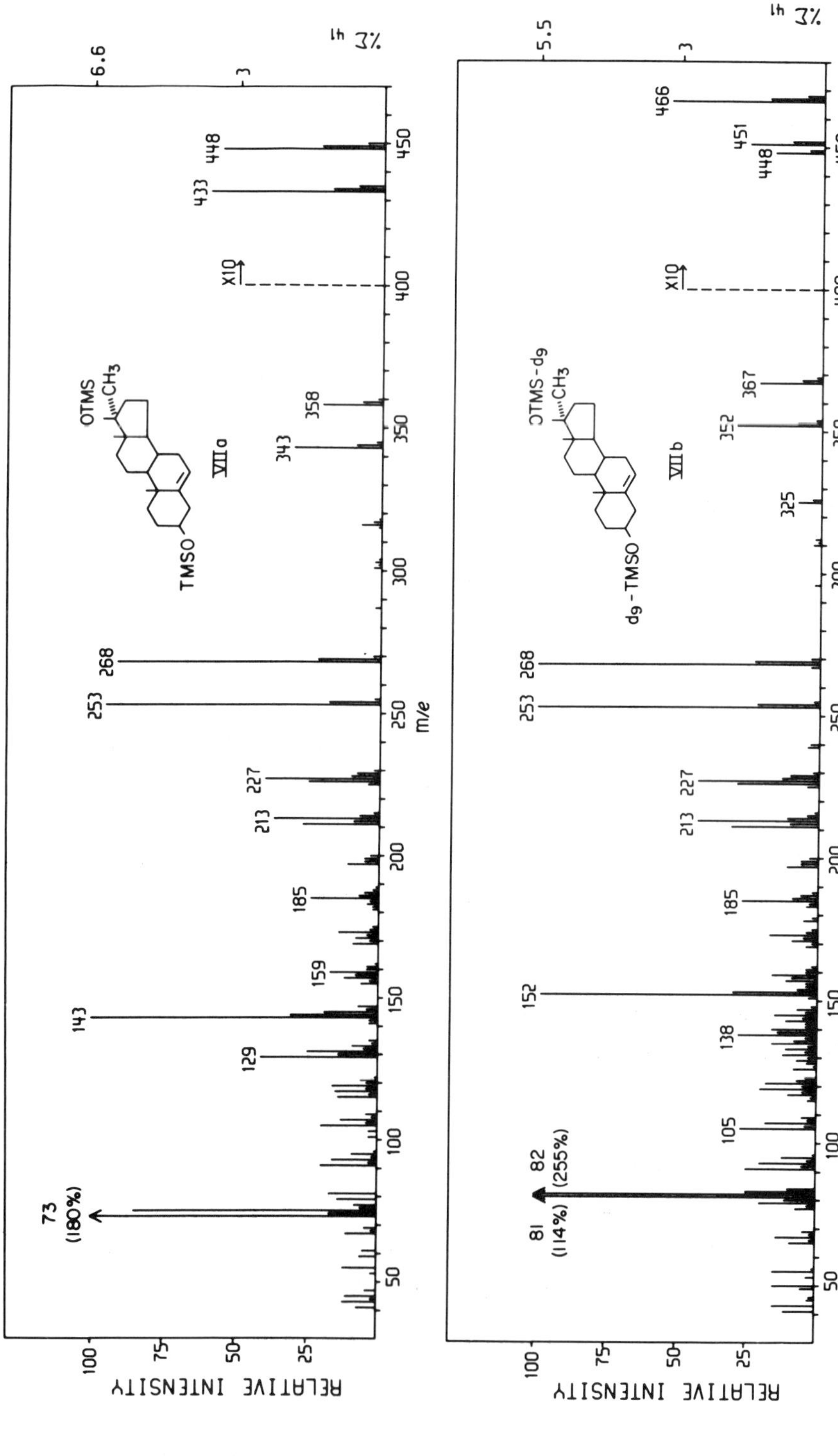

FIG. 6. Mass spectra of 3β,17β-bis(trimethylsilyloxy-d_0)-17 -methyl-5-androstene (top), 3β,17β-bis(trimethylsilyloxy-d_9)-17α-methyl-5-androstene (center), and 3β-(trimethylsilyloxy-d_0)-17β-(trimethylsilyloxy-d_9)-17α-methyl-5-androstene (bottom). (Reprinted from *Analytical Chemistry* with permission from the publisher.)

3β,17β-bis-TMS-d_9, and 3β-TMS-17β-TMS-d_9 derivatives of 17α-methyl-androst-5-ene-3β,17β-diol. The mixed derivative was prepared by initially reacting the unhindered 3β-hydroxy group with BSTFA in CH_3CN at 60° for 1 hr followed by removal of the excess solvent and reagent and reaction of the 17β-hydroxy group with TSIM-d_9 in cyclohexane. It is easy from the spectrum of the selectively labeled derivative to identify the origin of the TMS-containing ions. In addition, the spectrum of the TMS-d_0-TMS-d_9 derivative clearly shows that the $[M\text{-TMSiOH}]^{+\cdot}$ ion consists of two different species, each eliminating a methyl group with different relative preference.

The preparation of selectively labeled acetyl-d_0-acetyl-d_3 derivatives of glycals has also been described by Szafranek et al. [75]. Peracetyl-d_0 derivatives of, e.g., D-glucal, were prepared conventionally, followed by selective hydrolysis of the acetyl groups in methanol under controlled conditions. The partially acylated products were separated by column chromatography, their structures were identified by NMR, and they were finally treated with acetic anhydride-d_6 in pyridine to obtain a selectively labeled acetyl-d_0-acetyl-d_3 derivative. While permitting the elucidation of ion structures and fragmentation mechanisms, the preparation procedure is too cumbersome for routine use of mixed acetyl-d_0-acetyl-d_3 derivatives in the interpretation of mass spectra.

In a manner analogous to the on-column exchange of labile hydrogens, it is also possible to replace acidic trimethylsilyl groups with TMS-d_9 groups by conditioning the gas chromatographic column accordingly. This involves saturation of the GC column with a 20:1 mixture of BSA-d_{18}:TMCS-d_9 (typically 10 μl of the mixture for a 6-ft, 4-mm inside diameter, 1% SE-30 column) just prior to sample injection. This was followed by a second 10-μl injection of the labeled reagents 30 sec after introduction of the TMS derivative of the sample. The efficiency of exchange varied from 50% to as high as 90%. Phosphate trimethylsilyl groups are particularly susceptible to exchange under these conditions, and this technique has been utilized in the interpretation of the mass spectra of the TMS derivatives of nucleotides [76], glycerophosphoric acids [77], and steroidal phosphates [78]. Middleditch and Desiderio have further demonstrated that the ester-TMS groups in prostaglandin-TMS derivatives can also be selectively labeled in preference to ethereal-TMS groups during GC/MS, by on-column exchange. This facilitated significantly the interpretation of their mass spectra [79]. It is worth noting here that in estrogens the labile phenolic TMS groups were selectively exchanged for a dimethylethylsilyl group based on the same "sandwich" injection technique using dimethylethylsilylimidazole as a reagent [79a]. This provided for the recognition of phenolic OH groups as well as functional group labeling for mass spectral interpretation.

IV. MASS SPECTROMETRY OF SELECTED FUNCTIONAL DERIVATIVES

A. Steroids

1. *Silyl Derivatives*

a. The trimethylsilyl group. It is generally recognized that trimethylsilylation of hydroxy steroids affords the simplest and most practical derivative for the GC-MS analysis of these compounds [80, 81]. For the most part the mass spectra of TMS ethers of steroids exhibit easily discernible molecular ion peaks or, at least, indirect information about their molecular weight can be derived from the $[M-CH_3]^+$ ions that are invariably present in the spectra of TMS derivatives. The trimethylsilyl group strongly directs the fragmentation of the molecule and, as a result, the mass spectra are usually characterized by the presence of abundant TMS-containing ions. Ubiquitous is the ion of *m/e* 73 $[(CH_3)_3Si^+]$, which is extremely abundant in the 70-eV spectra and generally provides no structural information. Similarly an ion at *m/e* 75 $[(CH_3)_2\text{-Si-OH}]^+$ is always present in the spectra of O-TMS ethers.

Frequently the TMS-containing ions are of structural significance and the occurrence of specific ion peaks can be used to make structural assignments with a considerable degree of confidence. Formation of these ions can involve intramolecular migration of the TMS group or direct cleavage of adjacent C-C bonds. Some of the structurally significant ions in the mass spectra of TMS steroidal ethers are summarized below. It should be noted that many of these ions or rearrangement processes are also encountered in the TMS derivatives of other classes of compounds and will not be elaborated any further under those categories.

(1) Rearrangement ions and related processes:

Elimination of trimethylsilanol. The stepwise elimination of trimethylsilanol is a characteristic process in the spectra of TMS steroidal ethers and yields peaks corresponding to the Series $M^{+\cdot} \rightarrow [M-90]^{+\cdot} \rightarrow [M-2x90]^{+\cdot}$ . . . and $[M-15]^{+} \rightarrow [M-15-90]^{+} \rightarrow [M-15-2x90]^{+}$ . . ., etc. Trimethylsilanol elimination is dependent on the structural environment of the TMS group and, in principle, variations in the relative intensities of the corresponding spectral peaks can be used for structural identification on the basis of comparison with authentic spectra. The specificity of this sequential elimination is further demonstrated in Fig. 7, which shows the mass spectra of 5α-androstane-3β,16β-diol-TMS, 35, and androst-5-ene-3β,16β-diol-TMS, 36, each with an approximately 50%-^{18}O incorporation on the C-16 oxygen atom. It is apparent that the presence of the Δ^5-double bond provides the necessary site of stabilization for the positive charge to alter the sequence of trimethylsilanol elimination, and results in preferential elimination of the 3-substituent in compound 36 [82].

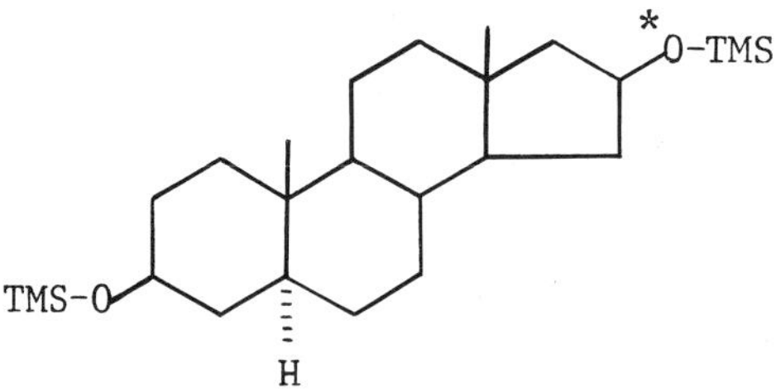

35

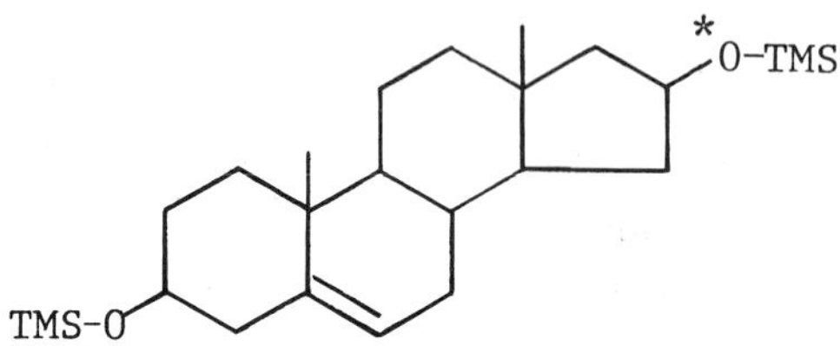

36

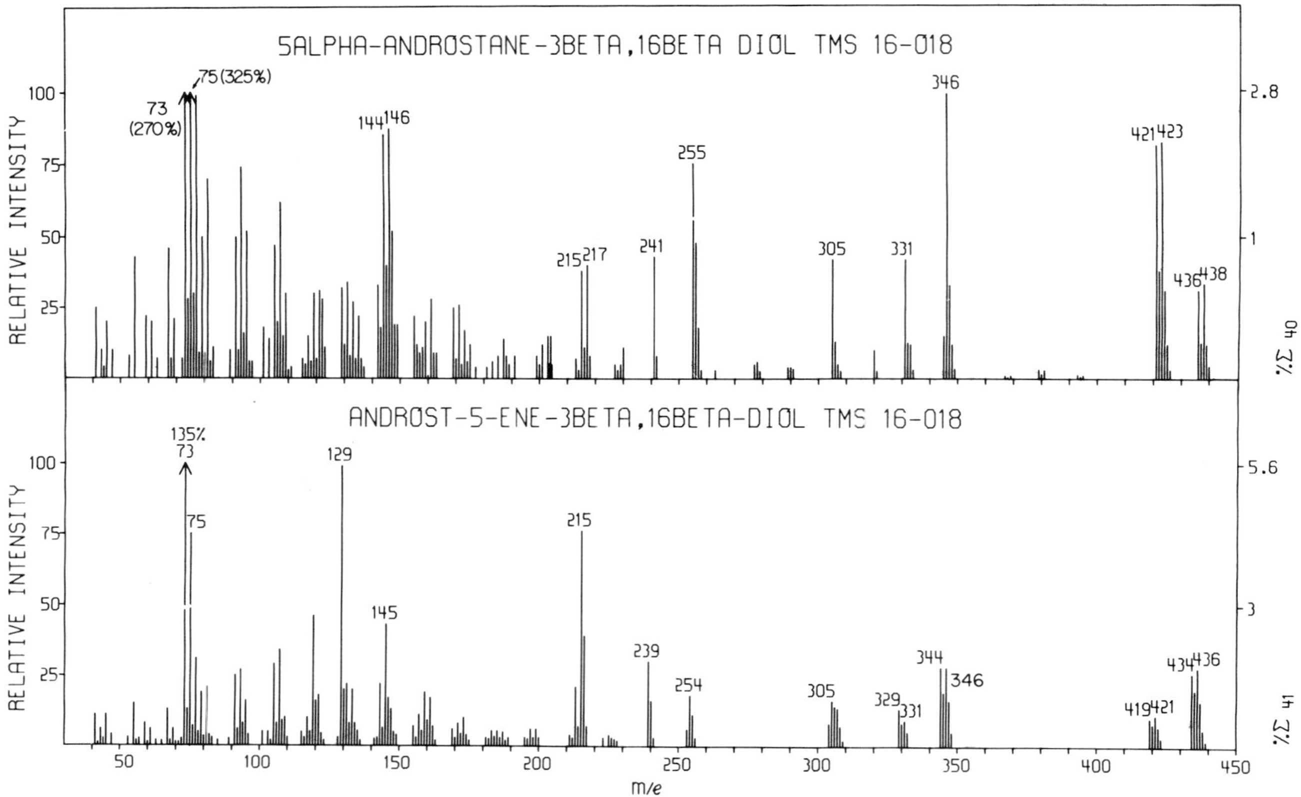

FIG. 7. Mass spectra of the trimethylsilyl derivatives of 5α-androstane-3β,16β-diol-16-^{18}O (top) and androst-5-ene-3β,16β-diol-16-^{18}O (bottom). (Reprinted from *Organic Mass Spectrometry* with permission from the publisher.)

m/e 147. The ion of *m/e* 147, a, is encountered in the mass spectra of compounds containing two or more trimethylsilyloxy groups. It has been proposed [83] that its formation is triggered by loss of a TMS methyl group and followed by reaction of the Si^+ center with the opposite oxygen (Scheme 13).

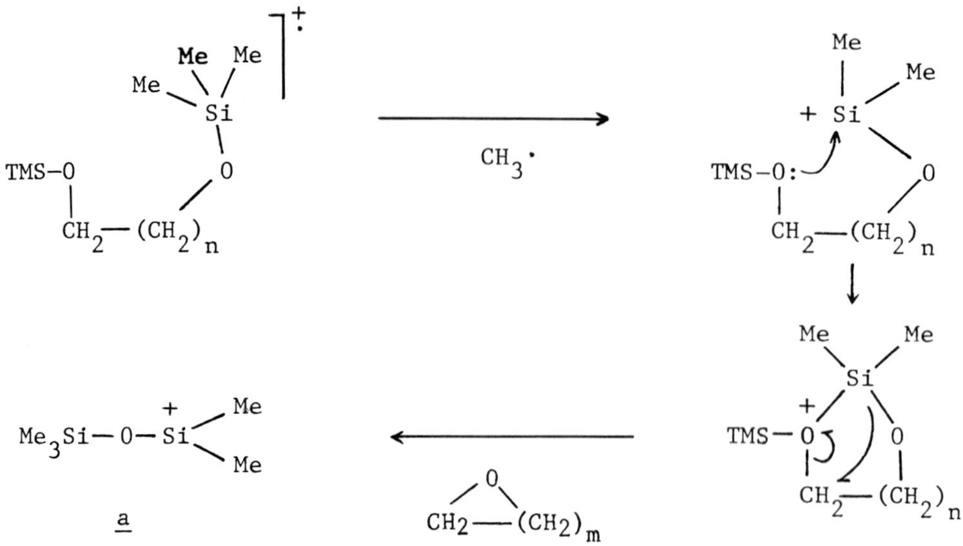

Scheme 13

While the *m/e* 147 ion occurs in significant abundance even in compounds where the TMSO groups are separated by a large number of carbons (e.g., n = 26 or higher) [84], it has been shown that in molecules of rigid conformation such as steroids its formation is strongly influenced by stereochemistry and the distance separating the interacting functionalities [85]. This is illustrated in Fig. 8, which shows the mass spectra of the TMS derivatives of the structural isomers androst-5-ene-3β,11β,17β-triol and androst-5-ene-3β,16α,17β-triol. Similarly, comparison of the *m/e* 147 ion intensities in the mass spectra of the TMS derivatives of the stereoisomers 5α-androstane-3β,11β,17β-triol and 5β-androstane-3α,11α,17β-triol revealed a ratio of ∿1:3 in the respective compounds. This difference can be rationalized in terms of the conformational transitions b ⇄ c and d ⇄ e (Schemes 14 and 15). Comparable effects have also been reported in

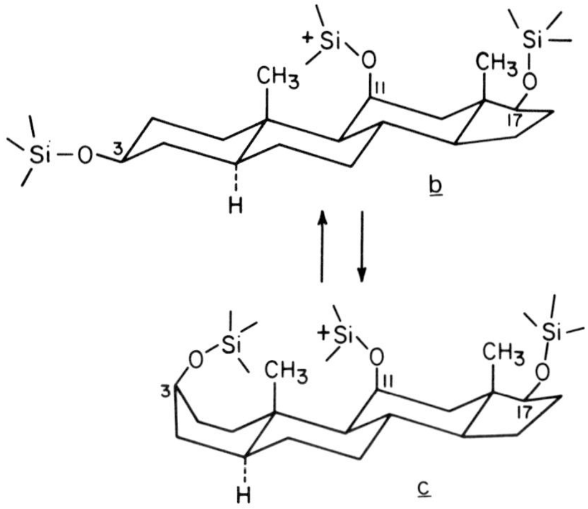

Scheme 14

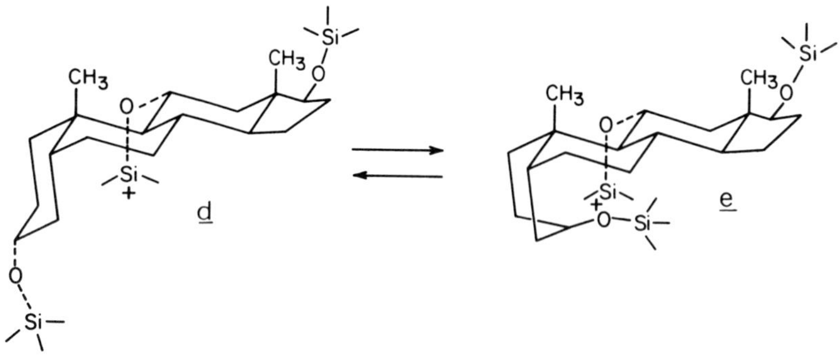

Scheme 15

the mass spectra of the TMS derivatives of cyclic sugars [86] and other related compounds [87].

m/e 191. The ion of *m/e* 191 contains two trimethylsilyloxy groups as indicated in structure f. Its mechanism of formation has been studied in detail with the aid of deuterium isotope labeling in

$$(CH_3)_3\text{-Si-O-CH=}\overset{+}{O}\text{-Si}(CH_3)_3$$

f, *m/e* 191

steroids containing TMSO groups on the D-ring and/or the 18-angular methyl group [88]. It has been postulated that it is formed by ring

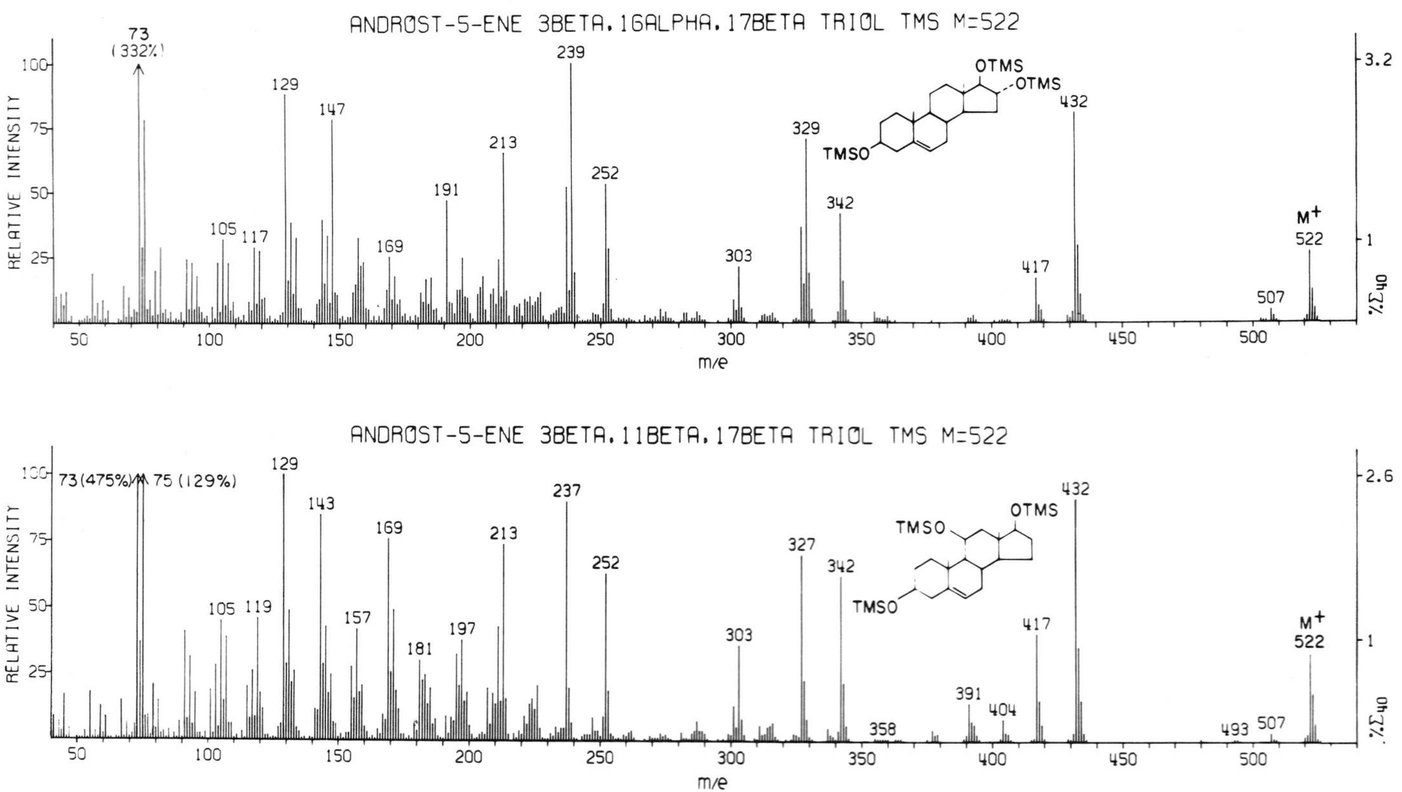

FIG. 8. Mass spectra of the trimethylsilyl derivatives of androst-5-ene-3β,16α,17β-triol (top) and androst-5-ene-3β,11β,17β-triol (bottom). (Reprinted from *Organic Mass Spectrometry* with permission from the publisher.)

opening and bond migration to the positively charged oxygen as shown in Scheme 15a for the case of a 17β,16α-bis-trimethylsilyloxy derivative. While the example of Scheme 15a involves the interaction of

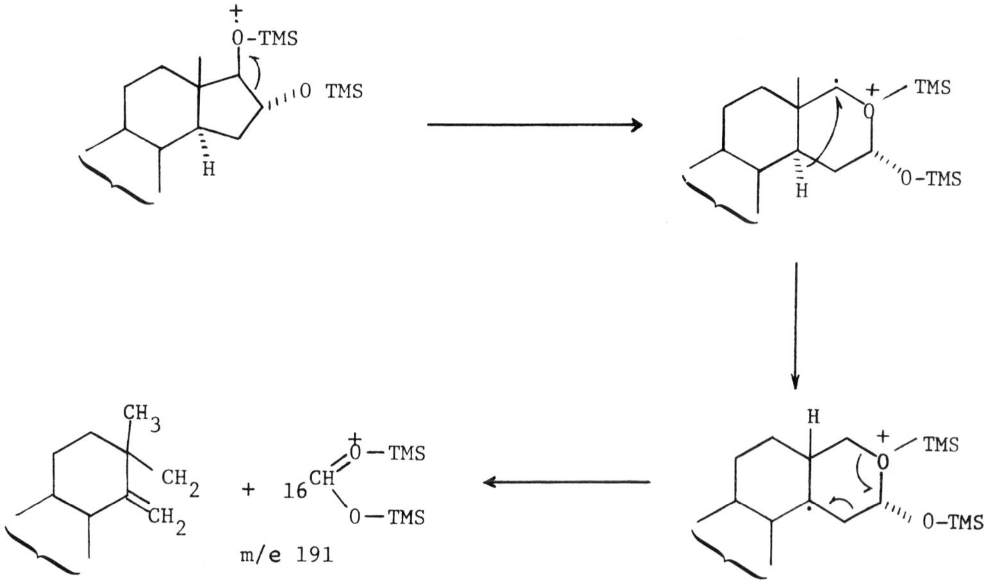

Scheme 15a

vicinal TMSO groups, the same ion was also observed in moderate abundance (∿ 20% RI) in the spectra of 11,17-trimethylsilyloxy steroids where the groups are well separated [89]. Selective deuterium labeling at C-17 and C-11 and ^{18}O-labeling of the C-17 oxygen suggested that formation of *m/e* 191 involved equally the C-11 and C-17 carbons and was initiated by cleavage of the C-9/C-11 bond and rotation about the C-12/C-13 and C-11/C-12 bonds.

Scheme 16 shows a plausible mechanism for the formation of the *m/e* 191 ion for the case in which the C-11 carbon is retained. Cleavage of the C-13/C-17 bond would have to be invoked to rationalize the formation of a *m/e* 191 ion comprising C-17 and the two TMSO groups.

[M-30]$^{+\cdot}$ *ions*. The ability of the TMS group to undergo a McLafferty rearrangement was first noted by Weber and co-workers [90]. The same process also occurs in 18-trimethylsilyloxy-17-oxo- and 19-trimethylsilyloxy-3-oxo-Δ^4-steroids, resulting in elimination of CH_2O (30 amu)

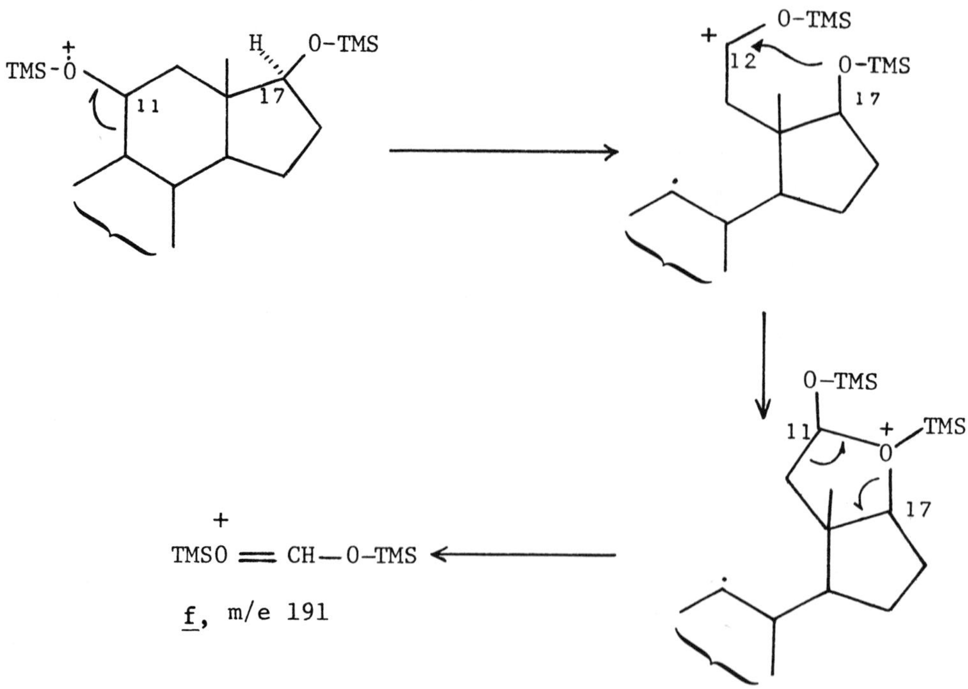

Scheme 16

from the molecular ion (Scheme 17). When considered in conjunction with other spectral data, this process can be of diagnostic value and has been used to identify the presence of steroids with those structures in biological mixtures [91].

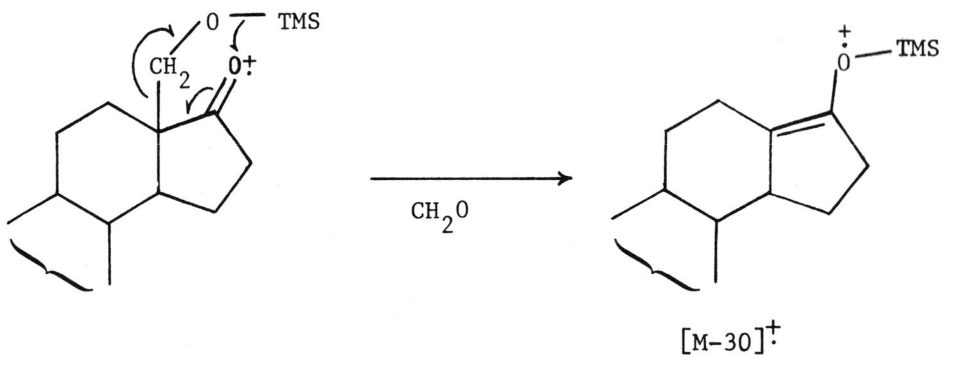

Scheme 17

[M-56]$^+$ ion. Formation of this ion is a classic example of the migratory properties of the TMS group and the confusion they can introduce in the interpretation of mass spectral data. The $[M-56]^+$ ion was originally observed, often as the base peak, in the mass spectra of the TMS derivatives of 5α-androstane-3α,11β-diol-17-one [92], androst-4-ene-6β,17β-diol-3-one [93], and pregn-4-ene-6β,17β-diol-3-one [94]. It was assumed that formation of $[M-56]^+$ involved loss of the elements of C_3H_4O, presumably by cleavage of the D-ring or the A-ring, depending on the site of the carbonyl group (17-keto or 3-keto). This did not explain, however, the presence of an abundant $[M-56]^+$ ion in the spectra of the TMS derivatives of pregn-5-ene-3β,17α-diol-20-one [95] or pregn-5,16-diene-3β-ol-20-one [96]. This ambiguity has since been clarified through the work of Brooks [97] and Sjövall [98] and co-workers. With the aid of ^{18}O, TMS-d_9,

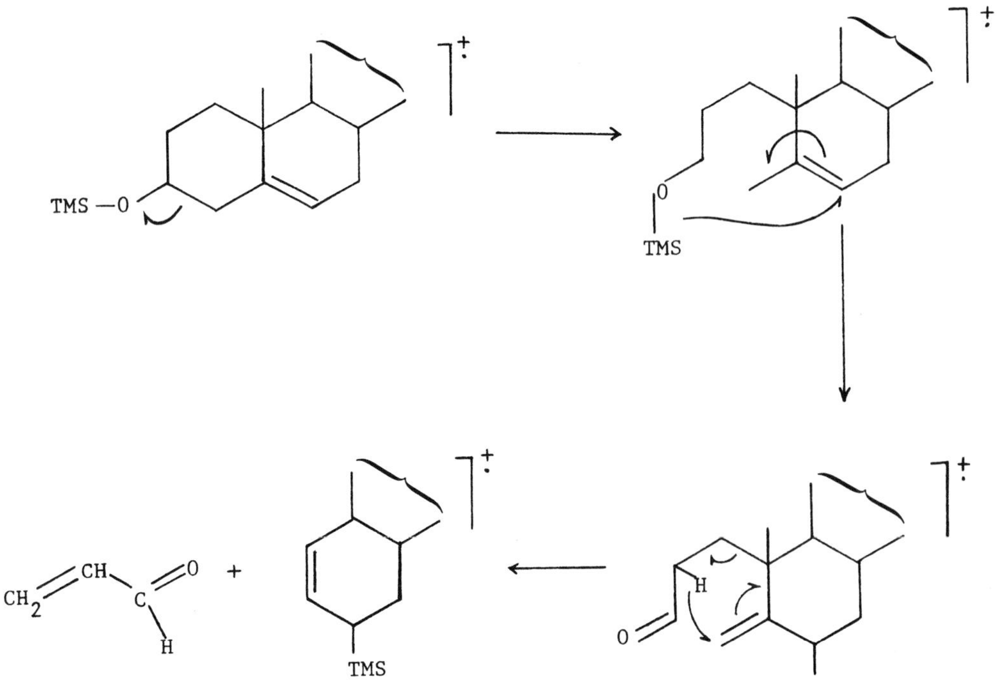

g

Scheme 18

and deuterium labeling of ring hydrogens, it was established that $[M-56]^+$ (g) was formed by intramolecular migration of the TMS group and elimination of C_3H_4O, the oxygen being that of the trimethylsilyloxy rather than that of the 17-keto function (Scheme 18).

(2) TMS-containing ions formed by direct cleavage:

A number of TMS-containing ions in steroid derivatives have come to be recognized as structurally diagnostic and although the formation of many of them does not involve intramolecular transfer of the TMS group itself, it does usually involve hydrogen and/or bond rearrangements. Some of the most important ones are summarized below.

m/e 129. The ion of *m/e* 129, h is characteristic of Δ^5-trimethylsilyl ethers of steroids and in early studies of the GC-MS of steroids it was used to identify the presence of such functionalities [99]. Its mechanism of formation was studied by Diekman and Djerassi [100] and is shown in Scheme 19. These authors pointed out that the same ion is also characteristic of 17-TMSO steroids and is formed by

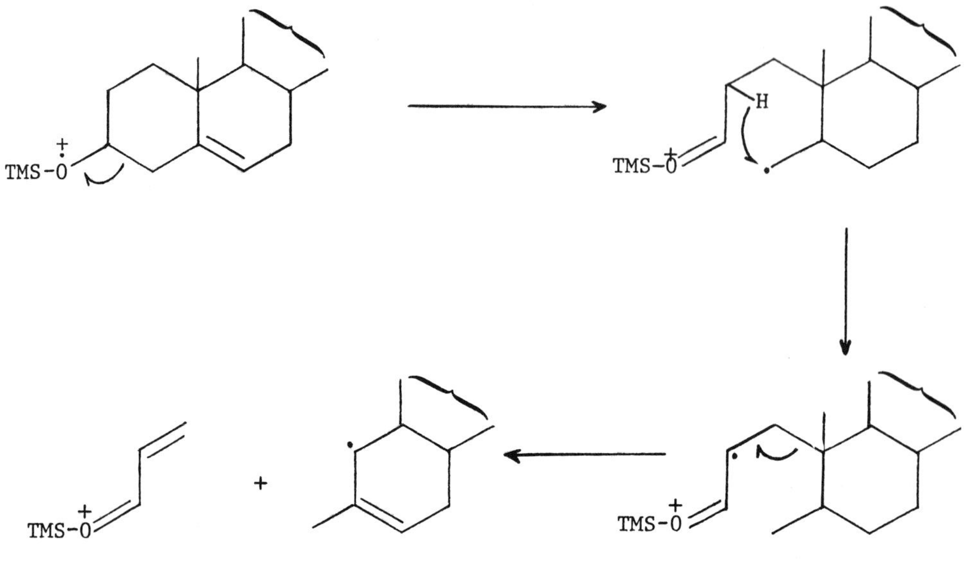

h, m/e 129

Scheme 19

scission of the C-17/C-13 bond and transfer of a 16-hydrogen to the steroid nucleus. It is interesting that the accompanying complementary ions in the case of the Δ^5-TMSO steroids occur at $[M-129]^+$ [101] while their 17-TMSO counterparts give rise to a peak corresponding to $[M-131]^+$. The value of these fragmentations from a structural identification viewpoint has been summarized in a more recent publication [102].

Typical ions produced by β-cleavage. Cleavage β- to the TMSO oxygen in D-ring substituted steroids yields the ions of *m/e* 103 and *m/e* 117 as indicated in structures i and j respectively [103,104].

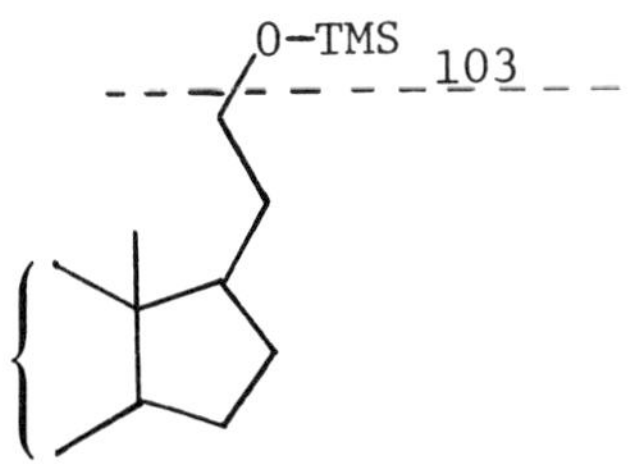

i

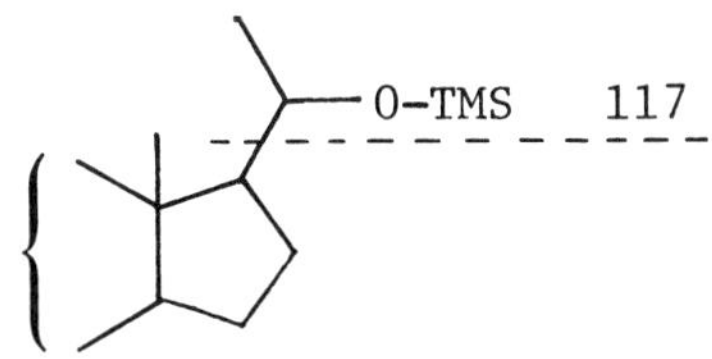

j

Their complementary ions $[M-103]^+$ and $[M-117]^+$ are usually also present, while *m/e* 103 and $[M-103]^+$ have also been observed in silyl ethers containing vicinal TMSO groups [93]. Of particular interest is the fragmentation of the bis-TMS derivative of the corticosteroid pregn-4-ene-3-one-17α,20α,21-triol in which partial silylation of the side chain yields a bis-trimethylsilyloxy compound, presumably with structure k. A major peak in the spectrum of this derivative occurs at $[M-133]^+$, rather than $[M-205]^+$, the fragmentation expected from the direct cleavage of the C-17/C-20 bond [105]. A similar fragmentation was observed in the spectrum of the tris-TMS-derivative of 5α-pregnane-3β,17α,20β,21-tetrol. Isotopic labeling [106] of the latter compound (C-20 oxygen with ^{18}O and the C-20 and C-21 hydrogens with ^2H) confirmed the loss of a rearranged side chain (Scheme 20). In many respects this is consistent with other intramolecular rearrangements and scramble of TMS groups located on the 17β- side chain of steroidal trimethylsilyl ethers [107].

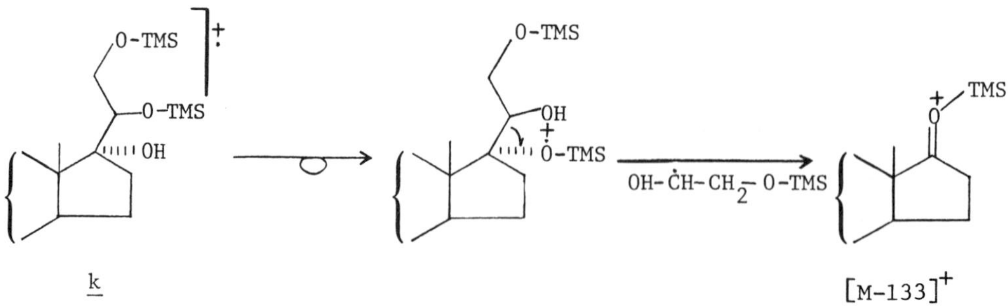

Scheme 20

b. Dimethylsiliconides. Dimethylsiliconide derivatives of steroids have been reported for certain corticosteroid diols. Dimethyldiacetoxysilane $[(AcO)_2Si(CH_3)_2]$ was shown to replace the C-17 and C-21 hydroxylic hydrogens and form a six-membered ring, with the silicon atom bridging the respective oxygens. A peak at $[M-28]^{+\cdot}$ in the spectra of several corticosteroids of dimethylsiliconides has been attributed to the elimination of CO [105,108]. It is noteworthy that unhindered hydroxyl groups also react to form dimethylacetoxysilyl derivatives whose mass spectrometric properties (e.g., molecular ion intensity) may not be as desirable as those of their corresponding TMS derivatives [109].

c. (Halomethyl)dimethylsilyl ethers. The spectra of chloromethyldimethylsilyl (CMDMS) ether derivatives of steroids closely resemble those of their trimethylsilyl counterparts, the chlorine isotopes providing a good label for mass spectral interpretation purposes. Specific effects on fragmentation are induced by this larger-than-TMS functionality and become more apparent when the derivatives of stereoisomers are compared. Such effects had been previously noted with the parent hydroxylic compounds or their TMS derivatives, but were not as significant as with the chloromethyldimethylsilyl analogs. For example, in a study of the mass spectra of the chloromethyldimethylsilyl derivatives of etiocholanolone, epietiocholanolone, androsterone, and epiandrosterone [110] the following features were noted.

1. Elimination of $ClCH_2$-$Si(CH_3)_2$-OH from $M^{\ddagger}$ to yield *m/e* 272 occurred in appreciable extent only from the 5β-isomers, presumably due to steric crowding of the hydrogens at C-2, C-4, and C-9.
2. A peak of significant intensity at *m/e* 271 occurred only in the spectra of the 3α-chloromethyldimethylsilyl ether isomers, and was derived in the sequence: $M^{\ddagger} \rightarrow [M\text{-}CH_2Cl]^{\ddagger}$ (*m/e* 347) → *m/e* 271.
3. In spite of the disadvantages of higher GC retention times of the CMDMS as compared with the TMS ethers, the occurrence of major peaks at high *m/e* values was potentially advantageous for improved detection limits by single-ion monitoring as background peaks are significantly reduced in intensity in that region of the spectrum.

d. Enol-trimethylsilyl derivatives. Under base-catalyzed conditions BSTFA was shown to react with carbonyl groups on steroids to produce enol derivatives. From a practical point of view this reaction should be most useful in the derivatization of the thermally unstable dihydroxyacetone side chain of steroids. A single C-20/C-21 enol derivative is formed in the latter case, while isomeric mixtures or partially derivatized starting material are encountered in the case of 3-keto-Δ^4-steroids or 17-ketosteroids [111]. Typical D-ring cleavages associated with the enolized dihydroxy acetone side chain are indicated on structure 1.

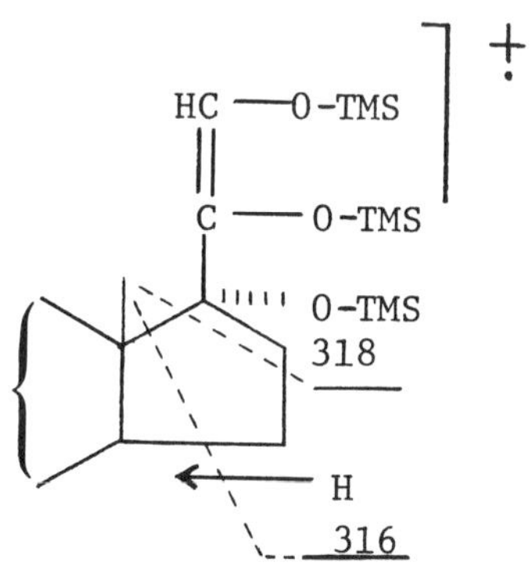

1

e. Other trialkylsilyl ethers. Trialkylsilyl derivatives with the general structure RX_2Si, where R = t-butyl or i-propyl and

X_2 = Me_2 or $-C_4H_8-$, have been used for the mass spectrometric characterization of steroids [112]. In comparison with their trimethylsilyl analogs, these sterically crowded trialkylsilyl derivatives offer certain advantages for GC-MS, most notably:

1. Higher resistance to hydrolysis
2. Higher GC retention properties, which can be utilized whenever desired
3. Less extensive fragmentation
4. Abundant ions in the important high mass region of the mass spectrum
5. Higher mass spectral sensitivity to the stereochemical environment of the functional group

In general, many fragmentations typical of the TMS derivatives of steroids (e.g., formation of ions equivalent to g or h, *vide supra*), are greatly suppressed in the spectra of the sterically crowded trialkylsilyl derivatives. The molecular ion peaks in the mass spectra of these derivatives are of low relative intensity, but molecular-weight information can be readily derived from the highly abundant $[M-R]^+$ ions, occurring at $[M-57]^+$ when R = t-butyl or $[M-43]^+$ when R = i-propyl. The high abundance of this ion can be explained by the relatively high stability of the t-butyl and i-propyl radicals compared with that of $CH_3^{\cdot}$, and, possibly, the relief of crowding in the silyl moiety. It is, in fact, this cleavage that initiates one of the most interesting and potentially most diagnostic processes in the spectra of these derivatives, as the loss of $R^{\cdot}$ is followed by elimination of HX_2SiOH. For the t-butyldimethylsilyl derivatives of 3α-hydroxy-5α-steroids, formation of the $[M\text{-t-butyl-}HSiMe_2OH]^+$ ion has been explained by the mechanisms indicated in Scheme 20a, which involve a double hydrogen transfer to oxygen and silicon respectively. Two transitions, $k_1 \rightarrow k_2$ and $k_3 \rightarrow k_4$, have been proposed as possible pathways for this fragmentation, both based on the chair conformation of a ring A from which HX_2SiOH can be eliminated by concerted bond rearrangements. In either case the indicated hydrogens can approach to within bonding distance of the Si and O atoms, which provides for an energetically favorable transition and

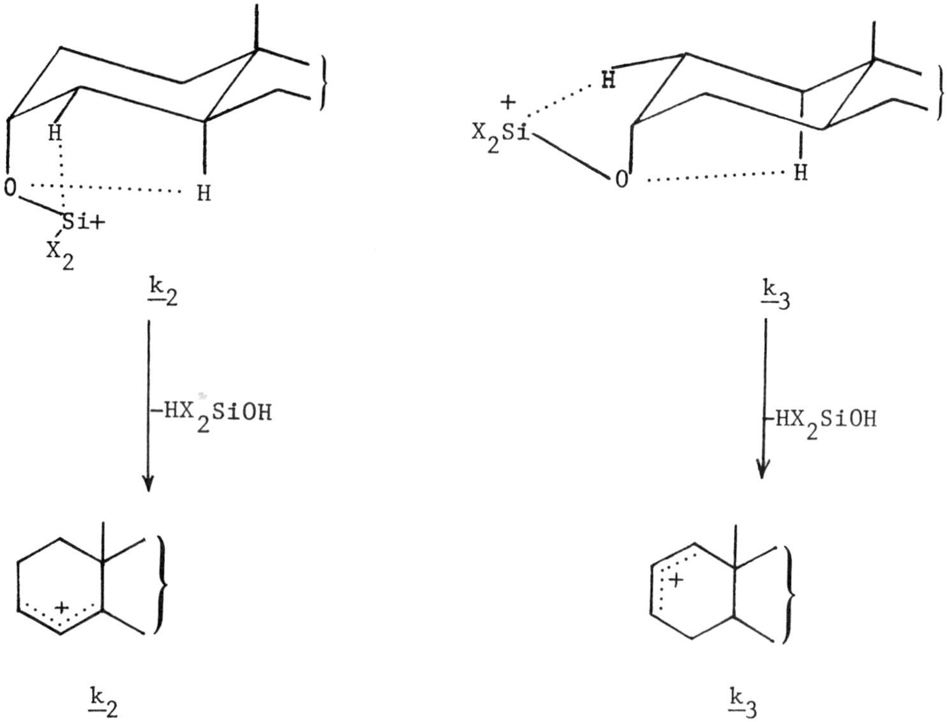

Scheme 20a

this is further enhanced by the stability of the resulting allyl carbonium ions $\underline{k}_2$ and $\underline{k}_4$. Analogous explanations are possible for other positions of silyloxy substitution. Several examples have been given [112] that confirm that the loss of HX_2SiOH is strongly dependent on the steric environment of the silyloxy group. It is conceivable that this dependence is influenced by the restrictions imposed by the multicenter nature of the postulated transition states and, if confirmed by isotopic labeling, this process can be of major diagnostic value.

2. *Methyl Derivatives*

Methyl ethers of sterols are noted for their resistance to hydrolysis and good gas chromatographic properties. In a study of the mass spectra of a number of methylated sterols by Idler et al. [113], the following characteristic features were noted.

- Mass increment introduced by permethylation is low as compared with other types of derivatization.
- Generally intense molecular ion peaks are observed.
- In an analogy to the elimination of H_2O from the molecular ions of free sterols, formation of $[M-CH_3OH]^{+\cdot}$ occurs mainly in the 5β-epimers.
- Δ^5-sterol methyl ethers exhibit intense $[M-CH_3OH]^{+\cdot}$ peaks.
- Δ^5-sterol methyl ethers show a $[M-71]^{+}$ ion (loss of CH_3O + C_3H_4), a fragmentation analogous to the formation of $[M-129]^{+}$ in TMS ethers.
- Much of the breakdown occurs from the steroid skeleton, indicating little or no charge localization on the methoxy group.

Despite the numerous gas chromatographic and mass spectrometric advantages offered by the permethyl derivatives, they have generally seen limited use in the GC-MS analysis of steroids. The limiting factor is actually the methylation procedure itself. In order to achieve the desired permethylation of all replaceable hydroxylic hydrogens, more drastic reaction conditions often have to be employed that result in isomerization of double bonds or other alterations in the structure of the parent molecule. For example, in permethylation of steroids using a methylsulphinyl carbanion as base, "overmethylation" of steroidal dienones or steroid glucuronides was observed [114,115]. This procedure has been used effectively in the 16,16-O-permethylation of 17-ketosteroids, but complications arose with 11-ketosteroids because of the production of a mixture of partially methylated compounds [116].

3. *Boronate Derivatives*

Cyclic boronates in the area of steroid analysis are most applicable for GC-MS studies of corticosteroids, where stable six- or five-membered ring diesters involving the C-17/C-21, C-20/C-21 or C-17/C-20 oxygens can be formed. Various alkyl- and phenylboronates have been investigated and the spectra of most of them exhibited intense molecular ion peaks [117]. Methylboronates are noted for their volatility and their use may often be desirable in spectrometry because of the small mass increment they introduce to the parent compound. Scission of the D-ring, Scheme 21, is a typical process of most boronate

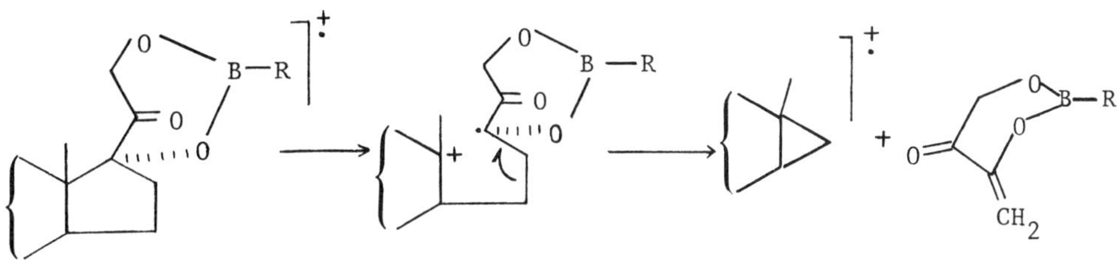

Scheme 21

derivatives of corticosteroids. In addition, the spectra of t-butyl boronates show an intense $[M-57]^+$ ion corresponding to the loss of the t-butyl group. In general, the utility of cyclic boronate derivatives is limited to corticosteroids containing no hydroxyl groups other than those in the dihydroxyacetone side chain, primarily because of complications associated with the derivatization reaction [105].

4. *Methyloxime (MO) Derivatives*

As discussed in Sec. III, methyloximes are the most commonly employed derivatives for carbonyl groups. From both a gas chromatographic and a mass spectrometric viewpoint, they are generally more desirable than benzyloximes or O-trimethylsilyloximes. The spectra of benzyloximes are dominated by the *m/e* 91 ion [3], while formation of O-trimethylsilyloximes requires a double derivatization step. Ethyloxime derivatives have been used mostly to provide group labeling and facilitate interpretation of the mass spectra of their methyloxime analogs.

The mass spectra of the methyloxime derivatives of steroids exhibit characteristic peaks of $[M-31]^+$, corresponding to a N-O bond cleavage and elimination of a methoxy radical. The same fragmentation was observed in the mass spectra of O-methyloximes of long-chain aliphatic aldehydes and ketones, although it was usually significantly suppressed in the shorter length (C-7 or less) compounds [118]. Cleavage of bonds β or γ to the C=N carbon, a process encountered in the spectra of methyloxime derivatives of long-chain compounds, is also of major significance in the MO derivatives of steroids. Typical ring fragmentations of 3- and 2-ketoxime steroids have been summarized by Brooks and Middleditch [81].

B. Prostaglandins

The principal types of chemical derivatives employed in the GC-MS analysis of prostaglandins were summarized in a recent review by Crain et al. [119]. The functionalities protected are the carboxylic group (methyl ester or trimethylsilyl ester), carbonyl group (methyl- or benzyloxime), and the hydroxyl group (trimethylsilyl, acetyl, trifluoroacetyl, phenylboronyl, or methyl ether). Typically, prostaglandins are analyzed as their methyl ester-trimethylsilyl ether-methyloxime (ME-TMS-MO) derivatives although the trimethylsilyl ester analogs have also been examined in detail.

Interesting solvent effects have been observed with the trimethylsilylation of prostaglandins of the E and A series. For example, reaction of PGE methyl esters with TSIM in piperidine [120] resulted in their quantitative conversion to the mono-TMS-PGB methyl esters, 38. This is shown in Scheme 22 for the case of the methylester of

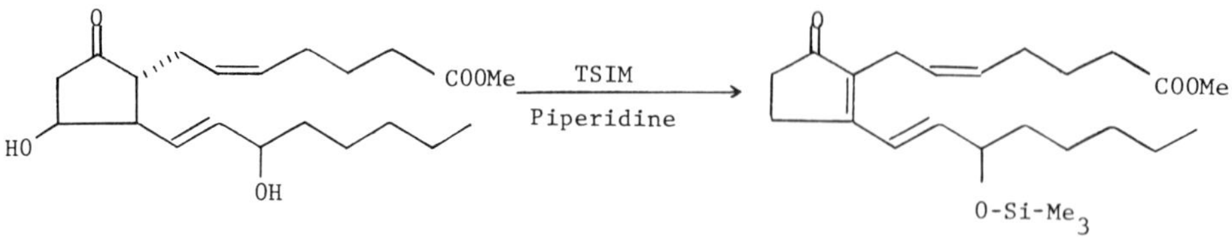

37 38

Scheme 22

PGE_2, 37. When the same reaction was carried out neat in TSIM several products were obtained, including PGA_2-TMS-ME and PGB_2-TMS-ME. It appears that piperidine was instrumental in providing the basic conditions needed for completion of the reaction and the yield of a single product as indicated in Scheme 22. In a related study, Rosello et al. [121] examined the effect of the mixture BSTFA-piperidine on the trimethylsilylation of prostaglandins of the A_1, A_2, B_1, B_2, E_1, E_2, $F_{1\alpha}$, and $F_{2\alpha}$ series. Prostaglandins F and B were converted to the tetrakis- and bis-TMS derivatives as anticipated, but the E prostaglandins were enolized to yield 9-enol-tetrakis (TMS) derivatives. The presence of piperidine was essential to ensure quantitative enolization at the 9-positions of the E prostaglandins, but the doubly conjugated 9-oxo group in the PGB series prevented the latter compounds from undergoing a similar enol reaction. Of particular significance was the observation of incorporation of nitrogen from the piperidine of the reaction mixture into the PGA prostaglandins to produce a unique and novel derivative corresponding to the TMS esters of 9-piperidyl-9,15-bis(trimethylsilyloxy)-PGA prostaglandins. Addition of a piperidyl group by reaction of the positively charged carbonyl center with the piperidine nitrogen to yield the derivative indicated by structure 39 was postulated [120]. While the structure, 39, proposed by the authors was consistent with

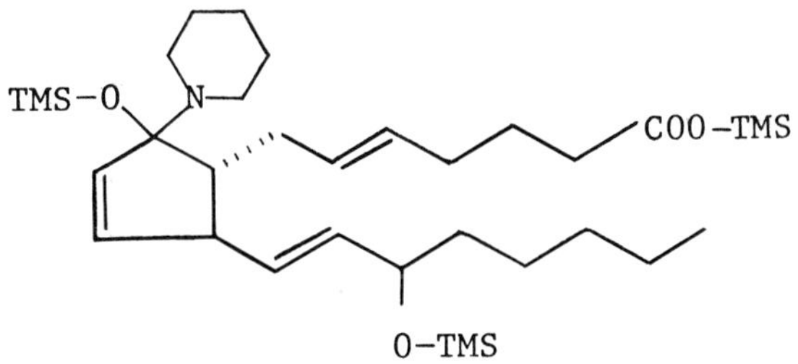

39

the molecular weight observed in the mass spectrum, a derivative, 40, formed by the classical Michael reaction (Scheme 23) should perhaps be considered. However, the mass spectral data alone are not sufficient to distinguish on a definitive basis between structures 39 and

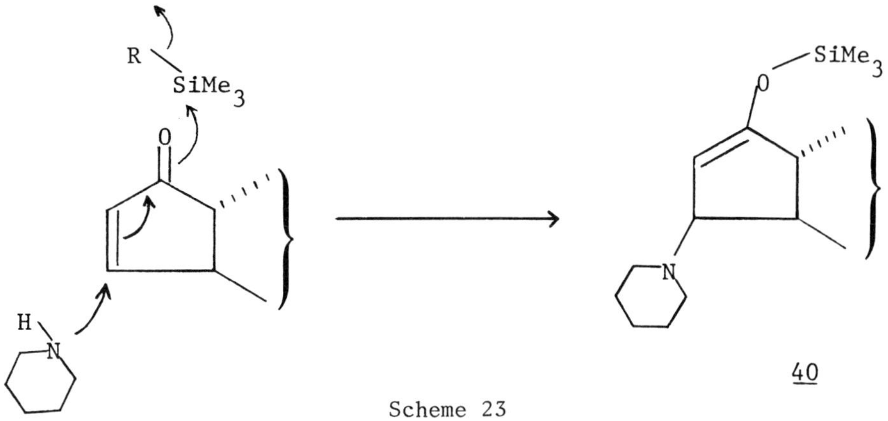

40

Scheme 23

40. It should be emphasized that in the absence of piperidine the expected bis-TMS derivatives of the PGA_1 and PGA_2 prostaglandins were obtained.

In view of the complications associated with the derivatization of oxo-prostaglandins, the formation of O-methyl oxime derivatives has been generally sought. The formation of syn and anti isomers can often be a nuisance, one known exception being the MO-TMS derivatives of prostaglandins of the B series where only one isomer was observed during GC [122]. A notable feature in the mass spectra of the MO-TMS derivatives of the PGB_1 and PGB_2 series is the stability of the ion formed by loss of the methoxy group of the oxime moiety, which carries well over 30% of the total ion current in the respective spectra. This contrasts with the spectra of their free keto-TMS ether counterparts where much more extensive fragmentation was observed.

In a series of papers, Middleditch and Desiderio [122-125] examined the mass spectra of the TMS-ester-TMS-ether-MO derivatives of prostaglandins of the A, B, F, and E series. With the aid of perdeuterio-trimethylsilyl and TMS-ether-d_9-TMS-ester derivatives, the identity of the trimethylsilyl groups involved in the formation of the various fragment ions was established. For example, ions produced due to remote interactions of trimethylsilyl groups were not found in the spectra of prostaglandins of the A and B series, but the ions at *m/e* 204 and *m/e* 217 in the spectra of derivatives of the E series (41) were found to contain an ethereal- and an ester-TMS group. Ion structures m and n (Scheme 24) are consistent with

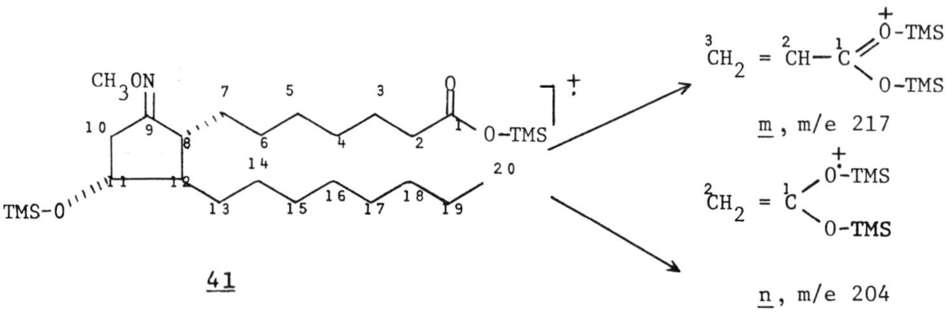

41

Scheme 24

the deuterium content of these ions in the spectrum of the MO-TMS derivative of 3,3,4,4-d_4-prostaglandin E_2. For the most part, however, the fragmentations in most of these derivatives appear to be initiated by direct β- cleavages or other fragmentation reactions typical of substituted alicyclic systems.

Prostaglandins of the F series as their TMS-ether-TMS-ester derivatives exhibit generally weak but recognizable molecular ion peaks. Sequential losses of 90 amu (trimethylsilanol) are more pronounced in the spectra of TMS derivatives of the F series than in those of the A or B series. Similarly, the spectra of the corresponding O-acetyl-ME derivatives of F series prostaglandins are dominated by ions corresponding to sequential eliminations of CH_3COOH. One notable difference is the virtual absence of molecular ion peaks [126]. The use of a stereospecific derivative-(9,11-butylboronate-15-trimethylsilyl-ether-) for nanogram-level analysis of $PGF_{2\alpha}$ has been reported by Kelly [43]. The cyclic derivative, 42, can provide for distinction between cis and trans diols and a peak at *m/e* 435 arising from cleavage of the C_{16}-C_{20} side chain, and which carried approximately 11% of the total ion current, was effectively used for selective ion monitoring.

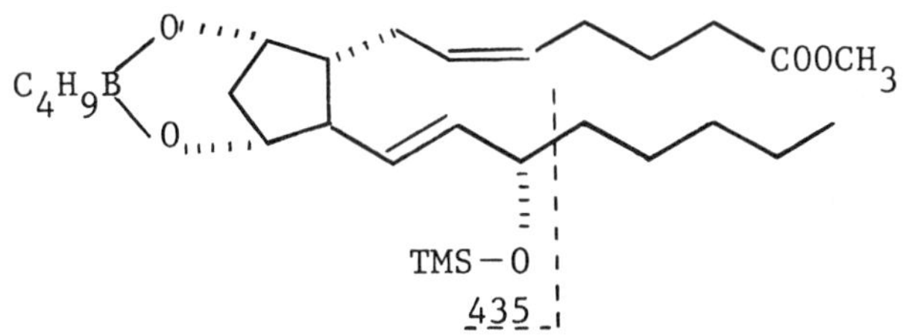

42

In a comparative study, Watson and Sweetman [127] evaluated different types of derivatives of isomeric prostaglandins A and B. In each case it was possible to distinguish between the PGA and PGB isomers on the basis of their different skeletal fragmentations. For example, both the PGA_2 and PGB_2 derivatives (43 and 44) exhibit characteristic peaks at $[M-71]^+$--i.e., $[M-C_5H_{11}]^+$, and $[M-99]^+$--i.e., $[M-C_5H_{11}CO]^+$, the latter ion occurring in much higher abundance in the PGB_2 isomers. The CO lost in this fragmentation originates from the ring carbonyl group. The spectra of the individual derivatives are further identified by specific fragmentations, characteristic of the functionalities involved. In the methyl ester derivatives expulsion of CH_3OH, from $[M-71]^+$, or from other fragment ions, was frequently observed. The spectra of the ME-t-Bu-DMS derivatives of PGA_2 and PGB_2 are noted for the abundant molecular ions and the intense peak at $[M-57]^+$ corresponding to loss of the t-butyl group. Formation of the latter ion is followed by elimination of methanol. Moreover, loss of t-butyldimethylsilanol, the equivalent of loss of trimethylsilanol (90 amu) in TMS derivatives, was a minor process in itself, but was followed by expulsion of a methoxy radical to yield a major ion at $[M-163]^+$ (*m/e* 299). Finally, the spectra of the ME-acetyl derivatives of PGA_1 and PGB_1 (45, 46) in addition to the typical elimination of CH_3COOH, $[M-60]^{+\cdot}$, also exhibit peaks corresponding to the loss of ketene, $CH_2=C=O$.

The high selectivity and sensitivity of combined GC-MS provides one of the ideal methods for the detection and quantitation of prostaglandins at the subnanogram level. This is usually performed with methyl ester-MO-TMS derivatives and is based on the use of multiple-ion-detection techniques [128]. Isotopically labeled analogs are used both as internal standards and carriers to extend the detection limits of the GC-MS systems [129,130], permitting reliable quantitations at the picogram level.

C. Amino Acids

Vapor phase analysis of amino acids is severely hampered because of their zwitterionic character. For MS introduction via the direct inlet probe it is normally sufficient to remove the polarity of the carboxylic acid function by formation of a methyl or ethyl ester. Structurally informative spectra of most amino acids were thus obtained in this fashion [131]. Further removal of the polarity associated with the amino group is necessary, however, for effective GC or GC-MS analysis of most amino acids.

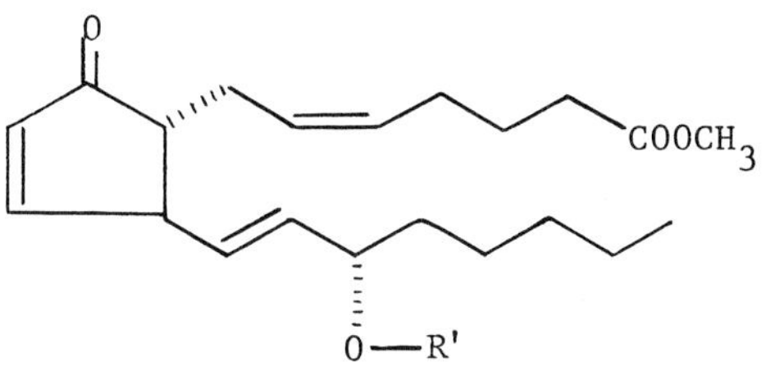

43, PGA_2

$R = CH_3$, $R' = Si(CH_3)_3$
$R = CH_3$, $R' = SiH(CH_3)_2$
$R = R' = Si(CH_3)_3$
$R = CH_3$, $R' = Si(CH_3)_2C(CH_3)_3$

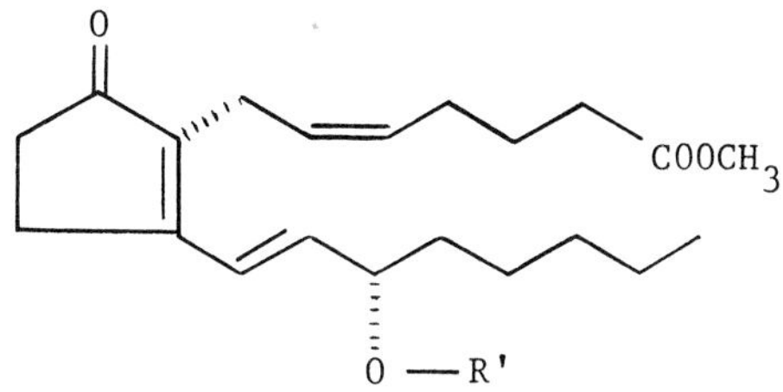

44, PGB_2

$R = CH_3$, $R' = Si(CH_3)_3$
$R = CH_3$, $R' = SiH(CH_3)_2$
$R = R' = Si(CH_3)_3$
$R = CH_3$, $R' = Si(CH_3)_2C(CH_3)_3$

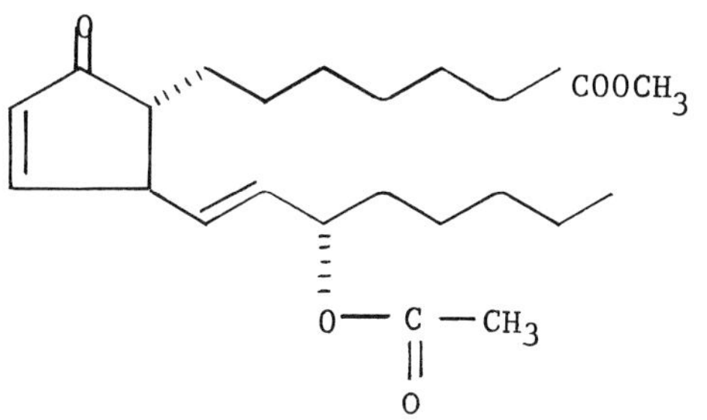

45, PGA_1

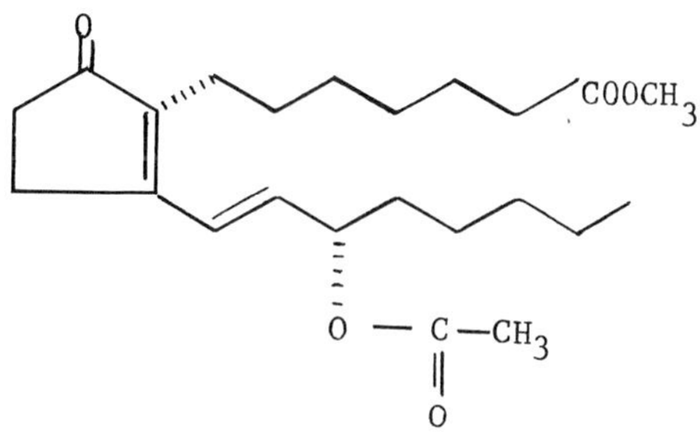

46, PGB_1

Derivatization of amino acids for GC-MS is most typically based on formation of their alkyl ester-N-acyl derivatives. N-Trifluoroacetyl (N-TFA)-n-butyl ester derivatives, first introduced in the early 1960s, are now commonly employed for GC-MS analysis of amino acids [132,133]. Variations of the reaction sequence leading to this derivative have been developed in an attempt to increase the yield and ensure the stability of products and intermediates [134]. In addition to N-TFA-n-butyl esters, other analogs have also been evaluated. For example, a recent communication reports on the preparation of HFB-i-amyl esters of amino acids via a three-step reaction as summarized in Scheme 25. It should be noted that in accordance

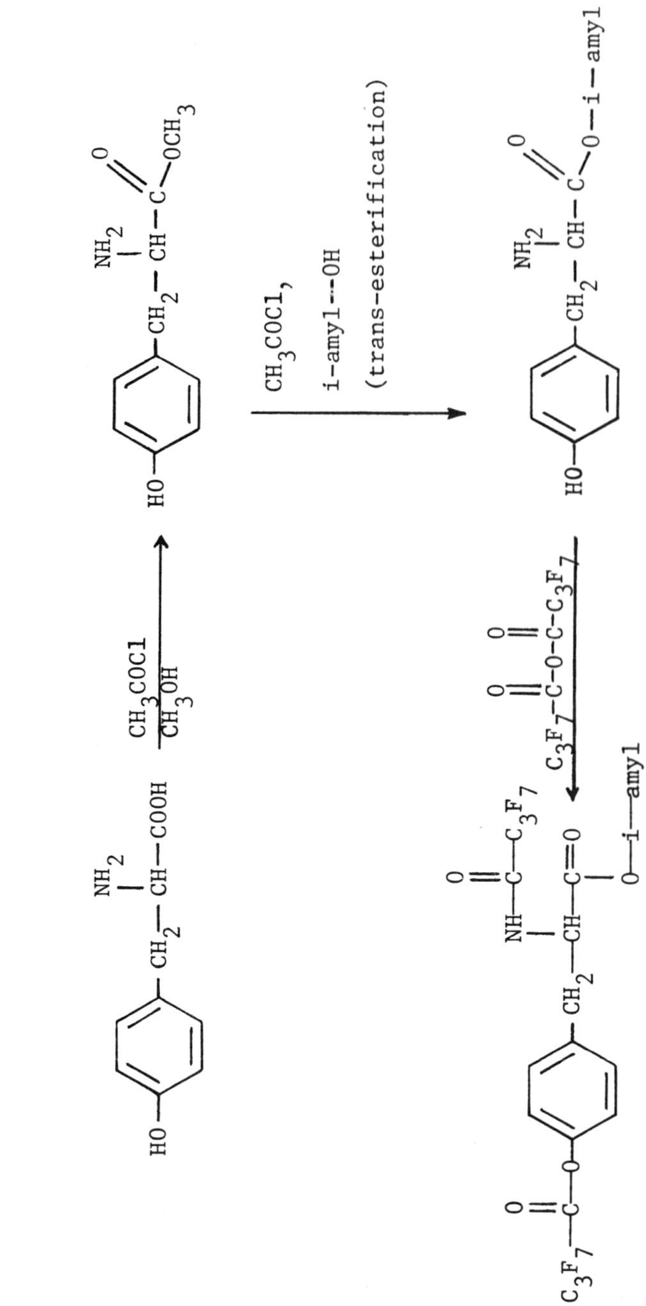

Scheme 25

with the methodology developed by Lamkin and Gehrke [134], direct esterification with i-amyl alcohol/HCl is not used in order to take advantage of the higher solubility of the amino acid in methanol and the high yield of the methyl ester formation particularly at microgram levels. Instead, a transesterification process following the rapid formation of a methyl ester is preferred [135].

Consistent with the fragmentation mode of the esters of amino acids [136], the principal cleavage of N-TFA-n-butyl ester derivatives also involves the carbon-carbon bonds next to the amino group. Thus, the base peak in the mass spectra of the N-TFA-n-butyl esters of α-amino acids such as alanine, sarcosine, N-methylalanine, and N-ethylglycine is formed according to the cleavage shown in Scheme 26 [137].

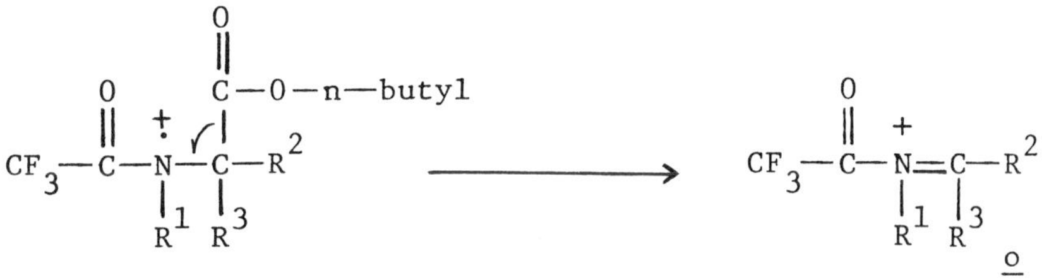

Scheme 26

While the peak arising from ion o dominates the spectra of N-TFA-n-butyl esters of α-amino acids, no single ion stands out as sharply in the spectra of the corresponding derivatives of β-amino acids. Accordingly, the fragmentation of N-methylalanine-N-TFA-n-butyl ester is indicated on structure 47 and its mass spectrum partially summarized in Table 5. Additional examples can be found in the study of Gelpi et al. [138], who examined the mass spectra of a variety of amino acids as their N-TFA-n-butyl ester derivatives.

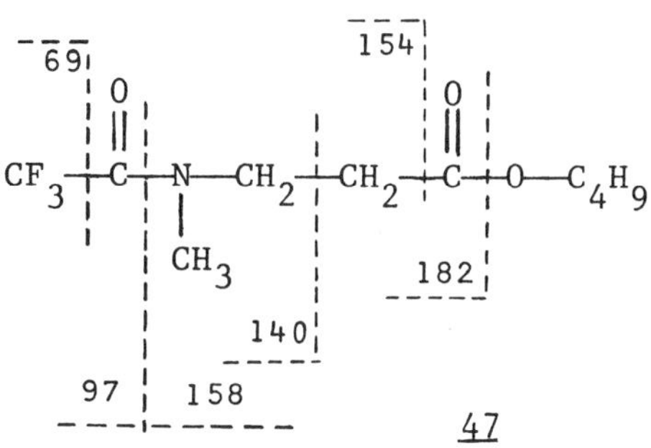

47

TABLE 5

Partial Mass Spectrum of N-TFA-n-Butyl Ester of N-Methylalanine

m/e	% RI
255, $M^{+\cdot}$	8
183	43
182	35
158	20
154	9[a]
153	82
140	100
97	5
69	16

[a]Corrected for isotopic contribution from *m/e* 153.

A two-step derivatization process of α-amino acids has been proposed by Nielsen and Solheim [139]. The N-benzoyl derivative, 48, produced by reaction of the amino acid with benzoyl chloride, was converted to 2-phenyl-4-methyl-oxazolin-5-one, 49, upon reaction with dicyclohexylcarbodiimide (Scheme 27). The mass spectra of the derived azlactones appear of little use for structural identification purposes

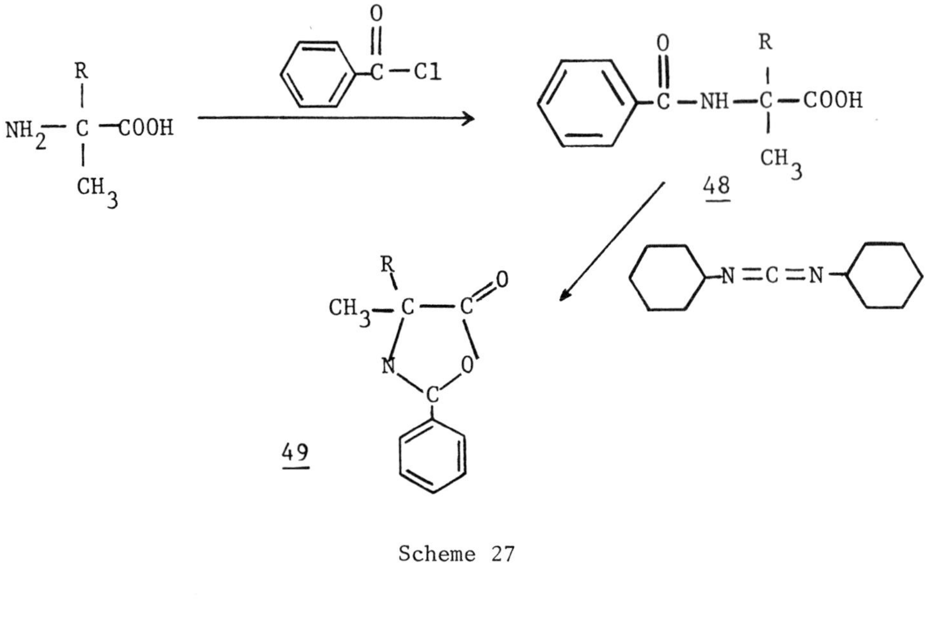

Scheme 27

as they exhibited very weak or no molecular ion peaks, and were generally dominated by the structurally uninformative ion of *m/e* 105 (C_6H_5-C≡O^+).

The principal drawback of all the aforementioned derivatization methods for amino acids is the requirement for a two- or three-step reaction for their conversion into a volatile entity suitable for GC-MS analysis. To overcome this problem a number of alternatives have been examined. Dimethylformamide dialkylacetals, 50, were shown to react with both carboxylic and amino groups to form N-dimethylaminomethylene alkyl esters [140]. The mass spectra of the N-dimethylaminomethylene-methyl ester (N-DMAM-ME) derivatives of several amino acids were examined. All of them gave recognizable molecular ion peaks and generally well-defined fragmentations. The flexibility of being able to prepare homologous alkyl esters provided an excellent means of group labeling and thus identifying the origin or relevant fragment ions. The mass spectrum of valine-N-DMAM-ME (MW 186) is as follows: *m/e* 186 [6], *m/e* 143 [100], *m/e* 127 [92], *m/e* 57 [22], *m/e* 44 [105], *m/e* 42 [60]. The cleavages involved in the formation of the major ions are indicated on 51. The dimethylformamide dialkyl acetal reagent has also been shown to react quantitatively with the guanidine group of arginine and the semicarbazone group of citrulline [141]. For hydroxy amino acids, however, a second step, preferably trimethylsilylation, is recommended to convert the N-dimethylaminomethylene alkyl ester into a derivative suitable for GC-MS analysis.

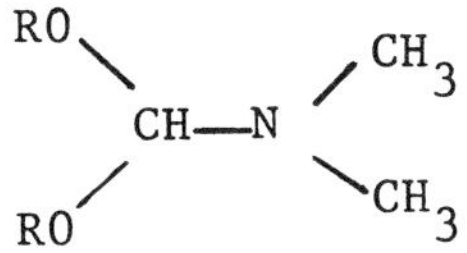

50

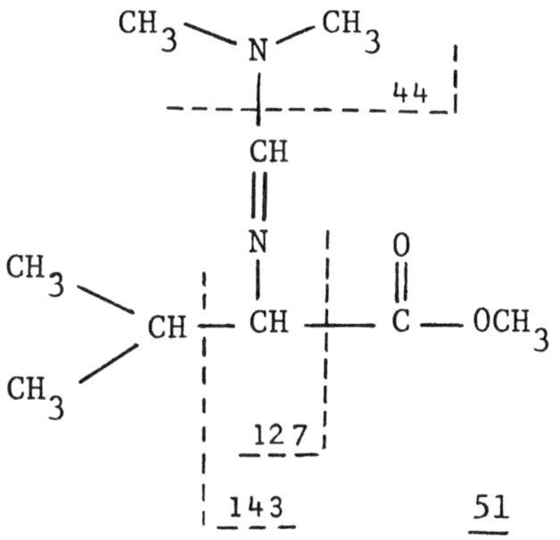

51

With the introduction of silylating reagents and their capability of replacing amino, carboxylic, and hydroxylic hydrogens, the objective of a single-step derivatization of amino acids has come closer to realization. Trimethylsilylation has been studied extensively vollowing its original use by Rühlmann and Giesecke [142]. Various silylating reagents have been evaluated, including HMDS and TMCS [142], BSA [143], TMSDEA [144], and BSTFA [145]. Gehrke has reported on the quantitative derivatization and GC separation of the 20 protein amino acids as their TMS derivatives formed by reacting with BSTFA [146,147]. Establishment of the correct conditions for quantitative derivatization is important in the trimethylsilylation of amino acids. Allowing adequate time for complete reaction in combination with the choice of proper reagent(s) minimizes the chances for formation of multiple derivatives for the same amino acid (e.g., N,N-bis-TMS and N-TMS). For example, this is the case in the derivatization of glycine and lysine when reacted with BSTFA under certain conditions [148]. Therefore, because of these potential complications the various parameters of reaction time, temperature, solvent, reagent, etc., have to be taken into consideration. Nonetheless, the feature of the one-step derivatization reaction and the added capability to use the reagent as a solvent make trimethylsilylation an attractive method for the GC-MS analysis of amino acids.

The mass spectra of the TMS derivatives of several amino acids have been determined and generally the fragmentations occur in a predictable fashion at positions β to the heteroatom. Partially (as compared with completely) silylated derivatives can be distinguished on the basis of the occurrence of ions at *m/e* 174 or *m/e* 102 as shown for glycine, 52, 52a, or the equivalent ions in other amino acids.

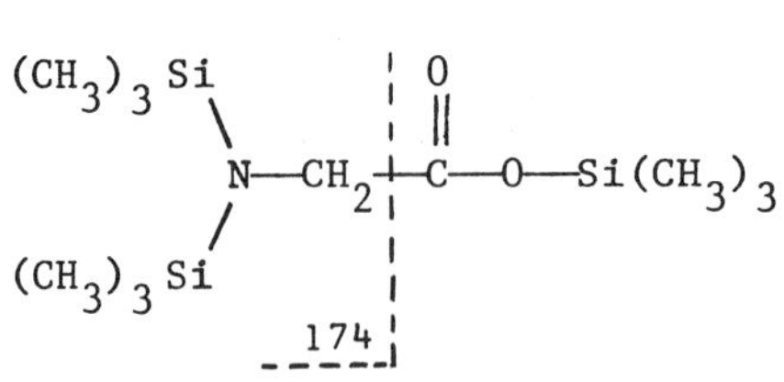

52

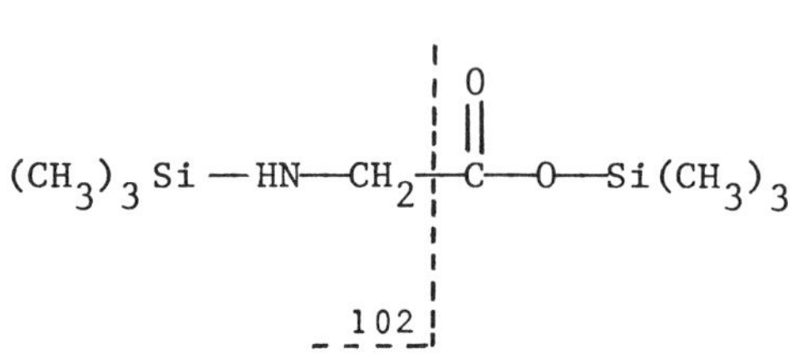

52a

The picture is somewhat more complex with lysine, however, where five different silylation sites are possible. Marked variations have been observed in the mass spectra of the tri-TMS derivative as a function of ion source temperature, presumably due to thermally induced intramolecular TMS arrangements. Preparation of the methyl ester-TMS derivative simplifies interpretation of specific data and suggests that it might be preferable to preparation of TMS esters.

A promising single-step derivatization method for amino acids involves an on-column methylation of their neopentylidene trimethylanilinium salts by pyrolysis at the injection port of the GC [149]. The amino acid was dissolved in a pivaldehyde-trimethylanilinium hydroxide (TMAH) solution and the trimethylanilinium salt of the resulting Schiff's base was converted to the methyl ester by pyrolysis (Scheme 28). Accordingly, tyrosine was converted to its O-methyl-neopentylidine-methyl ester and the ring nitrogens of tryptophane and

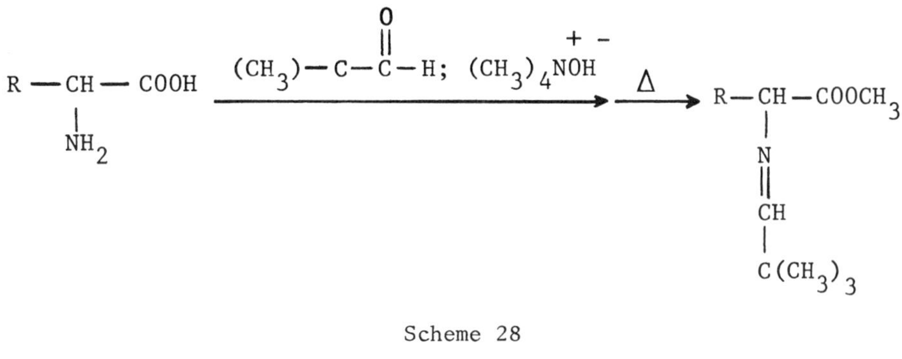

Scheme 28

proline were also methylated. The mass spectra of the 18 amino acids derivatized by this method showed very weak or no molecular ion peaks. The base peak usually consisted of ions of the type $[M-C(CH_3)_3]^+$ or $[M-COOCH_3]^+$. Deviations from this pattern were observed with the aromatic amino acids.

Identification of the N-terminal amino acid in a peptide chain by GC or GC-MS, following Edman degradation, has been the subject of numerous studies. As the degradation process with phenylisothiocyanate (Scheme 29) leads to the formation of a phenylthiohydantoin, 54, attention has been focused on the structural elucidation of the latter derivative. The phenylthiohydantoin (PTH) derivatives of all

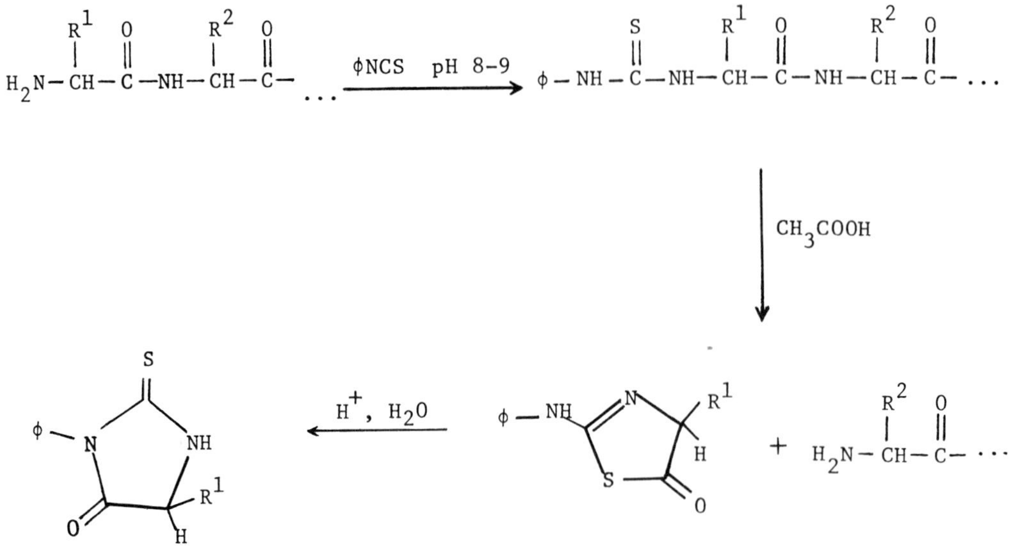

54, Phenylthiohydantoin

53, Thiazolinone

Scheme 29

protein amino acids, except arginine, were analyzed by GC [150,151]. Waterfield and Haber substituted methyl for phenyl isothiocyanate and the derived methylthiohydantoins were analyzed by GC [152]. In addition, unsubstituted thiohydantoins have been produced by reaction with NH_4CNS [153]. For GC-MS analysis the latter derivatives were further silylated to improve volatility [154]. Scheme 30 illustrates the reaction sequence for the conversion of glycine. The glycine thiohydantoin, 55, when treated with BSTFA:pyridine (1:1 v/v) yielded two products, a glycine-N,N'-bis-TMS-thiohydantoin, 56, and a glycine-N,N',O-tris-TMS-thiohydantoin, 57. The solvent used in this silylation reaction affects the number of products obtained [155]. For example, in dioxane only the N,N'-bis-TMS product was observed, whereas in pyridine the indicated mixture was found.

While the mass spectra of N-TFA-n-alkyl esters of amino acids generally exhibit weak molecular ion peaks, the opposite is true for their phenylthiohydantoin derivatives. In the latter, the molecular ion peak is invariably quite intense but the fragment ion peaks are

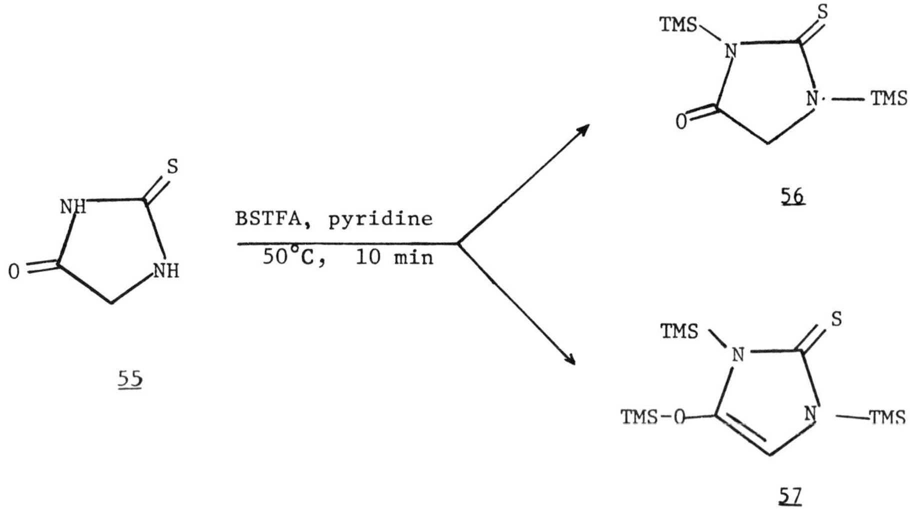

Scheme 30

usually representative of the thiohydantoin ring rather than the amino acid structure, and thus, in some respects, not quite as informative as might be desired. This is illustrated in the mass spectrum of the PTH derivative of isoleucine (Fig. 9). The ion at *m/e* 192 corresponds to phenylthiohydantoin, and is formed by elimination of i-butane from $M^{+\cdot}$. Furthermore, *m/e* 135 corresponds to phenylisothiocyanate and *m/e* 77 to $C_6H_5^+$ [156]. Similar spectra were obtained with methylthiohydantoin derivatives [157].

The mass spectra of the TMS derivatives of thiohydantoins exhibit easily recognizable molecular ion peaks accompanied by an abundant $[M-CH_3]^+$ ion, typical of compounds containing a TMS moiety. Another prominent peak also appears at *m/e* values corresponding to $[M-72]^+$ and attributed to the elimination of $(CH_3)_2$ $Si{=}CH_2$ [158]. The site of silylation in the trimethylsilyl derivative of PTH glycine and PTH alanine appears to be the NH function and is supported by the intense peaks at *m/e* 86 and *m/e* 100 respectively. Formation of these ions has been rationalized by the mechanism depicted in Scheme 31 [59].

Identification of the N-terminal amino acid by GC-MS of the thiohydantoin derivative proved somewhat of a problem with peptides containing asparagine or glutamine side-chain amides. The latter

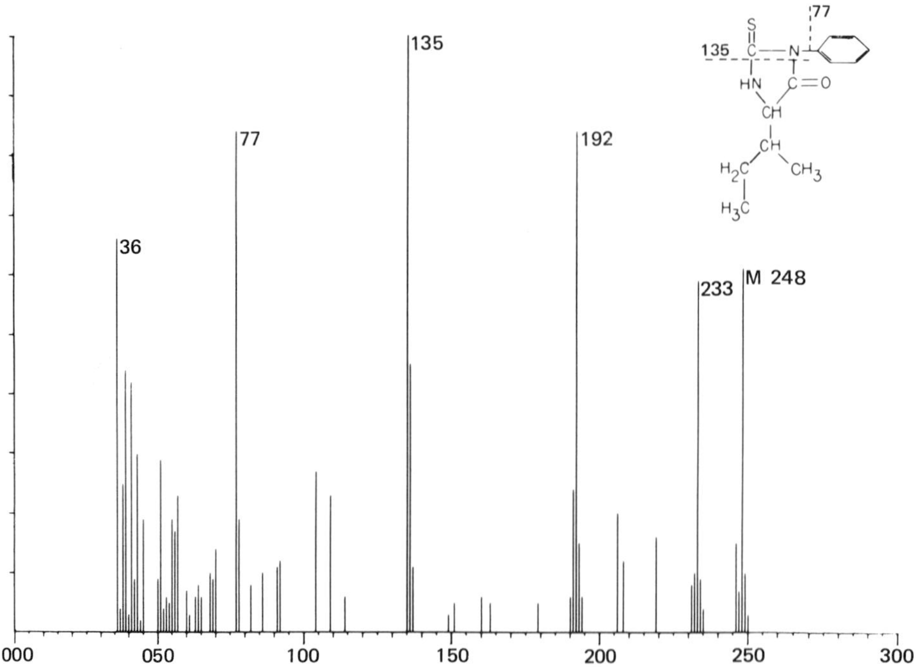

FIG. 9. Mass spectrum of the phenylthiohydantoin derivative of isoleucine.

hydrolyzed nonquantitatively to the corresponding carboxylic groups to yield aspartic and glutamic acid respectively. To circumvent this problem, Young and Desiderio [160] conducted GC-MS analysis of asparagine and glutamine by isolating their methylthiazolinones and converting them to their TMS counterparts. The mass spectra of these derivatives were remarkably simple and had intense molecular ion

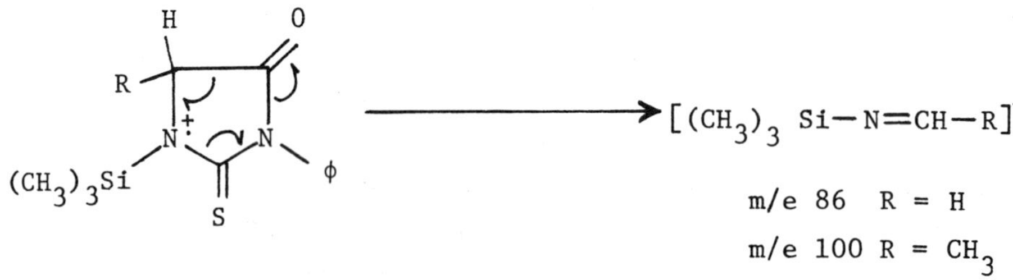

Scheme 31

peaks that provided for unequivocal identification of the amino acid.

Based on the general concept of a Schiff's base formation, Halpern et al. [161] obtained volatile derivatives of amino acids by reacting their methyl esters with CS_2. Primary amino groups were converted to isothiocyanate groups while secondary amines formed urethans. The authors report molecular ion peaks for the mass spectra of all the protein amino acids, except for arginine and cystine. For the simpler amino acids, fragmentation occurs primarily β to the nitrogen, 58, resulting in elimination of $\cdot COOCH_3$, in a manner similar to the free amino acid esters [131].

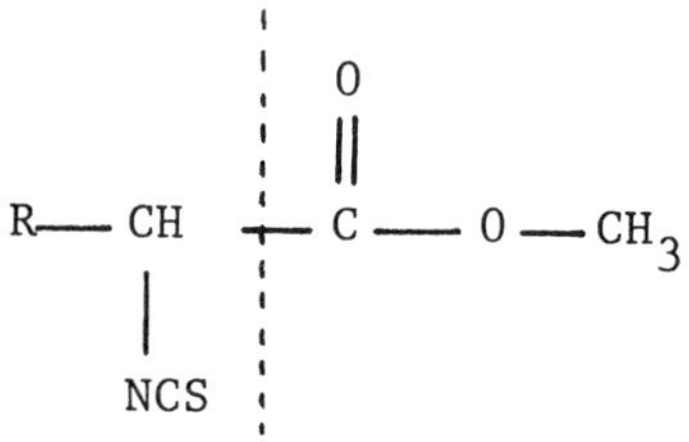

58

The two thyroid hormones 3,3',5,5'-tetraiodothyronine (T_4) and 3,5,3'-triodothyronine (T_3) are amino acids found in minute quantities (levels of pg/ml) in human serum. A variety of derivatives have been considered for their analysis by GC-MS, including N,O-dipivalyl-methyl esters. The mass spectra of the latter derivatives are characterized by relatively low-intensity ions in the high mass and molecular ion region [162]. A somewhat analogous situation has also been encountered with both the N,O-bis-trimethylsilyl-methyl ester, 59a and 60a, as well as the N,O-bis-trimethylsilyl-trimethylsilyl ester, 59b and 60b, derivatives of these compounds [163]. This contrasts sharply with the mass spectra of their corresponding N,O-bis-HFB-ME derivatives, 59c and 60c, which are dominated by intense peaks in the high mass region, and exhibit a high abundance of molecular ions. (See Fig. 3 for the spectrum of the N,O-bis-HFB-ME derivative of T_4.) Table 6 summarizes and compares the mass spectra of the N,O-bis-TMS-ME and N,O-bis-HFB-ME derivatives of the tetraiodothyronine,

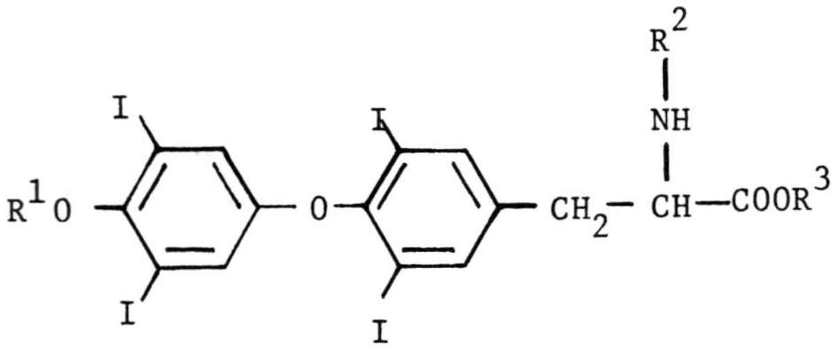

59

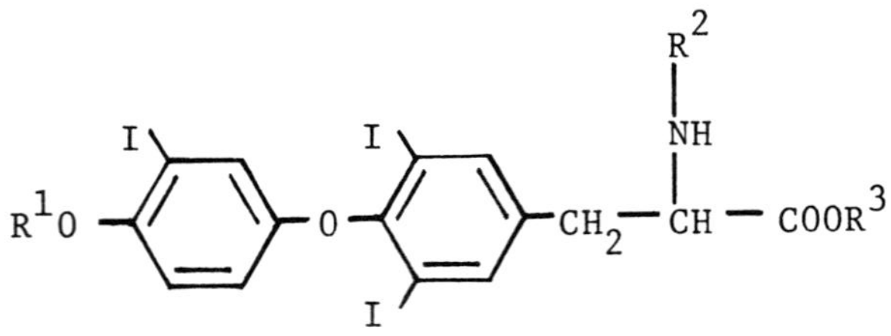

60

59a; $R^1 = R^2 = TMS$, $R^3 = CH_3$

59b; $R^1 = R^2 = R^3 = TMS$

59c; $R^1 = R^2 = C_3F_7CO$, $R^3 = CH_3$

60a; $R^1 = R^2 = TMS$, $R^3 = CH_3$

60b; $R^1 = R^2 = R^3 = TMS$

60c; $R^1 = R^2 = C_3F_7CO$, $R^3 = CH_3$

TABLE 6

Partial Mass Spectra of 59a[a] and 59c[b]

59a		59c	
m/e	% RI	*m/e*	% RI
935, $M^{+\cdot}$	<1	1183, $M^{+\cdot}$	60
920	2	970	100
892	<1	899	48
876	3	773	28
775	<1	716	36
748	<1	645	30
521	<1	519	8
430.5	2	448	21
160	100		

[a]Ref. 163.
[b]Ref. 9.

59. For a comparison of the spectra of the N,O-bis-TMS-TMS ester vs the N,O-bis-HFB-ME derivatives see Table 7, which summarizes the spectra of the corresponding analogs, 60b and 60c, of the triiodothyronine, 60. Elimination of a carboxy ester radical is common to both the methyl and the TMS esters yielding the ions at *m/e* 876 and 750 in the spectra of 59a and 60b, respectively, followed by a loss

TABLE 7

Partial Mass Spectra of 60b[a] and 60c[b]

60b		60c	
m/e	% RI	*m/e*	% RI
867, $M^{+\cdot}$	1	1057, $M^{+\cdot}$	67
852	4	844	100
750	17	773	67
649	1	647	25
622	1	590	22
496	1	519	35
218	100	393	6
		322	21

[a]Ref. 163.
[b]Ref. 9.

of HI in each case. Cleavage of the C-C bond beta to both the aromatic ring and the amino groups results in formation of an ion of structure $R^2\text{—}\overset{+}{N}H{=}CH\text{—}COOR^3$ corresponding to the base peak in the respective spectra (*m/e* 160 in 59a, *m/e* 218 in 60b). Weak peaks from the complementary ions of $[M-160]^+$ and $[M-218]^+$ are also observed, followed by loss of two iodine atoms in an apparently concerted mechanism. A notable feature in the spectrum of 59a, the N,O-bis-TMS-methyl ester derivative of the tetraiodothyronine, is the presence of a doubly charged ion with structure p at *m/e* 430.5.

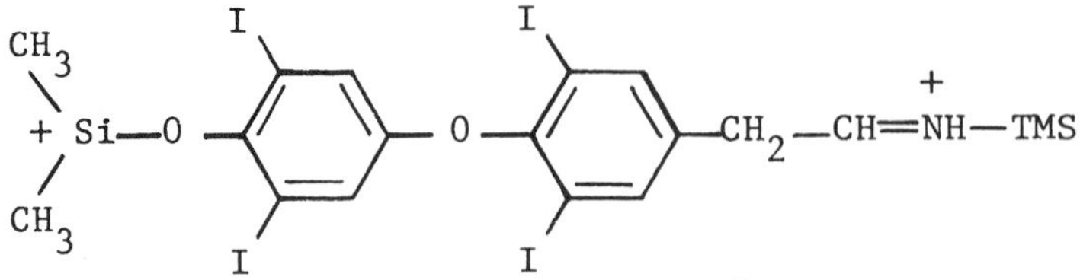

P

Its formation can be rationalized by the localization of the positive charges at the two opposite ends of the structure, a requirement for stabilization of doubly charged ions of this type [164,165].

The most dramatic feature in the spectra of the N,O-bis-HFB-ME derivatives of T_3 and T_4, as opposed to those at their TMS counterparts, is the remarkable intensity of the molecular ion peak and other peaks at the high mass region of the spectrum [9]. Fragmentations initiated from cleavage of the C-C bond beta to the aromatic system are also characteristic of the HFB-ME derivatives, paralleling the behavior of their TMS counterparts, 59a,b and 60a,b. However, in addition to these simple cleavages, the presence of the acyl group in the HFB-ME derivatives apparently initiates a McLafferty-type rearrangement to produce a conjugated ion of $[M-213]^{+\cdot}$ (Scheme 32). Further fragmentation associated with the latter ion produces additional ions of equivalent stability and hence high abundance in the mass spectra of the HFB-ME derivatives. This example demonstrates the effect that the selection of a derivative may have on the mass spectrometric behavior of a compound. In this specific case, the feature of intense peaks at the high mass region of the spectrum permits trace analysis studies to be conducted with a high degree of confidence, as

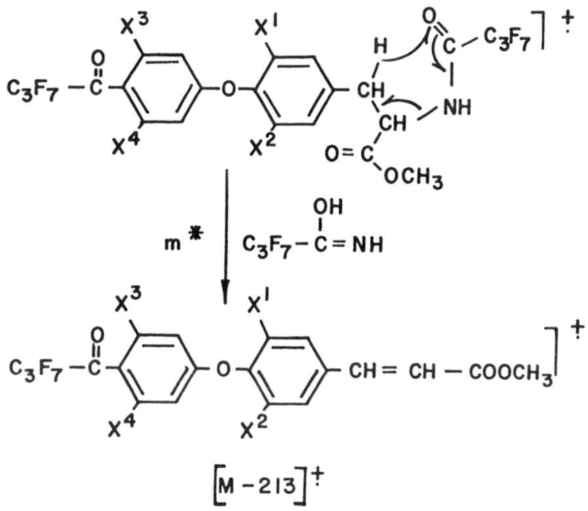

Scheme 32

that region of the spectra is free of interferences from ion peaks arising from column bleed, pump oil, or other background contributions.

D. Biogenic Amines and Related Compounds

1. Acyl Derivatives

N-Acylation of amino groups is a fast and simple reaction that can be carried out quantitatively at the picogram level. Acetyl or trifluoroacetyl groups are typical functionalities introduced in the derivatization of amines, and the following examples are representative of the mass spectrometry associated with these types of derivatives.

The mass spectrum of the N-acetyl derivative of phenylisopropylamine, 61, is summarized in Table 8. The fragmentations are generally predictive as they are controlled by the aromatic system and the amino nitrogen, thus producing the ions at *m/e* 91 and *m/e* 86. The somewhat low intensity of the *m/e* 91 ion can be, in part, rationalized by the tendency of the molecular ion to undergo a McLafferty rearrangement in the process of forming the conjugated ion of *m/e* 118 [166].

TABLE 8

Partial Mass Spectrum of Acetyl Derivative of Phenylisopropylamine

m/e	% RI	Type of ion
177	5	$M^{+\cdot}$
118	40	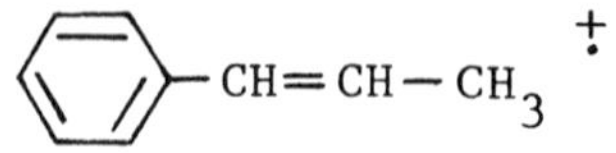
91	30	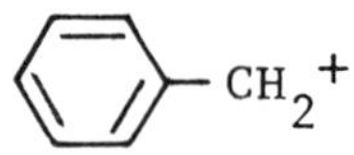
86	70	$CH_3—CH{=}NH—\overset{O}{\overset{\|}{C}}-CH_3$
44	100	$CH_3—CH{=}\overset{+}{N}H_2$

N-Trifluoroacetyl derivatives of phenylalkyl amines exhibit the same type of fragmentation, and the spectrum of the phentermine derivative, 62, has a distinct peak at *m/e* 132. However, the presence of a tertiary carbon alpha to the nitrogen results in the formation of a stable ion of *m/e* 154 [q, $(CH_3)_2C{=}\overset{+}{N}H—CO—CF_3$], which is the base peak in the mass spectrum of 62 [167]. In a related study, Zeman and Wirotama [168] compared the mass spectral characteristics of acetyl and TFA derivatives of amines of the general variety $R^1R^2N—H$. The following important differences were noted.

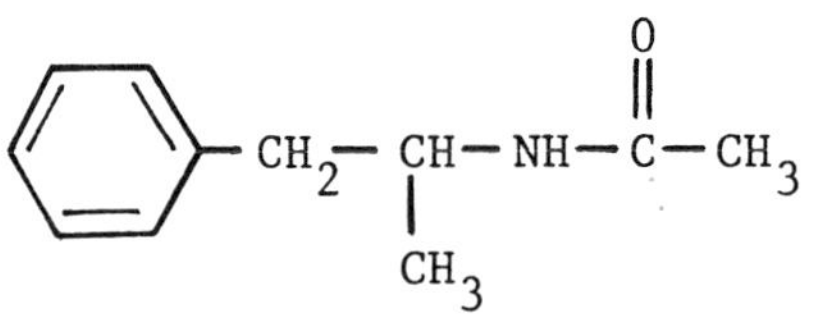

61

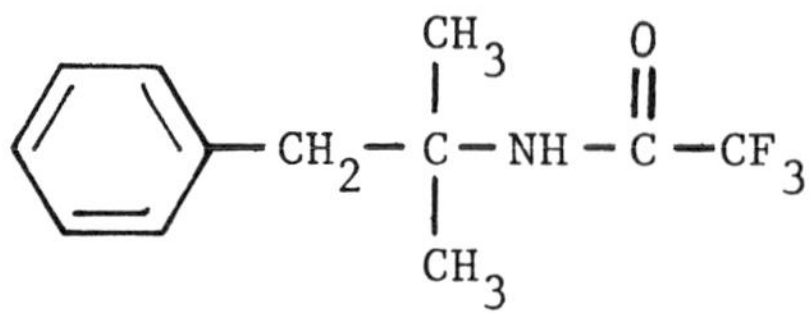

62

1. The $CH_3-\overset{+}{C}=O$ peak was quite intense in the spectra of the acetyl derivatives, but the analogous $CF_3-\overset{+}{C}=O$ ion did not stand out in the spectra of the TFA derivatives.
2. While elimination of ketene ($CH_2=C=O$) was observed frequently in the spectra of the acetyl derivatives, the corresponding elimination of $CF_2=C=O$ was not encountered in the spectra of TFA derivatives.

N,O-Trifluoroacetyl derivatives have been successfully prepared for N-hydroxy amines. For example, the psychotomimetic amine 1-(2,5-dimethoxy-4-methylphenyl)-2-aminopropane was N-hydroxylated by rabbit liver microsomes [169]. Treatment with excess trifluoroacetic anhydride at 25°C for 10 min produced the N,O-bis-TFA derivative, 63.

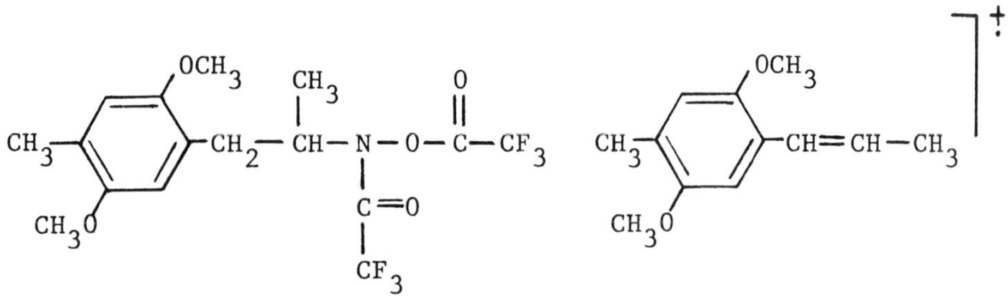

63 r, m/e 192

Its mass spectrum exhibits a relatively intense molecular ion peak (*m/e* 417, 25% RI) and the expected beta cleavage, as well as an ion produced by the McLafferty rearrangement (r, *m/e* 192, 22% RI).

2. *Cyclic Boronates*

Cyclic boronates derived from β-hydroxyphenethylamines have the general structure given by 64. For a large variety of compounds studied--e.g., ephedrine, metanephrine, normetanephrine, adrenaline, etc.--the molecular ion peak was of significant intensity [170]. Some general features in the mass spectra of these derivatives are the following.

1. The substituent (R) on the beta carbon can be readily identified on the basis of the facile cleavage to form an ion of structure s, whether R = H or R = alkyl group.

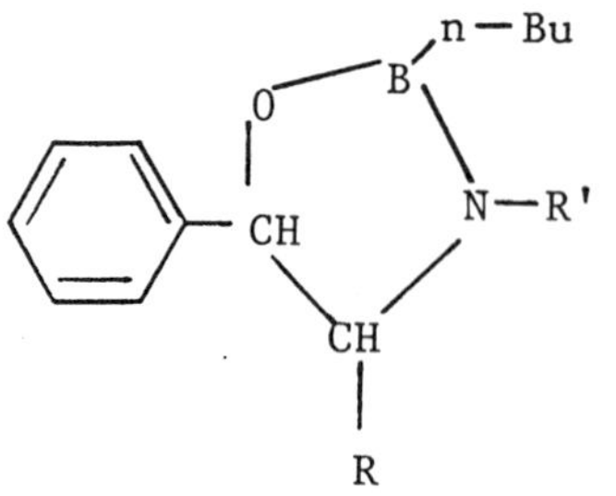

64

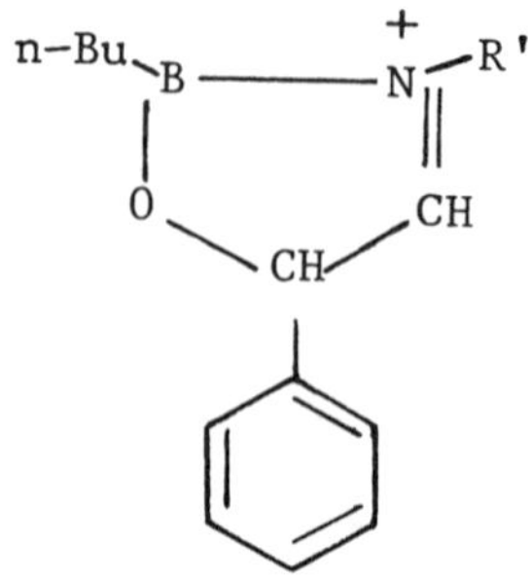

s

2. The substituents on the phenyl ring are usually retained, thus indicating the combined molecular weights in the prominent ions that are formed.
3. The nitrogen substituent R' can be usually identified on the basis of the transition t → u (Scheme 33), where ion t was formed from $M^{+\cdot}$ by elimination of n-Bu-B=O. It should be noted that isomeric compounds containing the same substituent on different ring positions could not be differentiated by mass spectrometry but were resolved by gas chromatography.

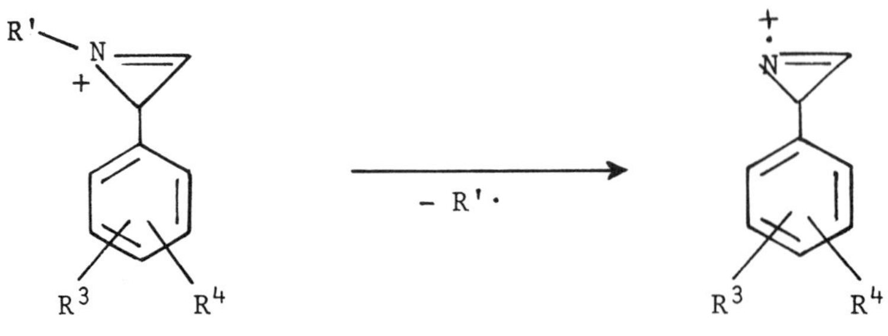

t u

Scheme 33

3. *Trimethylsilyl Derivatives*

The gas chromatographic properties of TMS derivatives of phenylalkylamines were first investigated by Sen and McGeer [171]. Trimethylsilylation of primary amines with BSA results in rapid introduction of a TMS group, but addition of a second TMS group proceeds more slowly. The presence of TMCS catalyst, however, facilitates the formation of N,N-di-TMS derivative [172,173]. The occurrence of mono- and di-TMS products in the trimethylsilylation of primary amines is, in fact, a point of concern and its possibility should be taken into consideration.

In an examination of the GC-MS properties of the TMS derivatives of biogenic amines, Abramson et al. [174] emphasized the importance of derivative selection in augmenting the sensitivity of the analytical method. By driving the silylation of the primary amino group to completion and forming a N,N-bis-TMS derivative, the authors were able to take advantage of the fragmentation directing properties of the phenyl and the amino groups that resulted in the formation of an ion of *m/e* 174, v (Scheme 34). This stable quaternary ammonium ion

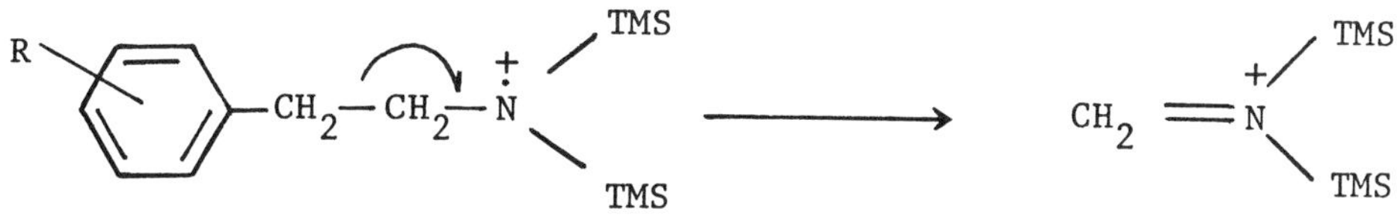

v, m/e 174

dominates the spectra of N,N-bis-TMS derivatives of phenylalkylamines and in some cases is responsible for more than 75% of the total ion current produced. As this ion is common to the mass spectra of a wide variety of biogenic amines, it was possible to use advantageously the inherent sensitivity of the mass spectrometer to improve the detection limits of GC analysis by selective ion monitoring.

4. *Isothiocyanate (NCS) Derivatives*

Reaction of carbon bisulfide with primary amines yields isothiocyanates or mustard oils [Eq. (8)]. This Schiff's base reaction was

$$R{-}NH_2 + S{=}C{=}S \longrightarrow R{-}N{=}C{=}S + H_2S \qquad (8)$$

originally utilized by Brandenberger for the separation of primary from secondary amines by gas chromatography [175]. More recently isothiocyanate derivatives have been employed effectively in the GC-MS analysis of amphetamines, biogenic amines, and other biologically active primary amines [2,176-178]. Some advantages of these derivatives for GC-MS are the following:

1. They have low polarity and hence high volatility for GC.
2. Low mass increment (42 amu) is introduced to the parent compound as opposed to 142 amu in the per-TMS derivatives.
3. The mass spectra usually exhibit reasonably intense molecular ion peaks.
4. Further derivatization (e.g., trimethylsilylation) of other polar groups such as aromatic hydroxy groups in catecholamines is possible following formation of the isothiocyanate.

In general the MS behavior of NCS derivatives of biogenic amines is quite predictive in that the principal fragmentation involves cleavage of the ethyl (or i-propyl) side chain and the formation of structurally diagnostic ions. Somewhat unique appears to be the fragmentation of the NCS derivative of the drug parnate, 65, in the spectrum of which the molecular ion eliminates a •NCS radical followed by a concerted loss of two hydrogen atoms (Scheme 35). The last fragmentation step is accompanied by an unusually intense metastable peak [176] and can be rationalized by the formation of the conjugated cyclic ion w.

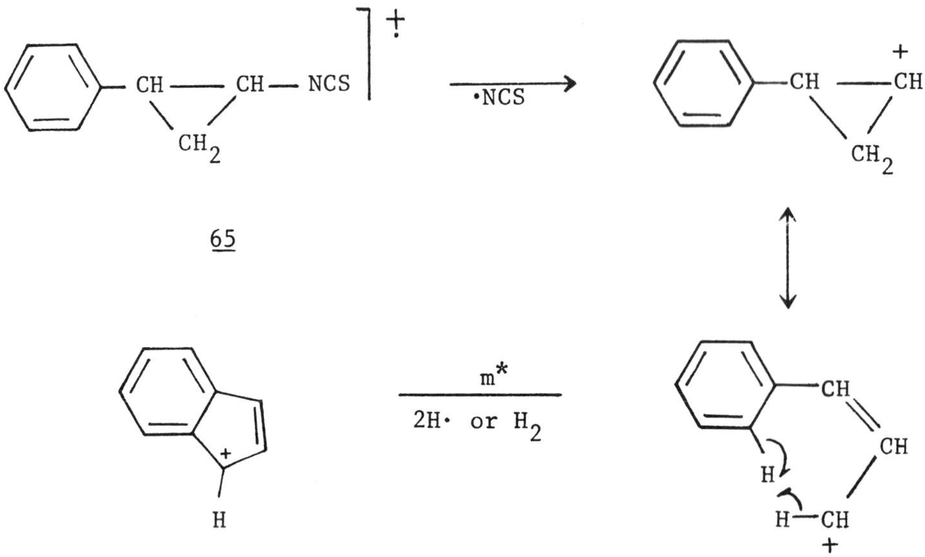

w, m/e 115

5. *Double Derivatives*

In catecholamines selective derivatization reactions can be carried out based on the different rates of reaction between the hydroxy and the amino groups with a specific reagent or, of course, the inherent selectivity of the derivatizing reagent. For example, in the case of the NCS derivatives discussed above, the CS_2 reacts specifically only with the amino group, thus permitting further derivatization of the hydroxylic groups. The selective derivatization of dopamine described by Donike [36,37] (Scheme 4) is a good example of the flexibility available to the analyst as far as the selection of the derivative is concerned. The most desirable derivative can thus be chosen depending on the information required from either the GC or the mass spectrum.

Secondary amino groups react relatively slowly to trimethylsilylating reagents while their acylation proceeds very readily. Horning et al. [35] have taken advantage of this and used sequentially trimethylsilylimidazole and heptafluorobutyrylimidazole to form N-TMS-O-HFB derivatives of catecholamines. This general approach was also employed in the derivatization of hydroxynoraporphines [179]. To achieve complete N,O-trimethylsilylation of 2,10-dimethoxy-11-hydroxynoraporphine, 66, relatively severe reaction conditions--e.g., well in excess of 100° overnight--were necessary. This had to be avoided in view of the previously demonstrated tendency of aporphines to undergo fission of ring B under certain alkylation conditions [180]. There was no difficulty associated with the formation of a bis-N,O-HFB derivative, although O-acylation proceeded at a distinctly slower rate than N-acylation. As it was desirable to retain an O-TMS functionality in the final derivative, a dual N-HFB-O-TMS derivative was prepared (Scheme 36) by reacting the noraporphine sequentially with heptafluorobutyric anhydride and BSTFA. The value of the O-TMS functionality is demonstrated from examination of the spectrum of this double derivative shown in Fig. 10. A major process associated with the trimethylsilyloxy group involves the loss of 30 amu from $M^{+\cdot}$ to

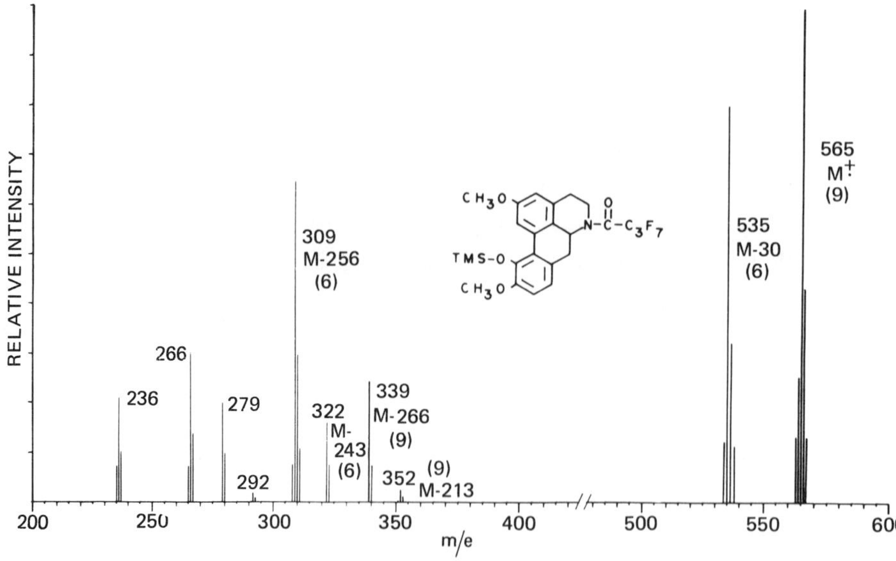

FIG. 10. Mass spectrum of N-heptafluorobutyryl-O-trimethylsilyl derivative of 2,10-dimethoxy-11-hydroxy-noraporphine.

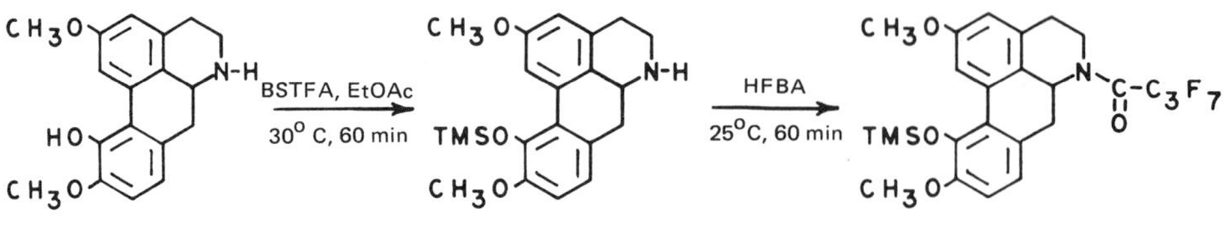

66

Scheme 36

form a cyclic silyldioxy ion (x). The structure of this ion was confirmed by its 6-amu shift in the spectrum of the perdeuteriotrimethylsilyl derivative. Monitoring for an ion of this type would help establish the relative disposition of a hydroxy and methoxy group (loss of 30 amu, C_2H_6) or two hydroxy groups (loss of 88 amu, $SiMe_4$) on the aromatic ring systems, something that would not be possible with the N,O-bis-HFB derivatives. Other fragmentations in the mass spectrum of Fig. 10 are associated with the N-heptafluorobutyryl group [179].

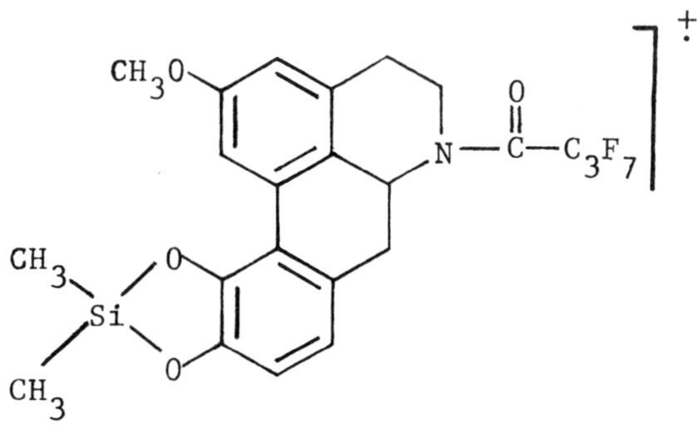

X

6. *Cyclic Derivatives*

Phenylalkylamines, tryptamines, and related compounds when reacted with formaldehyde form cyclic condensation products as indicated for tryptophan, 67, in Scheme 37. Upon trimethylsilylation of free hydroxylic or carboxylic groups these cyclic derivatives can be readily analyzed by GC-MS [181,182]. Owing to their partially aromatic and cyclic structure, the mass spectra of these derivatives exhibit only a few intense peaks, usually at the high mass region, and thus can be effectively utilized for low-level detection using single-ion monitoring. For example, the 70-eV mass spectrum of the N,N,O-tris-TMS derivative of the tryptophan condensation product, 68, exhibits only five major peaks [181]:

m/e 432, $M^{\ddagger}$ 45%
m/e 315 70%
m/e 216 73%
m/e 77 42%
m/e 73 100%

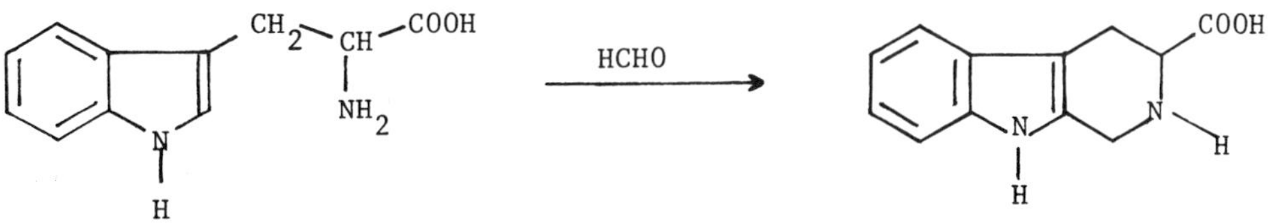

67 68

Scheme 37

E. Fatty Acids

The fundamental aspects of the mass spectrometry of fatty acid derivatives have been reviewed by McCloskey [183]. In this section we shall limit the discussion to some of the more recent developments in this area, a consideration of the mass spectrometric processes associated with specific types of derivatives and their utility in the elucidation of the structure of the parent compound. As the determination of the position of a double bond in a normal hydrocarbon chain is a problem frequently encountered in fatty acids, a discussion of the derivatization of olefinic bonds is included in this section.

1. *Saturated Normal Chain Acids*

a. Methyl esters. Saturated fatty acids are routinely analyzed as their methyl ester derivatives. The spectra of these derivatives have well-defined molecular ion peaks, thus permitting the unambiguous determination of their molecular weight. An ion of *m/e* 74 (y), produced by γ-hydrogen transfer via a McLafferty rearrangement characterizes the spectra of methyl ester derivatives of fatty acids bearing no C-2 substituent (Scheme 38). This ion usually carries a substantial fraction of the total ion current and can thus be used for the detection of low levels of acids in such a series by SIM.

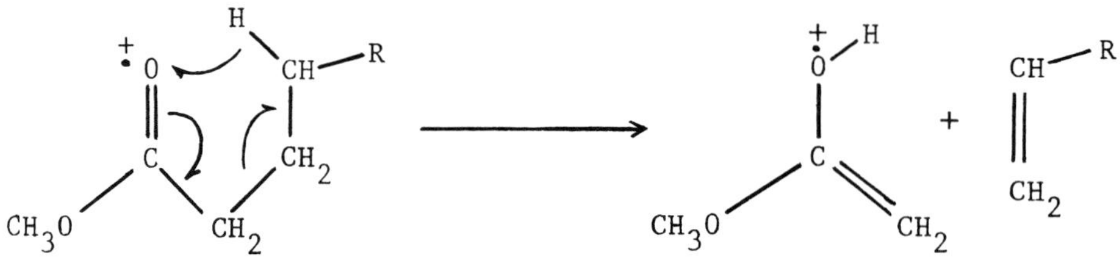

y

Scheme 38

Loss of a methoxy radical to give an ion of $[M-31]^+$ is another typical peak associated with the mass spectra of fatty acid methyl esters. Consideration of this process in conjunction with the McLafferty rearrangement shown in Scheme 38 can provide for the differentiation between two structural isomers such as an ethyl ester that may be the endogenous constituent of a mixture of fatty acids and the methyl ester of an α-methyl substituted fatty acid (see, for example, Ref. 15, page 13). In addition to $M^{+\cdot}$ and $[M\text{-}OCH_3]^+$ the molecular ion region also shows characteristic ions of $[M\text{-}29]^+$ and $[M\text{-}43]^+$ produced by extensive skeletal rearrangement of the chain and one hydrogen atom [183]. Other prominent ions correspond to the series $CH_3O—CO—(CH_2)_n^+$ beginning with *m/e* 87 (n = 2), which is also the ion of highest intensity in the series.

b. Trimethylsilyl esters. The mass spectra of trimethylsilyl esters of fatty acids bear many similarities to those of the methyl esters in that they exhibit the characteristic ions of $[M\text{-}29]^+$, $[M\text{-}43]^+$, and the McLafferty rearrangement ion, shifted by the expected 58 amu to *m/e* 132. Loss of a TMS methyl from *m/e* 132 yields an additional peak at *m/e* 117. Notable is the absence of a $[M\text{-}TMSO]^+$ ion, the equivalent of $[M\text{-}OCH_3]^+$ in methyl esters, while the series TMSO-CO-$(CH_2)_n^+$ is accompanied by a second series of ions 16 amu lower, produced by elimination of CH_4 (TMS methyl group + hydrogen) from the respective ions [184]. Trimethylsilyl derivatives are usually employed in the analysis of fatty acids containing additional functional groups with active hydrogens (see discussion below).

c. Benzyl esters. Long-chain fatty acids were derivatized by reaction with phenyldiazomethane to form benzyl esters [11]. The principal advantage of this form of esterification is realized in mixtures of fatty acids with other compounds containing long alkyl groups, where the fatty acids can be most readily distinguished either on the basis of GC retention differences [11a], or due to their characteristic mass spectra. In general benzyl esters of fatty acids in the series $R—CO—O—CH_2—\phi$ had the following mass spectral features.

1. In the range C_1-C_{12} the molecular ion intensities decreased with increasing alkyl residues.
2. At C_{12} the relative intensity of the molecular ion peak was 1.5%.

3. In the range C_{14}-C_{20} no molecular ion peaks were observed, but the molecular weight could be determined from the characteristic $[M-91]^+$ fragment ions.
4. The principal fragmentations are summarized in Scheme 39.

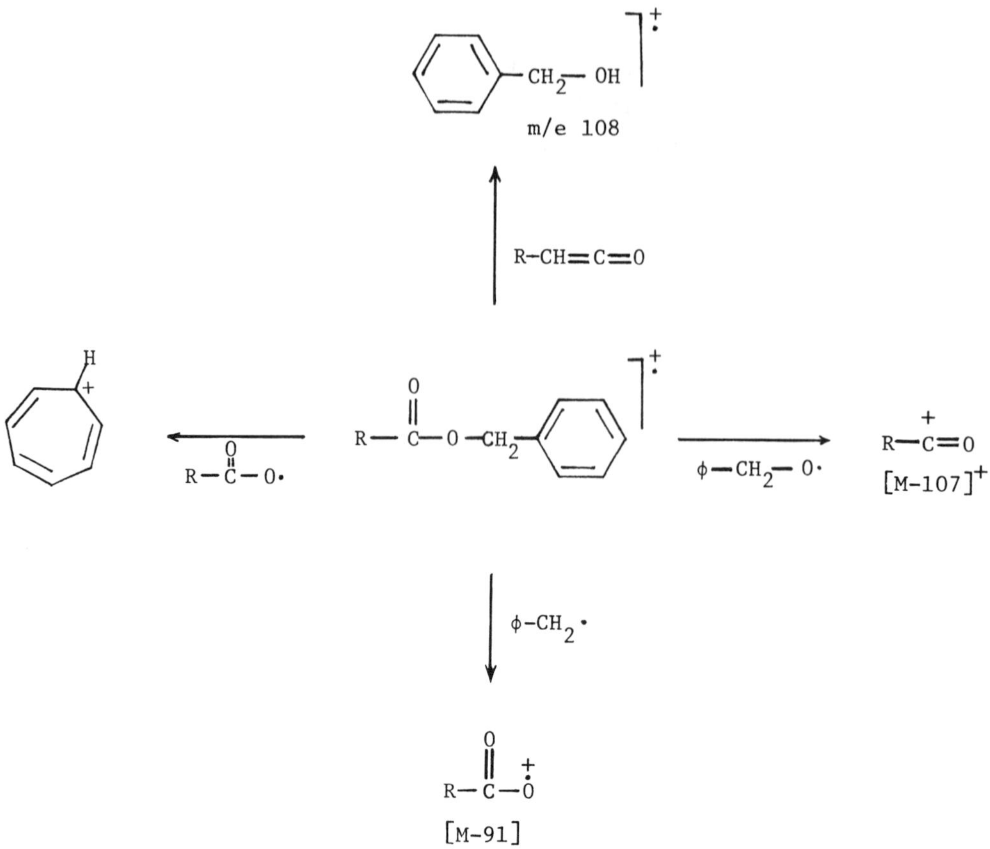

Scheme 39

d. Perfluoroalkyl esters. Carboxylic acids were esterified by reaction with excess α,α,ω-trihydroperfluoroalkohols under acidic conditions to produce esters with the general structure $CH_3(CH_2)_n$—$COOCH_2(CF_2)H$ [185]. The mass spectra of derivatives of this type parallel those of the methyl esters in many respects. For example, in the spectrum of $CH_3(CH_2)_8COOCH_2(CF_2)_2H$ abundant ions are observed at values corresponding to $M^{+\cdot}$, $[M-43]^+$, and $[M-112]^{+\cdot}$. The last ion is the equivalent of ion y, produced by a McLafferty rearrangement and has a structure corresponding to $[CH_2{=}C(OH){-}O{-}CH_2{-}(CF_2)_2H]^{+\cdot}$. In addition, an abundant ion with the structure $CH_3(CH_2)_8{-}\overset{+}{C}{=}O$ is also observed and corresponds to the $[M\text{-}OCH_3]^+$ ion encountered in the spectra of the methyl ester derivatives.

Analogous to the benzyl esters of long-chain acids, the fluoroester derivatives are particularly useful in cases where the carboxylic acids are in a mixture with hydrocarbons or alcohols of similar molecular weight. Under the latter circumstances, trimethylsilylation will also derivatize the alcohols, while methyl ester formation will add only 14 amu to the molecular weight, often an insufficient increment to distinguish or separate from other interfering components by GC.

e. Dicyanomethylene derivatives. Reaction of the acid chloride of a fatty acid with cyclopropene using pyridine as catalyst yielded a product which, upon treatment with diazomethane, gave a methyl ether of its dicyanomethelene (DCM) derivative; Eq. (9) [185]. The mass

$$R{-}CO{-}Cl \xrightarrow{\text{cyclopropene } (CH_2{-}CH{=}CH),\ C_5H_5N} \xrightarrow{CH_2N_2} R{-}(OCH_3)\ C{=}C(CN)_2 \qquad (9)$$

spectra of dicyanomethylene derivatives are characterized by a series of abundant fragment ions corresponding to $(CH_2)_n{-}(OCH_3)C{=}C(CN)_2^+$ and extending uniformly to the molecular ion. A distinct difference between the DCM and methyl ester derivatives is that in the former the fragmentations arising from C—C bond cleavage at a branching point are significantly enhanced. This is illustrated in Fig. 11, where the spectra of the DCM and methyl ester derivatives of 4,8,12-trimethyloctadecanoic acid are compared. Note that the peaks at *m/e* 303, 233, 163, and 135 indicating the branching site are as much as ten times more intense in terms of $\%\Sigma$ than the corresponding peaks in the spectrum of the methyl ester derivative.

It is apparent from the above that although from a practical point of view the DCM derivatives are not as versatile as the methyl esters due to the multistep derivatization procedure, in specific cases certain advantages may be derived from their application.

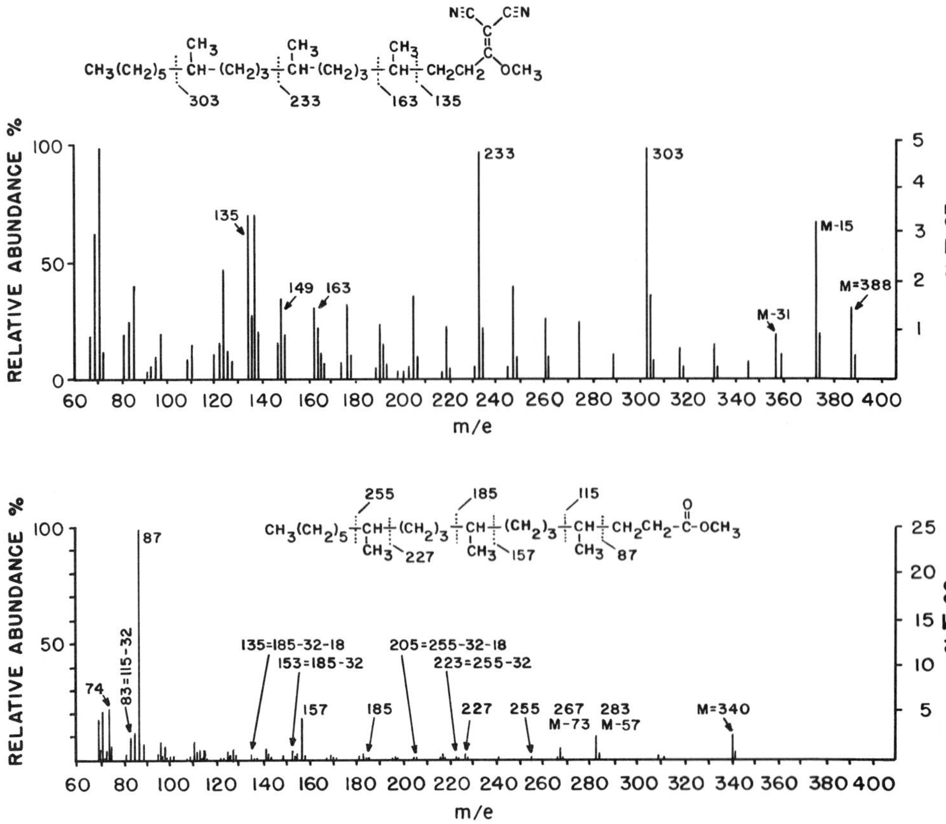

FIG. 11. Mass spectra of dicyanomethylene (top) and methyl ester (Bottom) derivatives of 4,8,12-trimethyloctadecanoid acid. (Reprinted from *Tetrahedron Letters* with permission from the publisher.)

2. *Hydroxy Acids*

a. Cyclic boronates. Reaction of α- or β-hydroxyacids with n-butyl-boronic acid (actually a mixture of the free acid and its anhydride) gave a cyclic boronate derivative; Eq. (10) [187]. In

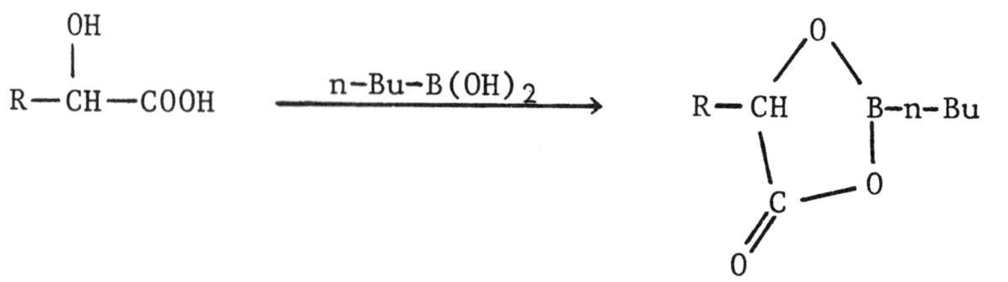

(10)

general the molecular ion peaks in the mass spectra of the boronate derivatives of fatty acids varied widely in intensity, but even in cases where they were extremely weak there was sufficient evidence from related fragment ions to determine the molecular weight of the compound. Loss of CO_2 to give an ion of structure z, Scheme 40, was a major process in the spectra of the α-hydroxyacid derivatives but rather insignificant in the β-hydroxyacids. In general, boronate derivatives of hydroxyacids have seen limited use compared with the methyl ester-TMS ethers or TMS-ethers-TMS-esters discussed below.

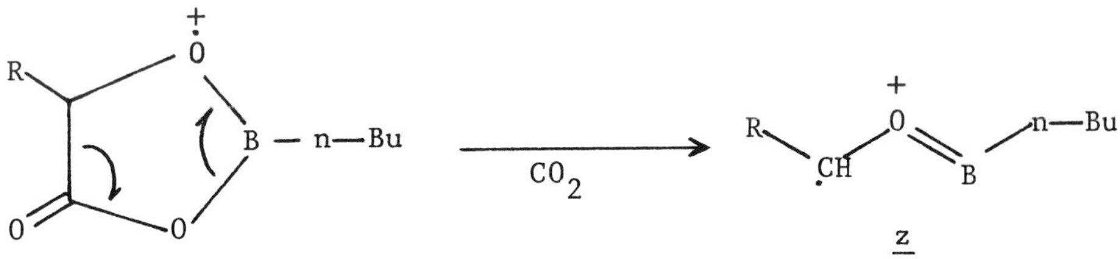

Scheme 40

b. Methyl- or TMS-ester-TMS-ether derivatives. Analysis of hydroxy fatty acids by GC-MS can be carried out by simple derivatization--i.e., methyl ester formation--of the carboxylic function. A discussion of the mass spectral characteristics of hydroxy-methyl ester derivatives can be found in the review by McCloskey [183]. For acids containing several hydroxy groups on the hydrocarbon chain further derivatization to form trimethylsilyl ethers is recommended. Presence of a trimethylsilyloxy group enhances the intensity of peaks produced by a direct beta cleavage, thus permitting easier confirmation of the position of substitution. An interesting fragmentation process occurs in the spectra of O-TMS-ethers-methyl esters of α-hydroxy acids and is apparently initiated by loss of a TMS methyl group and effective migration of the methoxy group to the positively charged siliconium center [188] (Scheme 41).

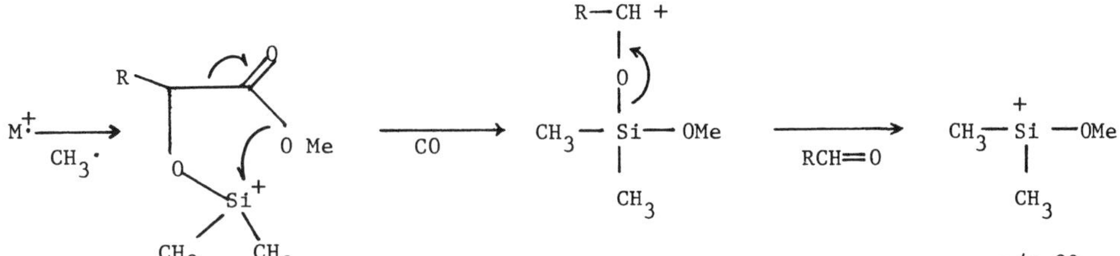

Scheme 41

Migrations of the trimethylsilyl group to the ester carbonyl oxygen are common in the TMS ethers of fatty acid esters [189]. A characteristic example is shown in Scheme 42 for the derivative of the C-42 unsaturated fatty acid $[C_{25}H_{49}CHOSi(CH_3)_3—C_{15}H_{30}COOCH_3]$ isolated from *Nocardia corallina* [190]. Transfer of the TMS group (69) yields a series of ions with structure aa formed by further C-C bond cleavages at alternate carbons from C-15 to C-13 [190].

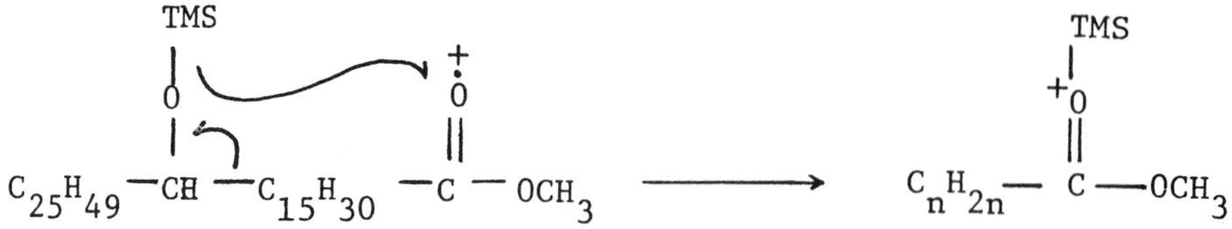

69

aa, where: n = 15 (m/e 342), n = 13 (m/e 314), n = 11 (m/e 286)

Scheme 42

Transfer of an intact TMS group to a radical carbonyl oxygen via a six-membered ring transition state (silyl McLafferty rearrangement) is a prominent process in the mass spectra of the TMS derivatives of hydroxy-carboxylic and dicarboxylic acids. In a series of papers, Petersson [191-194] examined in detail the mass spectra of hydroxy-acids or hydroxycarbonylacids derived from the alkaline degradation of carbohydrates. A summary of the fragmentations associated with this transfer and their diagnostic value is given in Scheme 43. The relative abundance of the silyl McLafferty rearrangement ion increased significantly in terms of %Σ at lower ionizing energies, indicating the low-frequency factor of the fragmentation. It is interesting that the compounds investigated by Petersson also have γ-hydrogens available for conventional McLafferty rearrangement [193]. Nevertheless, the ions produced by hydrogen rearrangement are considerably lower in abundance than those produced by the competing TMS rearrangement. In fact, on several occasions the hydrogen rearrangement ions were of insignificant abundance.

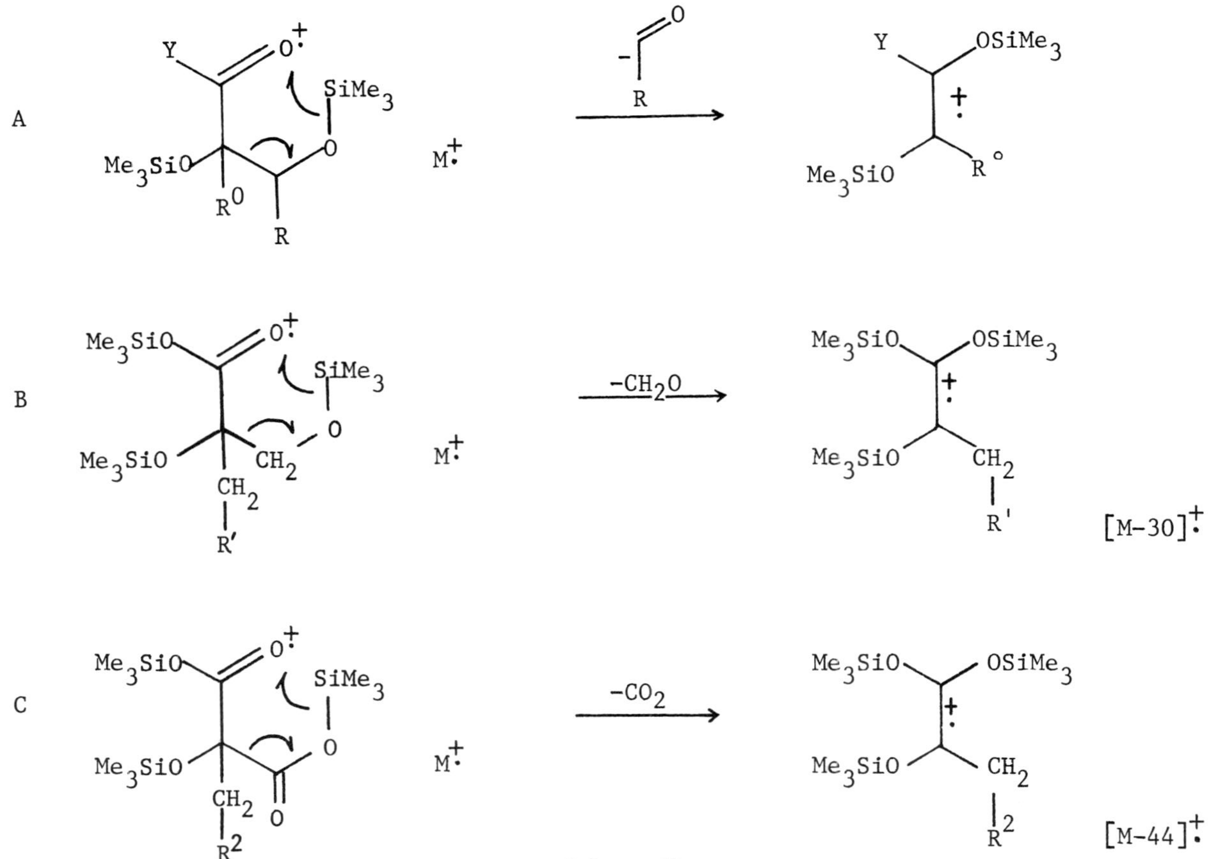

Scheme 43

McLafferty rearrangement of a TMS group in trimethylsilyl derivatives of hydroxy acids. A, general formulation; B, formulation for 2-substituted glyceric acids; C, formulation for 2-substituted tartronic acids.

3. Unsaturated Normal Chain Acids

Determination of the position of a double bond in a long-chain unbranched hydrocarbon chain by mass spectrometry represents a classic case where the employment of chemical derivatization procedures was early recognized as a necessity. It has been postulated [195] that electron-impact ionization of a double bond results in a radical induced hydrogen transfer along the length of the hydrocarbon chain and, in effect, a migration of the double bond. Thus, there is a loss of site specificity and to some extent a randomized fragmentation. While the mass spectra of isomeric olefins do exhibit quantitative differences, it is impossible to deduce on a predictive basis the site of unsaturation, or for that matter the stereochemistry (cis vs trans) at the olefinic junction.

With a few limited exceptions, derivatization of the double bond in a long-chain fatty acid involves saturation of the system and introduction of a heteroatom function that induces beta cleavage and stabilization of the positive charge as shown in Scheme 44. A problem with many of the derivatization procedures is the formation of mixtures

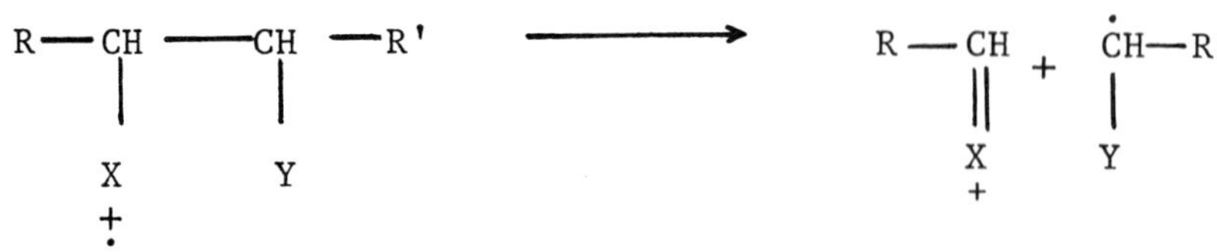

Scheme 44

of isomeric compounds (interchanged positions of X and Y), which may introduce confusion in the interpretation of the mass spectra or when the GC inlet is used. In either case, the difficulties are minimal when dealing with monoolefinic as compared with polyunsaturated systems.

Derivatization of a double bond has included the formation of a vicinal diol (X=Y=OH), N,N-dimethyl-aminoalcohol [X=$N(CH_3)_2$, Y=OH and X=OH, Y=$N(CH_3)_2$], vicinal dimethyl ether (X=Y=OCH_3), and vicinal bis-trimethylsilyl ether (X=Y=TMSO) [183]. Most useful among them appears to be the bis-trimethylsilyl ether produced by stereospecific oxidation to a diol and followed by trimethylsilylation; Eq. (11)

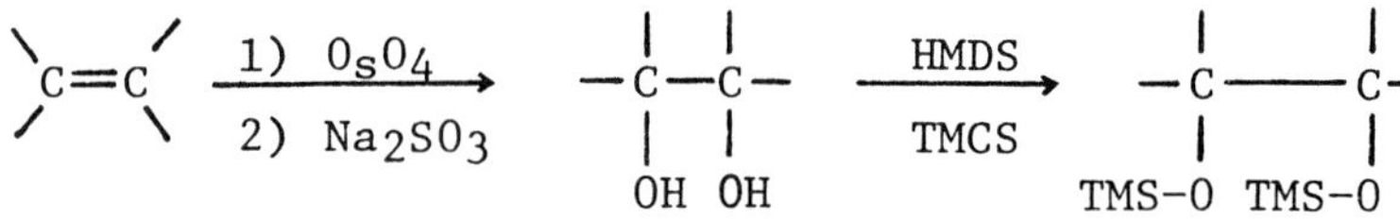

(11)

[196]. The advantages offered by the bis-trimethylsilyloxy derivative are apparent from Fig. 12, which shows the spectrum of the bis-OTMS derivative derived from methyl-oleate. The two ions at *m/e* 215 and *m/e* 259 formed by beta cleavage are distinct and the position of the double bond can be readily recognized. One of the disadvantages of these derivatives is the absence of a molecular ion peak, and certainly the inability to distinguish between cis and trans isomers. Wolff et al. [197] have taken advantage of the stereospecificity of the osmium tetroxide oxidation step [Eq. (11)] to further form an O-isopropylidene derivative [Eq. (12)] in which the characteristics of the double bond are retained by the relationship: cis → erythro and trans → threo.

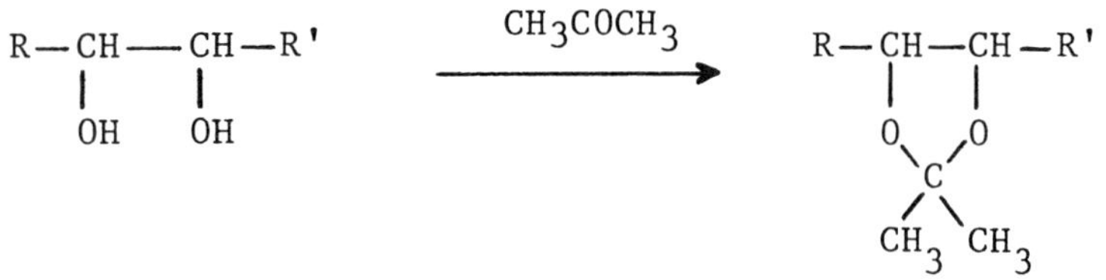

(12)

The spectra of O-isopropylidene derivatives are noted for the lack of molecular ion peaks, but the intense $[M-15]^+$ ions can be indirectly used for molecular-weight determination. The double-bond position can be recognized from the ions produced by direct beta cleavage (70), but a major drawback is the relatively low abundance of these ions due to other competing fragmentations. The stereochemical characteristics are reflected, however, in the intensity of an ion corresponding arithmetically to $[M-89]^+$ and formed in the sequence

$$M^{+\cdot} \xrightarrow[\cdot CH_3]{} [M-15]^+ \xrightarrow[CH_2=C=O]{} [M-57]^+ \xrightarrow[CH_3OH]{} [M-89]^+$$

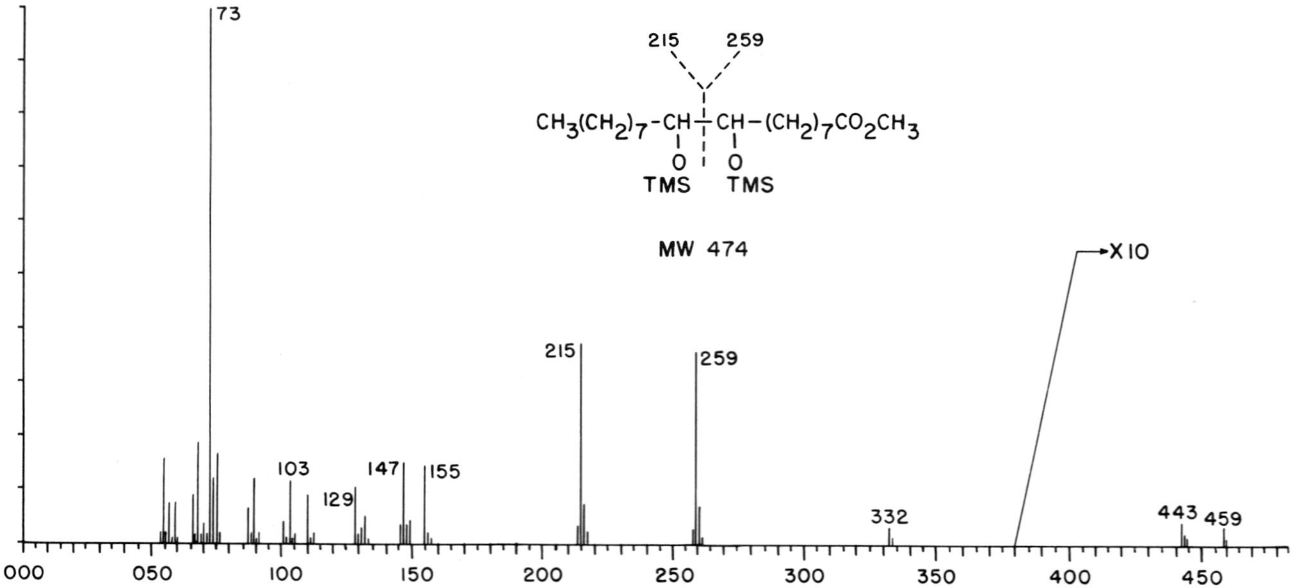

FIG. 12. Mass spectrum of the bis-trimethylsilyloxy-methyl ester derivative of the vicinal diol obtained by OsO_4 oxidation of oleic acid. (Reprinted from *Analytical Chemistry* with permission from the publisher.)

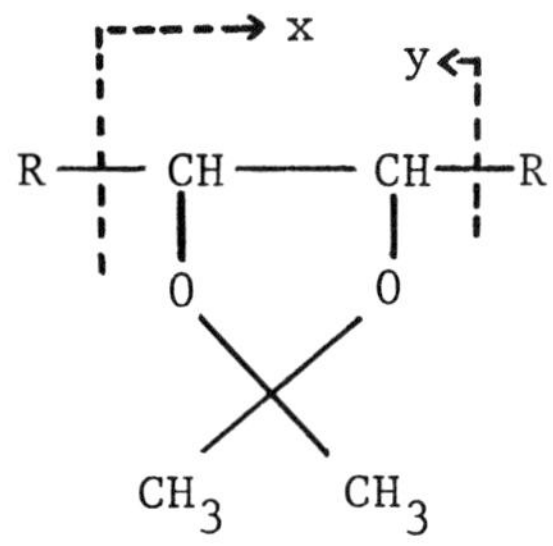

70

The 20-eV spectra of the erythro (cis) isomers show a distinctly higher intensity for this ion than the corresponding threo (trans) compounds. It is apparent that in certain cases double-bond stereochemistry can be determined even without reference spectra. More recently it has been proposed that chemical ionization (CI) mass spectrometry can be used for the same purpose [198]. The CI mass spectra of the bis-TMS ethers of the threo or erythro diols produced from the OsO_4 oxidation of the trans or cis olefins also exhibit quantitative differences in the intensities of certain peaks. These variations are not associated with the major peaks in the spectra, but they certainly appear sufficient for the differentiation of isomers with the aid of reference spectra if not on an absolute basis.

In addition to the derivatives discussed in the review of McCloskey [183] some of the derivatives suggested more recently for the GC-MS analysis of unsaturated fatty acid esters include hexafluoroisopropylidenes [Eq. (13)] [199], methoxy derivatives prepared by olefin reaction with mercuric acetate in methanol and demercuration of the methoxyacetoxymercuri product with sodium borohydride [Eq. (14)] [200], N-substituted amides [Eq. (15)] [201], and pyrrolidides [Eq. (16)] [202,203].

The hexafluoroisopropylidene derivatives are generally similar to the isopropylidenes as far as their mass spectrometric behavior is concerned. The spectra of N-substituted amides exhibit intense molecular ion peaks, but interpretation of the spectra is somewhat complex because of the multitude of isomeric products, especially

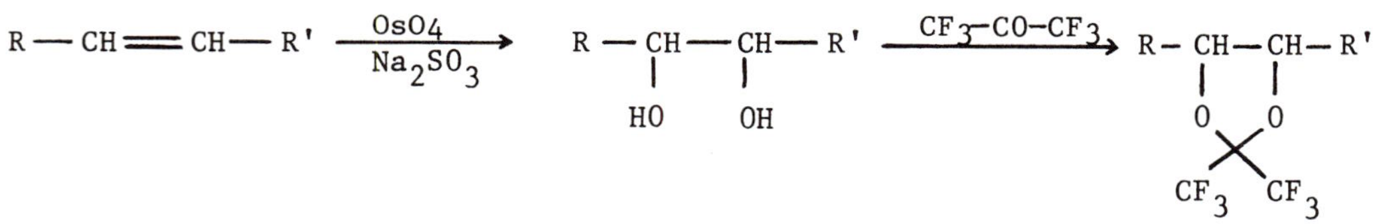

(13)

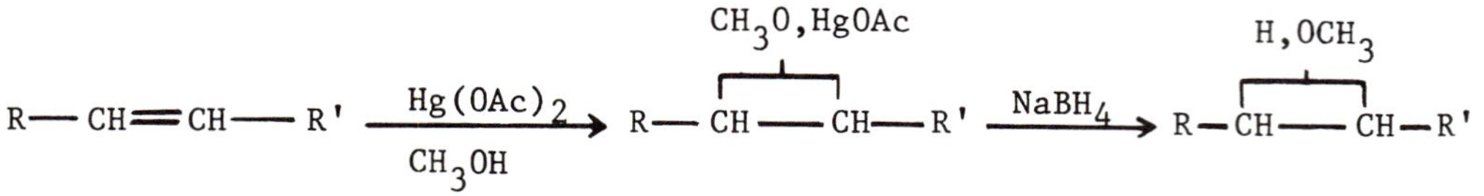

(14)

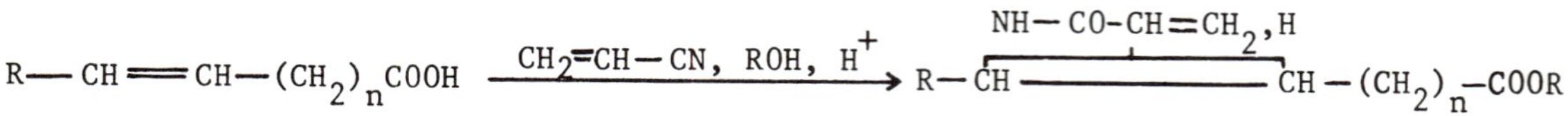

(15)

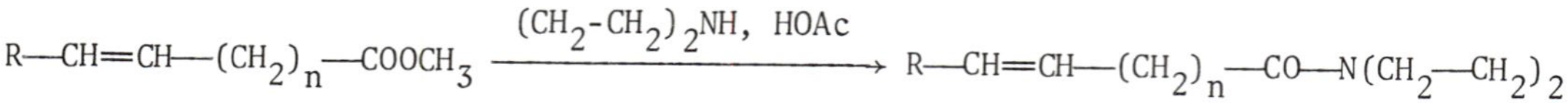

(16)

in the case of the derivatives of polyunsaturated compounds. One is faced with a similar problem in the evaluation of the GC-MS data of the methoxylated esters derived from the methoxymercuration-demercuration reaction. For example, consider the spectrum (Fig. 13) of the mixture of methoxy compounds obtained from the derivatization of methyl oleate. Fragmentation by beta cleavage yields two pairs of intense ions as depicted on 71. The set of ions at *m/e* 157, 171, 201, and 215

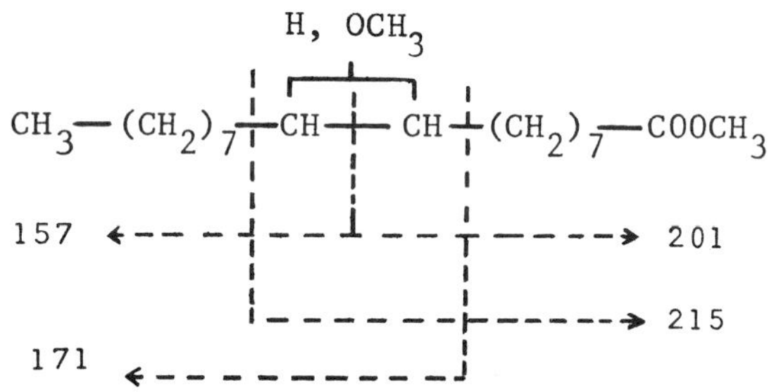

71

can certainly define the position of unsaturation, particularly since they are so distinct. One has to beware, however, of variations in the occurrence of the diagnostic ions in cases where the isomeric derivatives do not elute together. Nonetheless, the methoxy compounds appear indeed to have considerable potential as functional derivatives for the determination of the double-bond sites in long-chain unsaturated acids.

The pyrrolidides [202,203] in some respects occupy a rather unique position among chemical derivatives for olefinic systems. While all of the previously mentioned methods dealt with a direct chemical modification at the site of unsaturation itself, derivatization by pyrrolidide formation involves introduction of a functionality at the carboxylate site of the system. Recognition of the position of unsaturation is made possible by a distinct interruption in the spectrum of the 14-amu fragmentation pattern, which is typical of a long-chain saturated system. It has been postulated [202] that "occurrence of a 12-amu interval between the most intense peaks of two fragment clusters containing n and n - 1 carbon atoms of the acid moiety, is

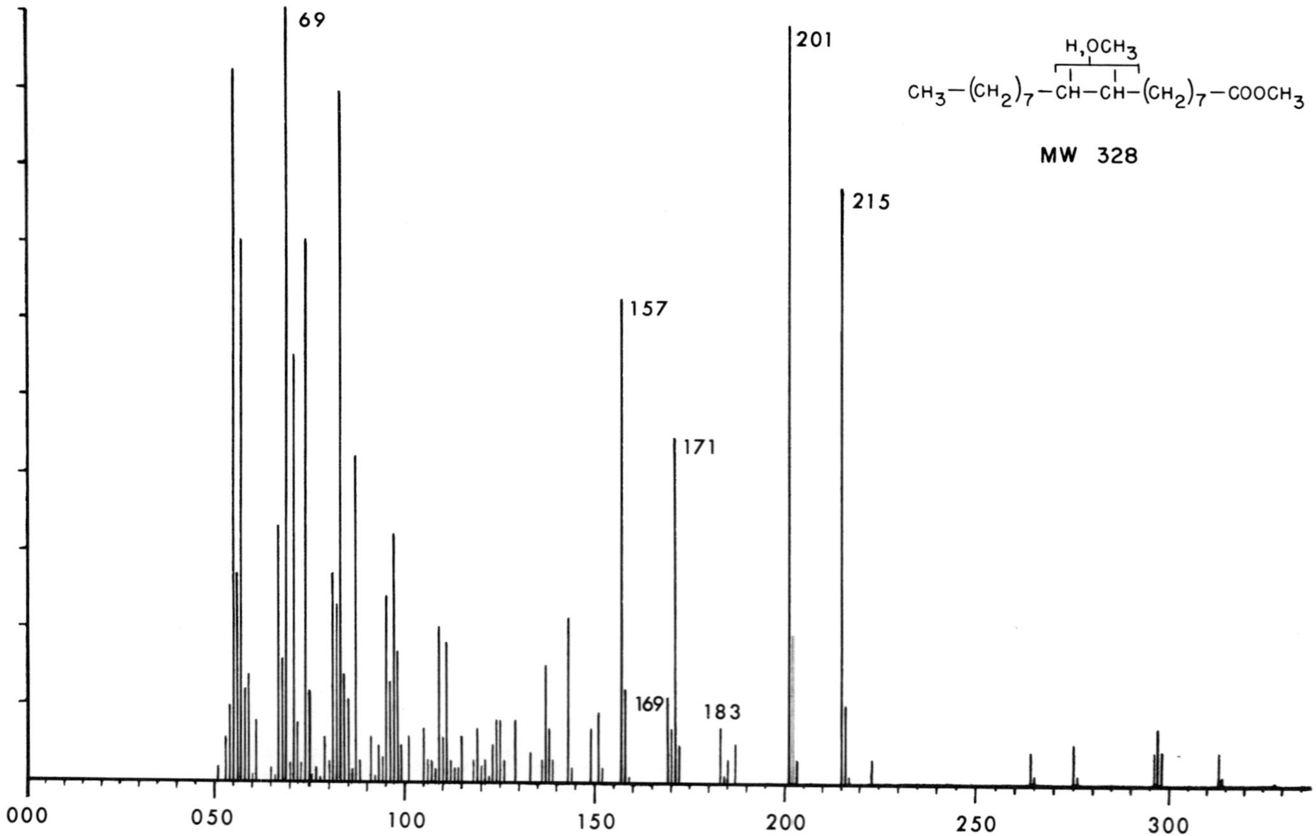

FIG. 13. Mass spectrum of methoxylated esters derived from methyl oleate by methoxymercuration-demercuration. (Reprinted from *Lipids* with permission from the publisher.)

indicative of a double bond between carbon n and n + 1 in the molecule." This is illustrated in Fig. 14, which compares the spectra of the pyrrolidine derivatives of a 6-18:1 and an 11-18:1 fatty acid. Note the 12-amu differences between the most intense peaks in the C_5 and C_6 ion clusters (*m/e* 154 - *m/e* 166; top spectrum) and in the C_{10}-C_{11} ion clusters (*m/e* 224 - *m/e* 236; bottom spectrum. In accordance with the postulated theory, these patterns are representative of C_6-C_7 and C_{11}-C_{12} unsaturation respectively. The cleavages leading to the diagnostic fragment ions in the spectra of the two derivatives are indicated in 72 and 73 respectively.

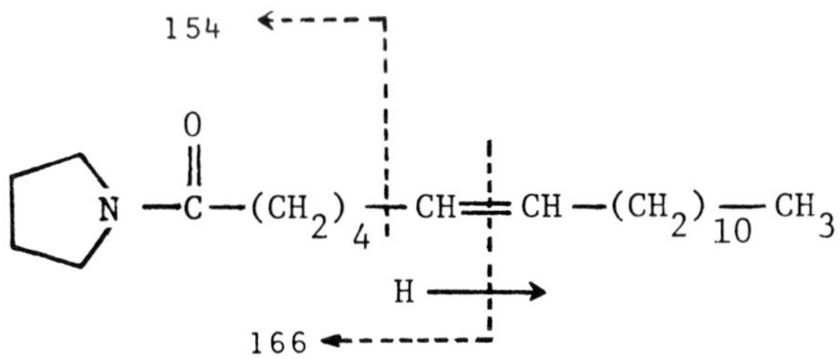

72

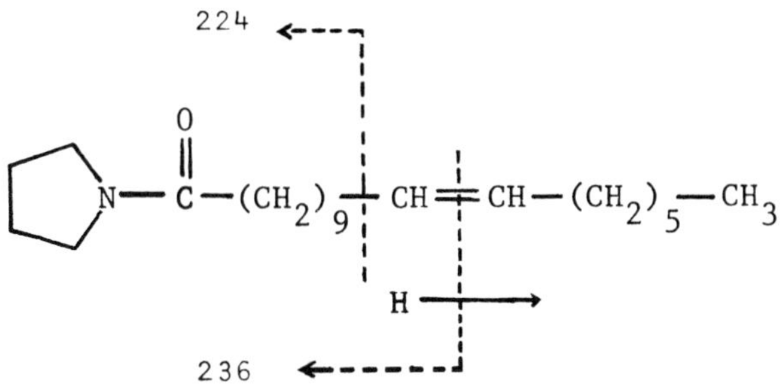

73

It is suggested that upon ionization and prior to fragmentation the double bond migrates toward the more polar pyrrolidine end of the molecular ion by one or more steps. This can explain the indicated cleavage across the olefinic system that results in formation of the *m/e* 166 and *m/e* 236 ions (see (72 and 73). In addition to the aforementioned structurally diagnostic fragment ions, typical ions in the

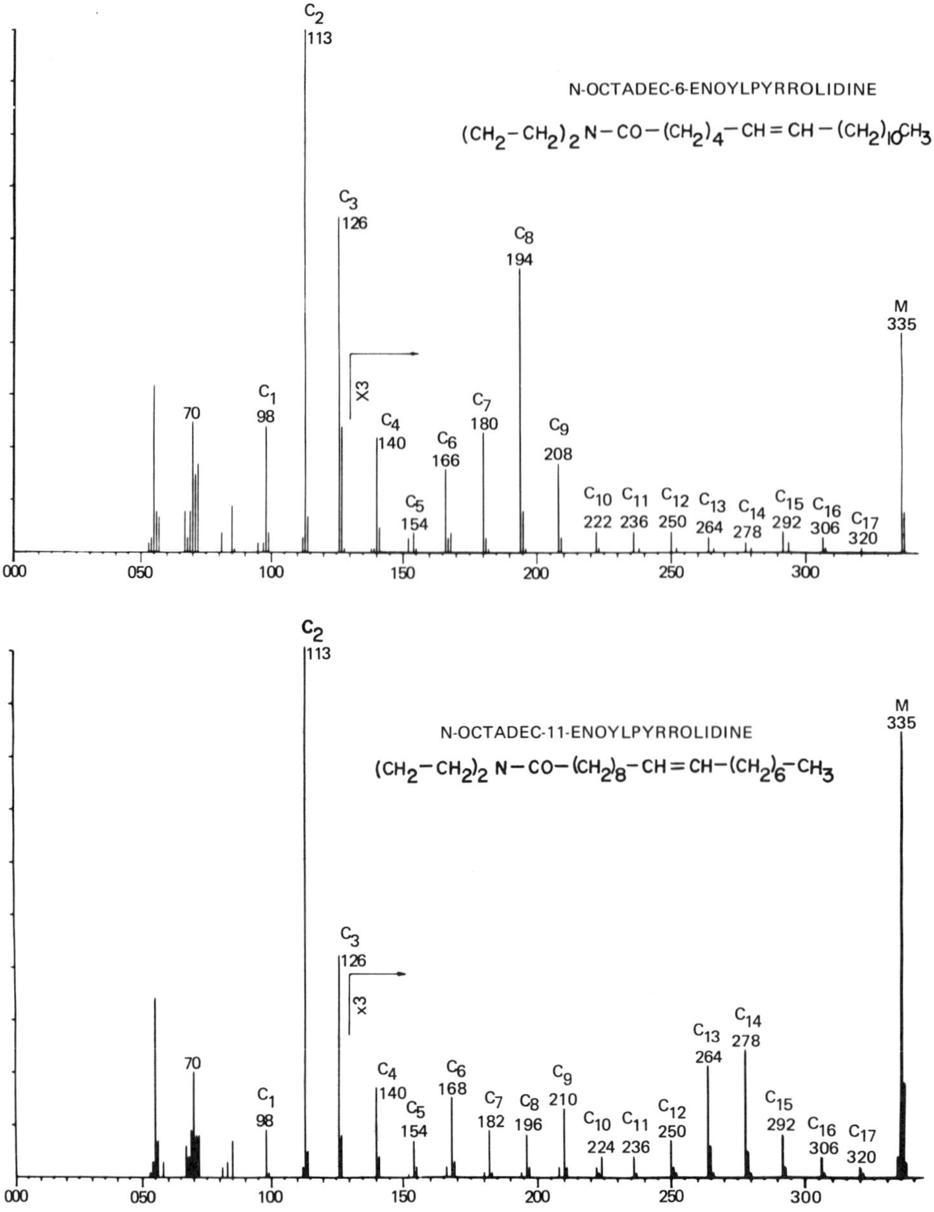

FIG. 14. Mass spectra of N-octadec-6-enoylpyrrolidine (top) and N-octadec-11-enoylpyrrolidine (bottom). (Reprinted from *Lipids* with permission from the publisher.)

spectra of these derivatives occur at *m/e* 70 $[(C_2H_2)_4NH]^+$, *m/e* 98 $[(C_2H_2)_4NCO]^+$, and *m/e* 113. The last ion is formed by γ-hydrogen rearrangement to the carbonyl oxygen of the enoylpyrrolidine moiety, analogous to the McLafferty rearrangement of Scheme 38.

The rules of interpretation of the mass spectrum of a pyrrolidide derivative of a monounsaturated fatty acid also apply to many dienoic and trienoic acids [203]. Although the spectra do not reflect the stereochemistry about the olefinic function this method of derivatization is potentially valuable because of the simplicity of the reaction and, in part, the uniqueness of the approach. More mass spectral data are needed, however, before its general applicability can be ascertained, especially for polyunsaturated systems.

F. Carbohydrates and Related Compounds

The mass spectrometry of carbohydrate derivatives has been reviewed in considerable detail by several investigators during the past decade. For a comprehensive analysis and comparison of the mass spectra of the various types of functional derivatives of sugars and sugar-containing natural products, the reader is referred to the monographs by Kochetkov and Chizhov [204], Radford and DeJongh [205], and Hanessian [206]. For a discussion of the mass spectra of other sugar-containing systems such as nucleosides and nucleotides the reader is referred to a more recent review by McCloskey [207]. In this section we summarize some of the major and/or more interesting mass spectrometric processes associated with specific functionalities, and their diagnostic value for the structural elucidation of simple sugars. As might be expected, not all of the observed fragmentations are applicable to polysaccharides, particularly beyond the trisaccharide level.

1. *Methyl Ethers*

Methyl ether derivatives of carbohydrates are well suited for GC-MS because of their volatility and ease of preparation. In that respect they are ideal for the GC-MS analysis of polysaccharides. It is

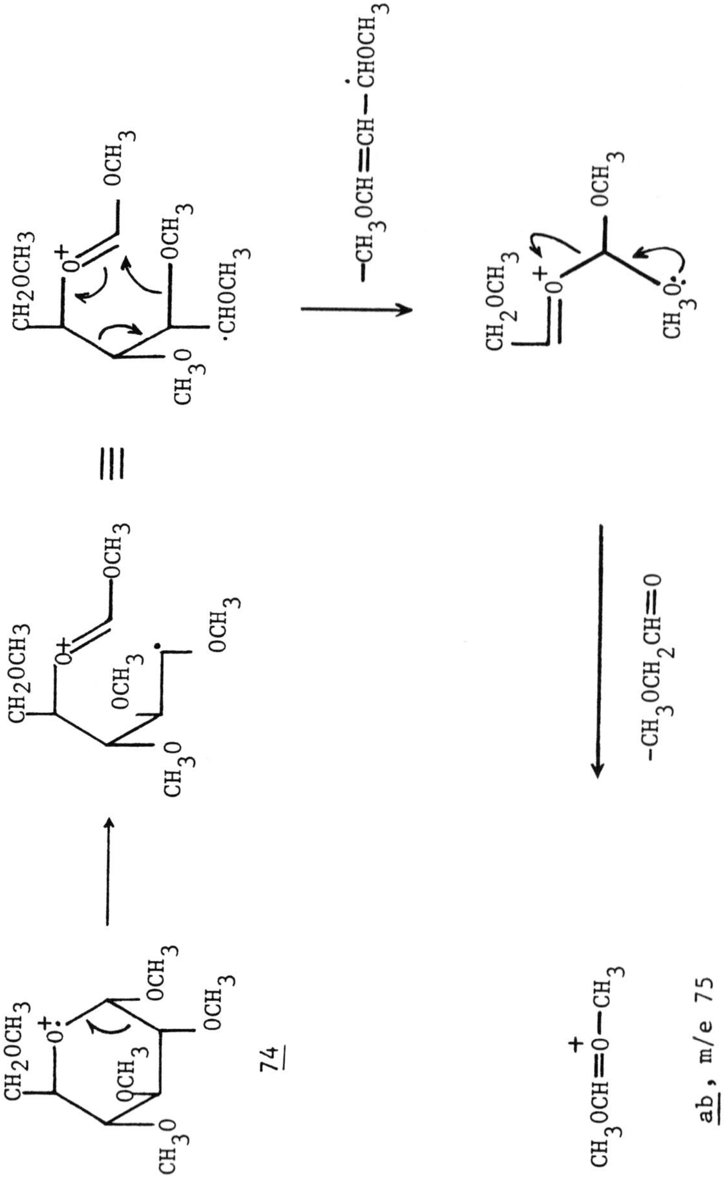
74
$-CH_3OCH{=}CH-\dot{C}HOCH_3$
$-CH_3OCH_2CH{=}O$
$CH_3OCH{=}\overset{+}{O}-CH_3$
ab, m/e 75

Scheme 45

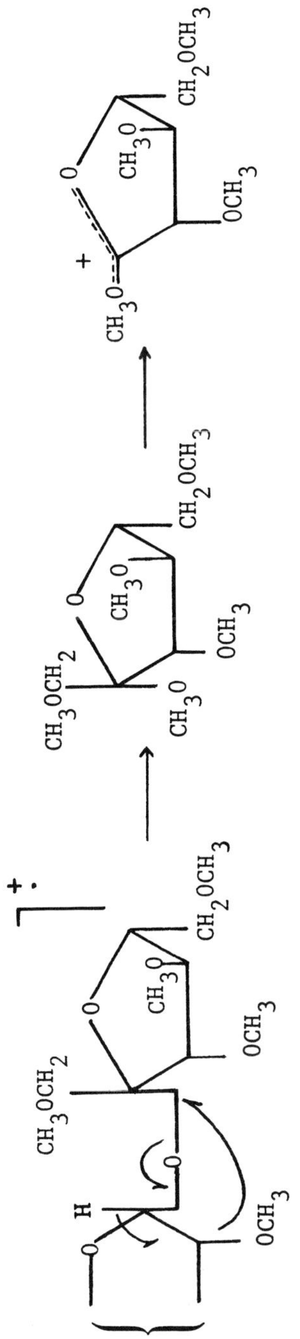

Scheme 46

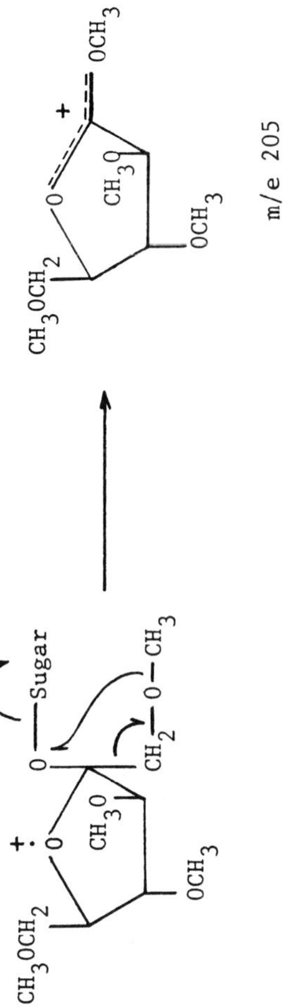

Scheme 47

unfortunate that in general the spectra of permethylated sugars do not exhibit molecular ion peaks. In cyclic sugars fragmentation of the pyranose or, for that matter, the furanose framework can be readily explained by assuming initial charge localization on the ring oxygen. This is followed by a series of competing β-cleavages and associated eliminations of CH_3OH or $CH_3O\cdot$. Of particular significance is the migration of a C-3 methoxyl group to C-1, which results in the formation of the ion of *m/e* 75 (ab). This is shown in Scheme 45 for the case of methyl 2,3,4,6-tetra-O-methyl-α-D-glucopyranoside, 74. A somewhat analogous process has been observed in the mass spectra of permethylated fructose-containing oligosaccharides [208].

In a comparative study of various oligosaccharides it was noted that fructofuranose units that are linked via C-2 to another sugar exhibit an intense *m/e* 205 peak. Formation of this fragment ion could be rationalized by postulation of either a methoxyl or methyl migration as depicted in Schemes 46 and 47, respectively.

Methoxyl group rearrangements in carbohydrate methyl ethers appear to be highly specific in nature and they can be useful for structural elucidation processes. For example, in disaccharides methoxyl migration via the mechanism shown in Scheme 45 results in the eventual formation of an abundant ion of *m/e* 353 characteristic of 1→6 linked disaccharides [209]. Oligosaccharides linked other than 1→6 show no or low-abundance ions of this mass. On the other hand, 1→2 and 1→4 glycosidic linkages are characterized by peaks at *m/e* 380 and 305. While a weak peak at *m/e* 380 arising from a different series of cleavages was also observed in 1→6 linked octamethyl disaccharides, the *m/e* 305 ion produced by a methoxyl migration in 1→4 linked saccharides (Scheme 48) could not be formed in the former case [209].

2. *Acetyl Derivatives*

O-Acetyl derivatives of sugars have good gas chromatographic properties and can be readily prepared on a microscale. However, their mass spectra usually do not exhibit molecular ion peaks, and the

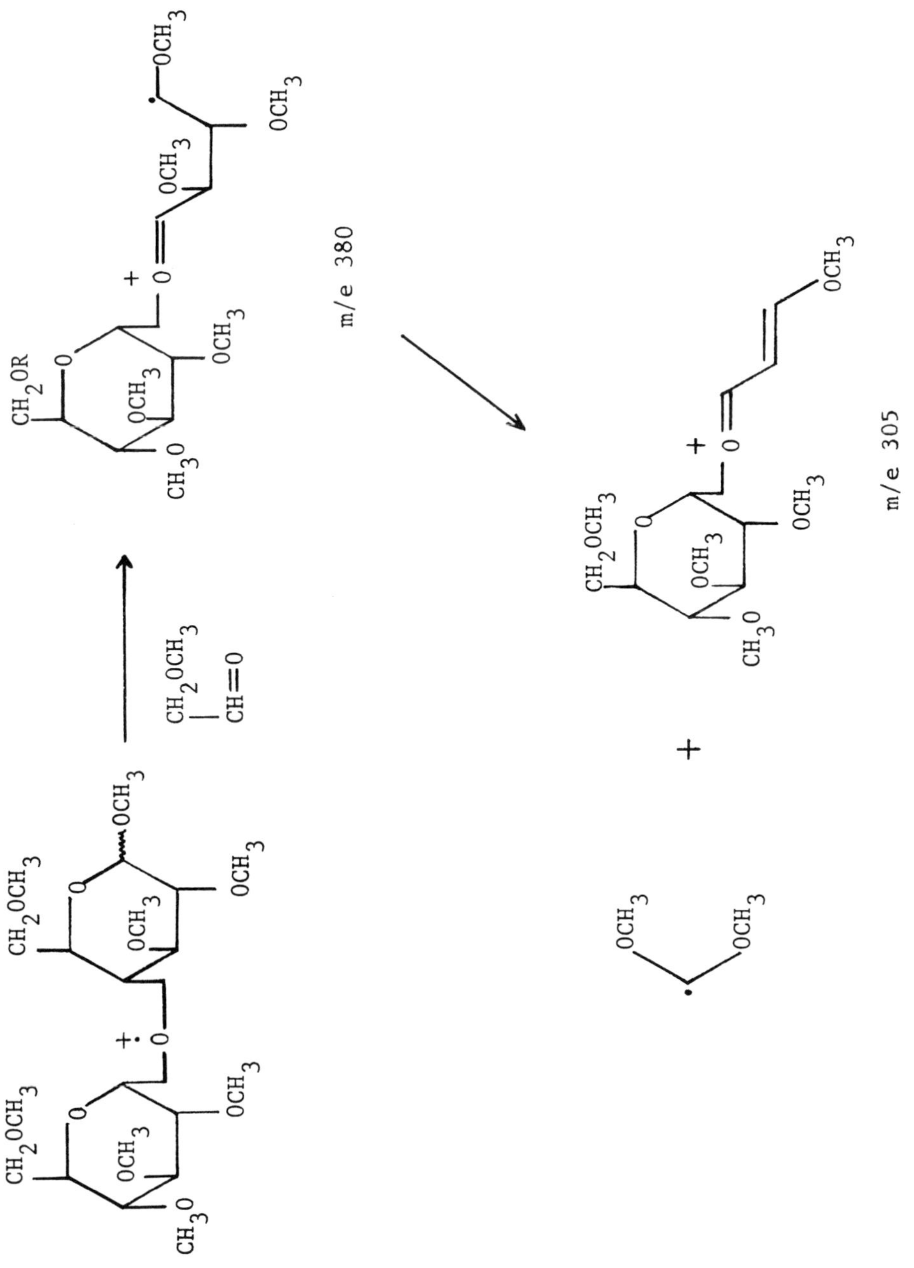

Scheme 48

acetate functionality can be lost in the form of CH_3COOH, $CH_3COO\cdot$, $CH_3CO\cdot$, $CH_2 = C = O$ or any combinations thereof. Interpretation of the mass spectra can be further complicated by the occurrence of peaks at $[M+43]^+$ formed by an ion-molecule reaction and corresponding to the attachment of a CH_3CO moiety to $M^{+\cdot}$. Three ions derived from the acetate moiety are ubiquitous to the spectra of acetyl derivatives and have been formulated as CH_3CO^+ (*m/e* 43), which is invariably the base peak, $CH_3CO—\overset{+}{O}(H)—COCH_3$ (*m/e* 103), and $CH_3CO—\overset{+}{O}(COCH_3)_2$ (*m/e* 145). Cleavage of the carbohydrate framework can be explained as with the methyl ether derivatives on the basis of initial charge localization on the ring oxygen. The resulting fragmentation pattern parallels that of the methyl ethers.

3. *Trimethylsilyl Ethers*

Trimethylsilyl derivatives are well suited for the GC-MS analysis of carbohydrates. They combine ease of preparation with high volatility, while a wealth of available information on the mass spectrometry of the trimethylsilyl group facilitates interpretation of their mass spectra. Molecular ions are of low abundance, but the presence of $[M-15]^+$ peaks is indicative of the molecular weight of the compound. Fragmentation processes and rearrangements associated with the TMS group closely parallel those described earlier for steroidal trimethylsilyl derivatives. The rearrangement ion of *m/e* 147 (a) can be used as a guide to identify stereochemical features [86] while that of *m/e* 191 (f) is produced by a TMSO rearrangement from C-3 to C-1, analogous to the methoxyl transfer described in Scheme 45, and can be used to deduce substitution on C-1 [210]. It should be kept in mind, however, that alternative sources for the derivation of this ion are also possible in such multi-TMS systems.

Cleavage of the sugar framework is again initiated, by the removal of a nonbonding electron from the ring oxygen, and, for the most part, the fragment ions formed are similar to those encountered in the spectra of methyl ethers. The distinction of pyranosides from furanosides poses no major difficulty, as the former are characterized by an intense peak at *m/e* 204, which retains carbons 2,3 and the substituents, 75, and the latter by an ion of *m/e* 205 produced by simple β cleavage, 76.

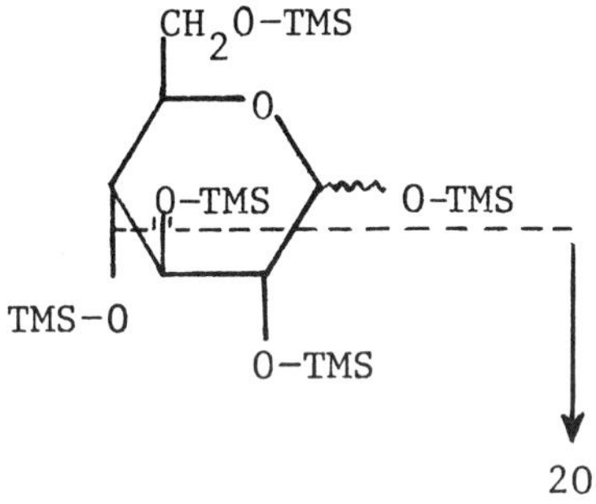

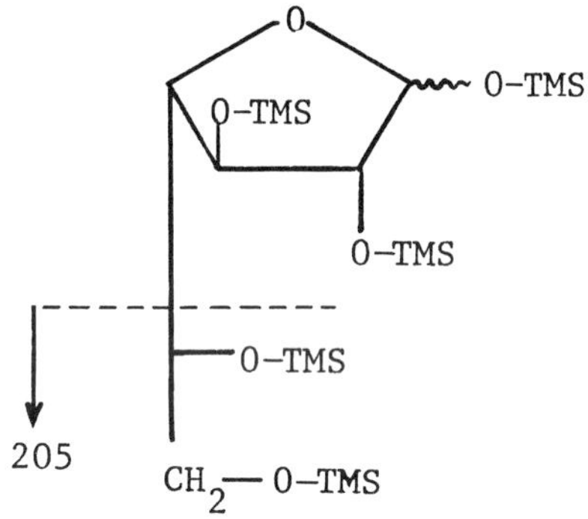

75 76

4. O-Isopropylidines

Isomeric sugars cannot be differentiated from the spectra of their methyl, acetyl, or TMS derivatives. This is possible, however, with their O-isopropylidene derivatives, since formation of these depends upon the spatial arrangement of the hydroxyl groups. Isopropylidene derivative formation enhances significantly the structural differences of isomeric sugars and results in markedly different mass spectral patterns. No molecular ions are observed in the spectra of isopropylidene derivatives of sugars due to the favorable elimination of an O-methyl group. Typical of O-isopropylidene derivatives in general, this peak helps define the molecular weight of the compound. Abundant fragment ions, however, are also formed directly from the molecular ion and occasionally the presence of metastable peaks provides indirect information about the molecular weight of the compound. For example, formation of the ion of *m/e* 101 in the spectrum of the 2,3-O-isopropylidene derivative of 1,6-anhydro-β-D-talopyranose, 77, is accompanied by an intense metastable peak at *m/e* 50.8 corresponding to the transition $M^{+\cdot}$ (*m/e* 202) → *m/e* 101 (Scheme 49) [211]. Isotopic labeling with deuterium was used to confirm the postulated hydrogen transfer.

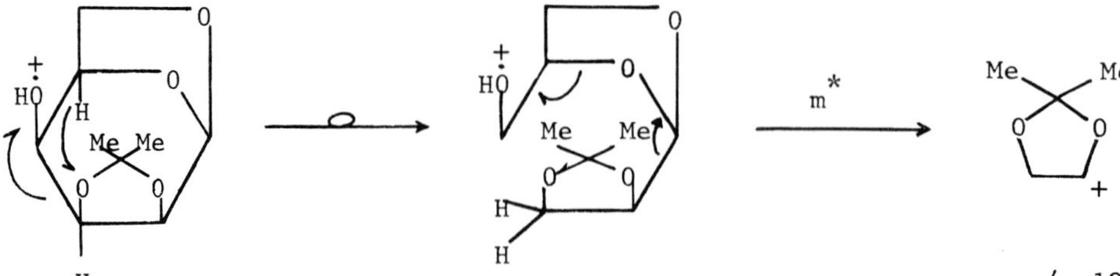

77, m/e 202 m/e 101

Scheme 49

The general applicability of O-isopropylidene derivatives in the GC-MS analysis of complex mixtures of monosaccharides is somewhat hampered by the long reaction times required for their quantitative preparation and the dependence of product-mixture composition on reaction conditions for some monosaccharides. A recent article [212] suggests the use of concentrated aldose solutions containing 1-2 mg of sample and reacting with 1% (v/v) sulfuric acid in acetone (1.5 ml) for 2 hr. Samples were injected into the GC-MS following neutralization with sodium bicarbonate.

5. *Cyclic Boronates*

Cyclic boronates provide the stereospecificity that also characterizes the O-isopropylidene derivatives of sugars. Hydroxylic groups with suitable stereochemical disposition form five- or six-membered rings, while the remaining free hydroxyl groups can be trimethylsilylated for further removal of the system's polarity. In a GC-MS study of such mixed boronate-trimethylsilyl derivatives of carbohydrates it was shown that it is possible to identify the isomeric structures of pentoses, hexoses, 6-deoxyhexoses, and 2-acetamido-2-deoxyhexoses [38]. For example, borosilylation of D-galactose, D-mannose, and D-glucose gave a single derivative for each compound represented by 78, 79, and 80 respectively. The relative intensities of TMS-containing

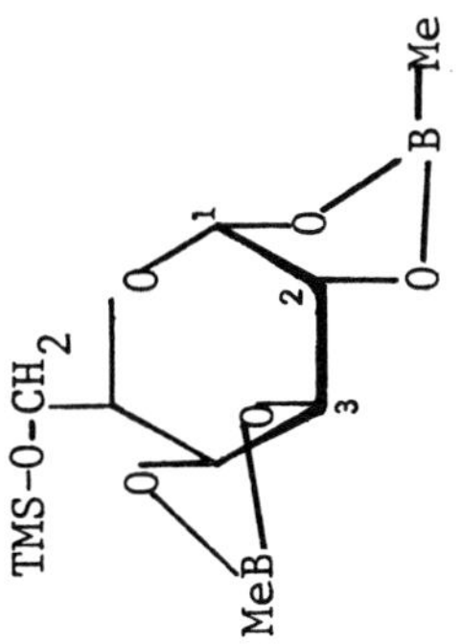

78

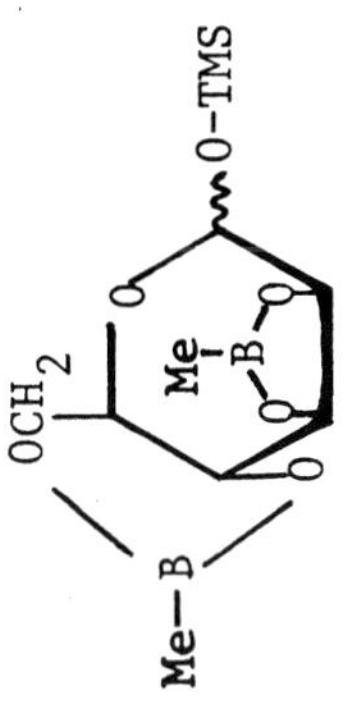

79

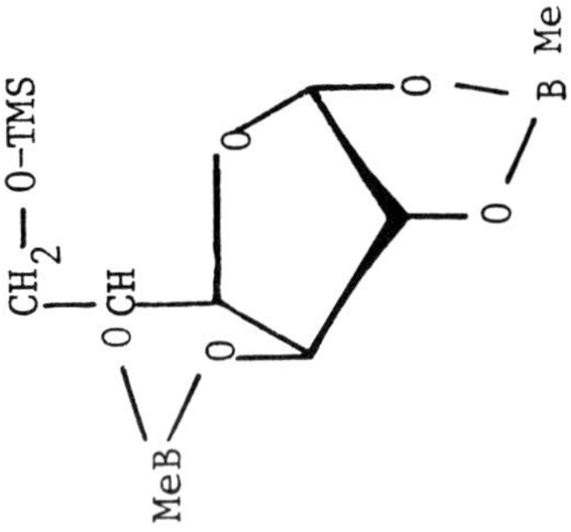

80

$(CH_3)_2-\overset{+}{Si}-O-CH_2-CHO$

ac, m/e 117

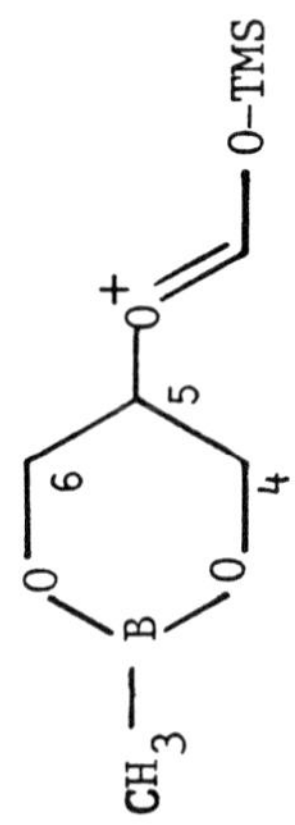

ad, m/e 215

$CH_2=\overset{+}{O}-TMS$

ae, m/e 103

ions were used to identify the linkage of the trimethylsilyloxy group. A peak at *m/e* 117 (ac) in the spectrum of 78 indicated a TMSO linkage at C-6, an ion of *m/e* 215 (ad) in the spectrum of 79 was indicative of a cyclic 2,3:4,6 diboronate structure with a TMS group at C-6, and the ion of *m/e* 103 (ae) in the spectrum of 80 suggested the presence of a furan ring and two 1,2:3,5 boronate groups.

6. *Acyclic Derivatives*

Most prominent among the acyclic derivatives are the thioacetals whose mass spectrometric properties have been reviewed previously [204-206]. In general, acyclic derivatives fragment in a predictive fashion by simple β-cleavages, which permits determination of the position of specific groupings. The spectra of thioacetals are noted for their well-defined molecular ion peaks, as opposed to the various derivatives of cyclic sugars discussed above whose spectra usually exhibit no molecular ions.

In a recent publication, Laine and Sweeley [213] reported on the preparation of open-chain methyl oxime-trimethylsilyl derivatives of sugars. Methyl oxime derivatives of aldoses and hexoses were prepared by reaction with methoxylamine hydrochloride in pyridine at 80° for 2 hr, followed by trimethylsilylation of the free hydroxylic groups with BSTFA. For the most part, the mass spectra of these acyclic derivatives exhibited very weak molecular ion peaks. Consideration of structure 81, the MO-TMS derivative of D-glucose, exemplifies the type of fragmentation encountered in the mass spectra of these derivatives. Mass values on the left indicate fragment ions containing the methyl oxime end of the molecule and their complementary ions produced from the same cleavage are shown on the right. Values in parentheses indicate the percent relative intensities (RI) of the respective ions. Secondary ions produced by loss of trimethylsilanol (90 amu) may often be of significant abundance. For example, the base peak in the spectrum of the MO-TMS derivative of D-glucose, 81, occurs at *m/e* 319, which is derived from *m/e* 409 by elimination of trimethylsilanol.

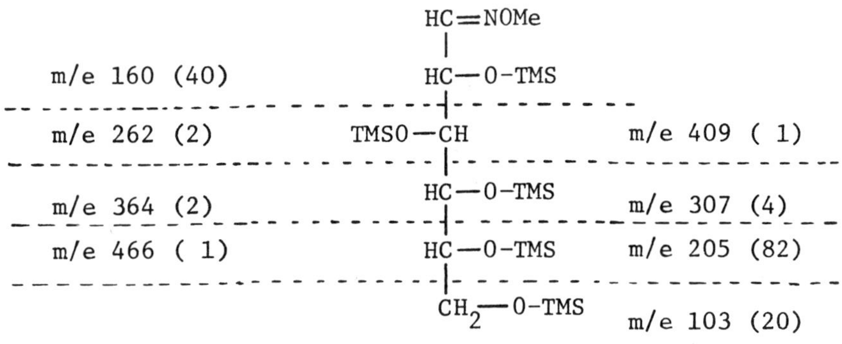

81

In view of their simple fragmentation pattern and the absence of rearrangement ions, the MO-TMS derivatives of sugars appear ideally suited for the location of substituents on the various carbon atoms. Structures 82 and 83, the MO-TMS derivatives of 2-O-methyl-D-ribose and 3-O-methyl-D-ribose, respectively, illustrate this point. Determination of the position of methyl substitution is made possible from the fragmentations leading to the formation of the ions at *m/e* 102 and 307 in 82 vs *m/e* 160 and 249 in 83.

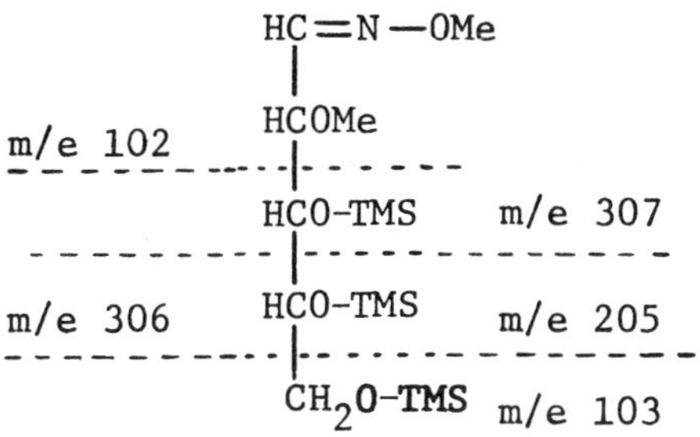

82

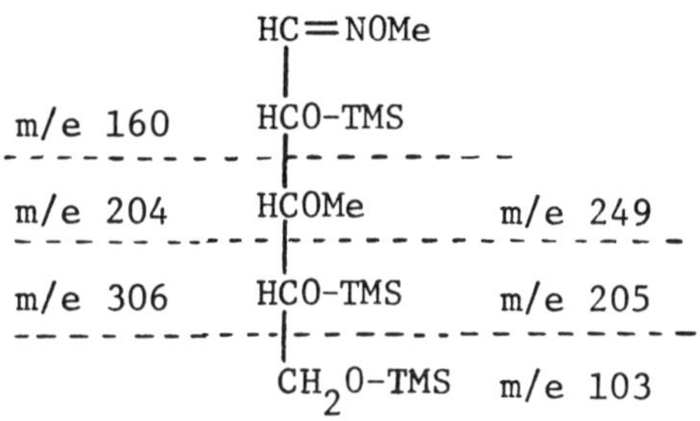

83

REFERENCES

1. W. L. Gardiner and E. C. Horning, *Biochim. Biophys. Acta, 115,* 524 (1966).
2. N. Narasimhachari and P. Vouros, *J. Chromatog., 70,* 135 (1972).
3. P. Devaux and E. C. Horning, *Anal. Biochem., 41,* 70 (1971).
4. J. Drozd, *J. Chromatog., 113,* 303 (1975).
5. J. Karliner, *Tetrahedron Lett., 1968,* 3545.
6. A. N. H. Yeo and D. H. Williams, *J. Chem. Soc. (C), 1968,* 2666.
7. H. Budzikiewicz, D. H. Williams, and C. Djerassi, *Structure Elucidation of Natural Products by Mass Spectrometry,* vol. 2, Holden-Day, San Francisco, 1964, p. 76.
8. *Ibid.*, p. 26.
9. B. A. Petersen and P. Vouros, *Anal. Chem., 49,* 1304 (1977).
10. K. Biemann, *Mass Spectrometry: Organic Chemical Applications,* McGraw-Hill, New York, 1962, pp. 83-84.
11. U. Hintze, H. Röper, and G. Gerchen, *J. Chromatog., 87,* 481 (1973).

11a. D. Karashina, A. Shigematsu, T. Furukawa, T. Magayoshi, and I. Matsumoto, *J. Chromatog., 130,* 77 (1977).

12. E. W. Robb and J. J. Westbrook, *Anal. Chem., 35,* 1644 (1963).
13. B. S. Middleditch and D. M. Desiderio, *Anal. Lett., 5,* 605 (1972).
14. N. H. Andersen, *J. Lipid Res., 10,* 320 (1969).
15. J. T. Watson, *Introduction to Mass Spectrometry,* Raven Press, New York, 1976, p. 146.
16. E. M. Chambaz, G. Defaye, and C. Madani, *Anal. Chem., 45,* 1090 (1973).
17. J. R. G. Challis and R. B. Heap, *J. Chromatog., 50,* 228 (1970).
18. L. Dehennin, A. Reiffsteck, and R. Scholler, *J. Chromatog. Sci., 10,* 224 (1972).
19. P. G. Devaux and E. C. Horning, *Anal. Lett., 2,* 637 (1969).
20. J. W. Vogh, *Anal. Chem., 43,* 1618 (1971).
21. W. J. A. VandenHeuvel and E. C. Horning, *Biochim. Biophys. Acta, 74,* 560 (1963).
22. H. M. Fales and T. Luukkainen, *Anal. Chem., 37,* 955 (1965).
23. T. A. Baillie, C. J. W. Brooks, and E. C. Horning, *Anal. Lett., 5,* 351 (1972).
24. A. E. Pierce, *Silylation of Organic Compounds,* Pierce Chemical Company, Rockford, Ill., 1968.

25. Initially introduced by: A. G. Sharkey, Jr., R. A. Friedel, and S. H. Langer, *Anal. Chem., 29,* 770 (1957).

26. C. J. W. Brooks and B. S. Middleditch, "Gas Chromatography--Mass Spectrometry," in *Specialist Periodical Reports, Vol. 2* (D. H. Williams, Senior Reporter), The Chemical Society, London, 1973, p. 306.

27. D. J. Harvey and W. D. M. Paton, *J. Chromatog., 109,* 73 (1975).

28. M. A. Quiliam and J. B. Westmore, Paper No. E-3, ASMS, 24th Annual Conference on Mass Spectrometry, San Diego, Calif., May 1976.

29. K. K. Ogilvie and D. J. Iwacha, *Tetrahedron Lett., 1973,* 317.

30. G. Philippou, D. A. Gibham, and R. F. Seamark, *Lipids, 10,* 714 (1975).

31. J. T. Watson, *Introduction to Mass Spectrometry,* Raven Press, New York, 1976, pp. 10, 11, 48, 147.

32. C. Zomzely, G. Marco, and E. Emery, *Anal. Chem., 34,* 1414 (1962).

33. P. I. Jaakonamaki and J. E. Stouffer, *J. Gas Chromatog., 5,* 303 (1967).

34. B. A. Petersen, R. N. Hanson, R. W. Giese, and B. L. Karger, *J. Chromatog., 126,* 503 (1976).

35. M. G. Horning, A. M. Moss, E. A. Boucher, and E. C. Horning, *Anal. Lett., 1,* 311 (1968).

36. M. Donike, *Chromatographia, 7,* 651 (1974).

37. M. Donike, *J. Chromatog., 78,* 273 (1973).

38. V. N. Reinhold, F. Wirtz-Peitz, and K. Biemann, *Carbohydrate Res., 37,* 202 (1974).

39. J. Wiecko and W. R. Sherman, *Proceedings, ASMA,* 22nd Annual Conference on Mass Spectrometry and Allied Topics, Philadelphia, Pa., 1974, p. 323.

40. C. J. W. Brooks and J. Watson, *Chem. Commun., 1967,* 952.

41. C. J. W. Brooks and I. MacLean, *J. Chromatog. Sci., 9,* 18 (1971).

42. G. M. Anthony, C. J. W. Brooks, I. MacLean, and I. Sangster, *J. Chromatog. Sci., 7,* 623 (1969).

43. R. W. Kelly, *Anal. Chem., 45,* 2079 (1973).

44. a) M. Beroza and M. N. Inscoe, "Precolumn Reactions for Structure Determination," in *Ancillary Techniques of Gas Chromatography* (L. S. Ettre and W. H. McFadden, Eds.), Wiley-Interscience, New York, 1969. b) W. C. Kossa, J. MacGee, S. Kamachandran, and A. J. Webber, *J. Chromatog. Sci., 17,* 177 (1979).

45. M. Widman, I. M. Nilsson, J. L. G. Nilsson, S. Agurell, and K. Leander, *Life Sci.,* Part II, *10,* 157 (1971).

46. H. J. Kupferberg, *Clin. Chim. Acta, 29,* 283 (1970).

47. R. H. Hammer, B. J. Wilder, R. R. Streiff, and A. Mayersdarf, *J. Pharm. Sci., 60,* 327 (1971).

48. E. Brochmann-Hanssen and T. O. Oke, *J. Pharm. Sci., 58,* 371 (1969).

49. I. Gan, J. Korth, and B. Halpern, *J. Chromatog., 92,* 435 (1974).

50. M. J. Barrett, *Clin. Chem. Newsletter, 3,* 1 (1971).

51. R. Osiewicz, V. Aggarwal, R. M. Young, and I. Sunshine, *J. Chromatog., 88,* 157 (1974).

52. L. K. Wong, C. E. Costello, and K. Biemann, *J. Chromatog., 116,* 321 (1976).

53. W. F. Bryant and P. D. Klein, *Anal. Biochem., 65,* 73 (1975).

54. W. J. A. VandenHeuvel, R. N. Stillwell, W. L. Gardiner, S. Wikstrom, and E. C. Horning, *J. Chromatog., 19,* 22 (1965).

55. W. J. A. VandenHeuvel, *J. Chromatog., 26,* 396 (1967).

56. W. J. A. VandenHeuvel, *J. Chromatog., 43,* 215 (1969).

57. P. R. Vagelos, W. J. A. VandenHeuvel, and M. G. Horning, *Anal. Biochem., 2,* 50 (1961).

58. K. Biemann, *Mass Spectrometry: Organic Chemical Applications,* McGraw-Hill, New York, 1962, pp. 170-172.

59. J. Vessman and G. Rietz, *J. Chromatog., 100,* 153 (1974).

60. W. J. A. VandenHeuvel, J. L. Smith, R. A. Firestone, and J. L. Beck, *Analyt. Lett., 5,* 285 (1972).

61. R. M. Thompson and E. C. Horning, *Steroids Lipids Res., 4,* 135 (1973).

62. W. J. A. VandenHeuvel, V. B. Gruber, and R. W. Walker, *J. Chromatog., 87,* 341 (1973).

63. D. L. von Minden, R. N. Stillwell, W. A. Koenig, K. J. Lyman, and J. A. McCloskey, *Anal. Biochem., 50,* 110 (1972).

64. T. S. Ro-Choi, Y. C. Choi, D. Henning, J. A. McCloskey, and H. Busch, *J. Biol. Chem., 250,* 3921 (1975).

65. R. D. Coffin and R. M. Thompson, *J. Chromatogr., 138,* 223 (1977).

66. A. Wu, *Clin. Chem., 20,* 630 (1974).

67. R. M. Caprioli, "Use of Stable Isotopes," in *Biochemical Applications of Mass Spectrometry* (G. R. Waller, Ed.), Wiley-Interscience, New York, 1972.

68. H. Budzikiewicz, C. Djerassi, and D. H. Williams, *Structure Elucidation of Natural Products by Mass Spectrometry, Volume 1: Alkaloids,* Holden-Day, Inc., San Francisco, 1964, p. 17.

69. M. Senn, W. J. Richter, and A. L. Burlingame, *J. Am. Chem. Soc.*, *87*, 680 (1965).

70. P. Vouros, *J. Org. Chem.*, *38*, 3555 (1973).

71. G. J. Kallos and L. B. Westover, *Tetrahedron Lett.*, *1967*, 1223.

72. G. M. Anthony and C. J. W. Brooks, *J. Chem. Soc. (D)*, *1970*, 200.

73. J. A. McCloskey, R. N. Stillwell, and A. M. Lawson, *Anal. Chem.*, *40*, 233 (1968).

74. P. Vouros and D. J. Harvey, *Anal. Chem.*, *45*, 7 (1973).

75. J. Szafranek, A. Wisniewski, and P. Vouros, *Carbohydrate Res.*, *56*, 11 (1977).

76. A. M. Lawson, R. N. Stillwell, M. M. Tacker, K. Tsuboyama, and J. A. McCloskey, *J. Am. Chem. Soc.*, *93*, 1014 (1972).

77. D. J. Harvey, M. G. Horning, and P. Vouros, *J. Chem. Soc., Perkin Trans. I.*, *1972*, 1074.

78. D. J. Harvey, M. G. Horning, and P. Vouros, *Tetrahedron*, *27*, 4231 (1971).

79. B. S. Middleditch and D. M. Desiderio, *J. Org. Chem.*, *38*, 2204 (1973).

79a. H. Miyazraki, M. Ishibashi, M. Itoh, K. Yamashita, and T. Nambara, *J. Chromatog.*, *133*, 311 (1977).

80. W. J. A. VandenHeuvel, J. L. Smith, G. Albers-Schonberg, B. Plazonnet, and P. Belanger, "Derivatization and Gas Chromatography in the Mass Spectrometry of Steroids," in *Modern Methods of Steroid Analysis* (E. Heftman, Ed.), Academic Press, New York, 1973.

81. C. J. W. Brooks and B. S. Middleditch, "Some Aspects of Mass Spectrometry in Steroid Analysis," in *Modern Methods of Steroid Analysis* (E. Heftman, Ed.), Academic Press, New York, 1973.

82. P. Vouros and D. J. Harvey, *J. Chem. Soc., Chem. Comm.*, *1972*, 765.

83. J. Diekman, J. B. Thomson, and C. Djerassi, *J. Org. Chem.*, *34*, 3147 (1969).

84. G. H. Draffan, R. N. Stillwell, and J. A. McCloskey, *Org. Mass Spectrom.*, *1*, 669 (1968).

85. S. Sloan, D. J. Harvey, and P. Vouros, *Org. Mass Spectrom.*, *5*, 789 (1971).

86. S. C. Havlicek, M. R. Brennan, and P. J. Scheuer, *Org. Mass Spectrom.*, *5*, 1273 (1971).

87. R. T. Gray, J. Diekman, G. L. Larson, W. K. Musker, and C. Djerassi, *Org. Mass Spectrom.*, *3*, 973 (1970).

88. J.-Å. Gustafsson, R. Ryhage, J. Sjövall, and R. M. Moriarty, *J. Am. Chem. Soc.*, *91*, 1234 (1969).

89. P. Vouros and D. J. Harvey, *J. Chem. Soc.*, *Perkin Trans. I*, *1973*, 727.

90. W. P. Weber, R. A. Felix, and A. K. Willard, *J. Am. Chem. Soc.*, *92*, 1420 (1970).

91. J.-Å. Gustafsson and B. P. Lisboa, *Steroids*, *15*, 723 (1970).

92. J.-Å. Gustafsson and J. Sjövall, *Eur. J. Biochem.*, *6*, 227 (1968).

93. B. P. Lisboa, J.-Å. Gustafsson, and J. Sjövall, *Eur. J. Biochem.*, *4*, 496 (1968).

94. J.-Å. Gustafsson and B. P. Lisboa, *Steroids*, *11*, 555 (1968).

95. H. Erikson, J.-Å. Gustafsson, and J. Sjövall, *Eur. J. Biochem.*, *6*, 219 (1968).

96. J.-Å. Gustafsson, C. H. L. Shackleton, and J. Sjövall, *Eur. J. Biochem.*, *10*, 302 (1969).

97. C. J. W. Brooks, D. J. Harvey, and B. S. Middleditch, *J. Org. Chem.*, *37*, 3365 (1972).

98. J. Bjorkhem, J.-Å. Gustafsson and J. Sjövall, *Org. Mass Spectrom.*, *7*, 277 (1973).

99. P. Eneroth, K. Hellström, and R. Ryhage, *J. Lipid Res.*, *5*, 245 (1964).

100. J. Diekman and C. Djerassi, *J. Org. Chem.*, *32*, 1005 (1967).

101. C. J. W. Brooks, E. M. Chambaz, W. L. Gardiner, and E. C. Horning, *Proceedings*, 2nd Int. Congr. Horm. Steroids, 1967, p. 366.

102. B. S. Middleditch, P. Vouros, and C. J. W. Brooks, *J. Pharm. Pharmacol.*, *25*, 143 (1973).

103. J.-Å. Gustafsson and J. Sjövall, *Eur. J. Biochem.*, *6*, 236 (1968).

104. J. Sjövall and R. Vihko, *Steroids*, *7*, 447 (1966).

105. T. A. Baillie, C. J. W. Brooks, and B. S. Middleditch, *Anal. Chem.*, *44*, 30 (1972).

106. P. Vouros and D. J. Harvey (unpublished data).

107. P. Vouros, *J. Org. Chem.*, *38*, 3555 (1973).

108. R. W. Kelly, *Steroids*, *13*, 507 (1969).

109. R. W. Kelly, *J. Chromatog.*, *43*, 229 (1969).

110. J. R. Chapman and E. Bailey, *Anal. Chem.*, *45*, 1636 (1973).

111. E. M. Chambaz, G. Defaye, and C. Madani, *Anal. Chem.*, *45*, 1090 (1973).

112. M. A. Quilliam and J. B. Westmore, *Steroids*, *29*, 579 (1977).

113. D. R. Idler, L. M. Safe, and S. H. Safe, *Steroids, 16,* 251 (1970).

114. S. Danishefsky, P. Solomon, L. Crawley, M. Sax, S. C. Yoo, E. Abola, and J. Pletcher, *Tetrahedron Lett., 10,* 961 (1972).

115. R. M. Thompson and D. M. Desiderio, *Biochem. Biophys. Res. Commun., 48,* 1303 (1972).

116. P. A. Leclercq, *Biomed. Mass Spectrom., 1,* 109 (1974).

117. C. J. W. Brooks and D. J. Harvey, *J. Chromatog., 54,* 193 (1971).

118. B. S. Middleditch and B. A. Knights, *Org. Mass Spectrom., 6,* 179 (1972).

119. P. F. Crain, D. M. Desiderio, Jr., and J. A. McCloskey, "Mass Spectrometry of Prostaglandins," in *Methods in Enzymology* (J. M. Lowenstein, Ed.), Academic Press, New York, 1975.

120. S. Nicosia and G. Galli, *Anal. Biochem., 61,* 192 (1974).

121. J. Rosello, J. Tusello, and E. Gelpi, *J. Chromatog., 130,* 65 (1977).

122. B. S. Middleditch and D. M. Desiderio, *Lipids, 8,* 267 (1973).

123. B. S. Middleditch and D. M. Desiderio, *Prostaglandins, 4,* 31 (1973).

124. B. S. Middleditch and D. M. Desiderio, *Anal. Biochem., 55,* 509 (1973).

125. B. S. Middleditch and D. M. Desiderio, *J. Org. Chem., 38,* 2204 (1973).

126. U. Axen, K. Green, D. Horlin, and B. Samuelsson, *Biochem. Biophys. Res. Commun., 45,* 519 (1971).

127. J. T. Watson and B. J. Sweetman, *Org. Mass Spectrom., 9,* 39 (1974).

128. F. C. Falkner, B. J. Sweetman, and J. T. Watson, *Appl. Spectrosc. Rev., 10,* 51 (1975) and references cited therein.

129. B. Samuelsson, M. Hamberg, and C. Sweeley, *Anal. Biochem., 38,* 301 (1970).

130. J. T. Watson, D. R. Pelster, B. J. Sweetman, J. C. Frolich, and J. A. Oates, *Anal. Chem., 45,* 2071 (1973).

131. K. Biemann, "Mass Spectrometry," in *Organic Chemical Applications,* McGraw-Hill, New York, 1962, p. 260 ff.

132. C. Zomzely, G. Marco, and E. Emery, *Anal. Chem., 34,* 1414 (1962).

133. D. E. Johnson, S. J. Scott, and A. Meister, *Anal. Chem., 33,* 669 (1961).

134. W. H. Lamkin and C. W. Gehrke, *Anal. Chem., 37,* 383 (1965).

135. P. Felker and R. S. Bandurski, *Anal. Biochem., 67,* 245 (1975).

136. W. Vetter, "Amino Acids," in *Biomedical Applications of Mass Spectrometry* (G. Waller, Ed.), Wiley-Interscience, New York, 1972.

137. J. G. Lawless and M. S. Chadha, *Anal. Biochem., 44,* 473 (1971).

138. E. Gelpi, W. A. Koenig, J. Gilbert, and J. Oro, *J. Chromatog. Sci., 7,* 604 (1969).

139. O. G. Nielsen and E. Solheim, *Anal. Chem., 47,* 333 (1975).

140. J. P. Thenot and E. C. Horning, *Anal. Lett., 5,* 519 (1972).

141. I. Horman and F. J. Hesford, *Biomed. Mass Spectrom., 1,* 115 (1974).

142. K. Rühlmann and W. Giesecke, *Angew. Chem., 73,* 113 (1961).

143. J. Klebe, H. Finkbeiner, and D. White, *J. Am. Chem. Soc., 88,* 2290 (1966).

144. E. D. Smith and H. Sheppard, Jr., *Nature, 208,* 878 (1965).

145. D. L. Stalling, C. W. Gehrke, and R. W. Zumwalt, *Biochem. Biophys. Res. Commun., 21,* 616 (1968).

146. C. W. Gehrke, H. Nakamoto, and R. W. Zumwalt, *J. Chromatog., 45,* 24 (1969).

147. C. W. Gehrke, K. Kuo, and R. W. Zumwalt, *J. Chromatog., 57,* 209 (1971).

148. K. Bergström, J. Gürther, and R. Bromstrand, *Anal. Biochem., 34,* 74 (1970).

149. K. M. Williams and B. Halpern, *Anal. Lett., 6,* 839 (1973).

150. J. J. Pisano and T. J. Bronzert, *J. Biol. Chem., 224,* 5597 (1969).

151. B. W. Melvas, *Acta Chem. Scand., 23,* 1679 (1969).

152. M. Waterfield and E. Haber, *Biochemistry, 9,* 832 (1970).

153. C. R. Stork, *Biochemistry, 7,* 1976 (1968).

154. M. Rangarajan, R. E. Ardrey, and A. Darbre, *J. Chromatog., 87,* 499 (1973).

155. C. W. Gehrke and K. Leimer, *J. Chromatog., 53,* 201 (1970).

156. B. W. Melvas, *Acta Chem. Scand., 23,* 1679 (1969).

157. T. Fairwell, W. T. Barnes, F. F. Richards, and R. E. Lovins, *Biochemistry, 9,* 2134 (1970).

158. M. Rangarajan, R. E. Ardrey, and A. Darbre, *J. Chromatog., 87,* 499 (1973).

159. R. E. Harman, J. L. Patterson, and W. J. A. VandenHeuvel, *Anal. Biochem., 25,* 452 (1968).

160. M. A. Young and D. M. Desiderio, *Anal. Lett., 8,* 1 (1975).

161. B. Halpern, V. A. Close, A. Wegmann, and J. W. Westley, *Tetrahedron Lett., 1968,* 3119.

162. R. Docter and G. Hennemann, *Clin. Chim. Acta, 34,* 297 (1971).

163. R. Hoffenberg, A. M. Lawson, D. B. Ramsden, and P. J. Raw, "The Application of Mass Spectrometry to the Study of Thyroxine and Related Compounds in Biological Fluids," in *Mass Spectrometry in Biochemistry and Medicine* (A. Frigerio and N. Castagnoli, Eds.), Raven Press, New York, 1974.

164. P. Vouros and K. Biemann, *Org. Mass Spectrom., 2,* 375 (1969).

165. J. L. Smith, J. L. Black, and W. J. A. VandenHeuvel, *Org. Mass Spectrom., 5,* 473 (1971).

166. E. Anggard and A. Hankey, *Acta Chem. Scand., 23,* 3110 (1969).

167. A. K. Cho, B. Lindeke, and D. J. Jenden, "Gas Chromatography-Mass Spectrometry in the Distribution and Metabolism of Phentermine," in *Mass Spectrometry in Biochemistry and Medicine* (A. Frigerio and N. Castagnoli, Eds.), Raven Press, New York, 1974.

168. A. Zeman and I. P. G. Wirotama, *Z. Anal. Chem., 247,* 158 (1969).

169. J. Gal, L. D. Gruenke, and N. Castagnoli, Jr., *J. Med. Chem., 18,* 683 (1975).

170. G. M. Anthony, C. J. W. Brooks, and B. S. Middleditch, *J. Pharm. Pharmacol., 22,* 205 (1970).

171. N. P. Sen and P. L. McGeer, *Biochem. Biophys. Res. Commun., 13,* 390 (1963).

172. M. G. Horning, A. M. Moss, and E. C. Horning, *Biochim. Biophys. Acta, 148,* 597 (1967).

173. W. J. A. VandenHeuvel, *J. Chromatog., 36,* 354 (1968).

174. F. P. Abramson, M. W. McCaman, and R. E. McCaman, *Anal. Biochem., 57,* 482 (1974).

175. H. Brandenberger and E. Hellback, *Helv. Chim. Acta, 50,* 958 (1967).

176. N. Narasimhachari and P. Vouros, *Anal. Biochem., 45,* 154 (1972).

177. H. Brandenberger and D. Schnyder, *Z. Anal. Chem., 261,* 297 (1972).

178. N. Narasimhachari and P. Vouros, *Biomed. Mass Spectrom., 1,* 367 (1974).

179. J. V. Evans and P. Vouros, *Anal. Biochem., 90,* 389 (1978).

180. F. E. Granchelli, A. H. Soloway, J. L. Neumeyer, and C. N. Filer, *J. Org. Chem., 42,* 2014 (1977).

181. B. S. Middleditch, *Anal. Lett., 8,* 397 (1975).

182. B. S. Middleditch, *J. Chromatog., 126,* 581 (1976).

183. J. A. McCloskey, "Mass Spectrometry of Fatty Acid Derivatives," in *Topics in Lipid Chemistry* (F. D. Gunstone, Ed.), vol. I, Logos Press, Ltd., London, 1970.

184. G. H. Draffan, R. N. Stillwell, and J. A. McCloskey, *Org. Mass Spectrom., 1,* 669 (1968).

185. R. M. Teeter, *Anal. Chem., 39,* 1742 (1967).

186. R. E. Wolff, J. T. Watson, and B. J. Sweetman, *Tetrahedron Lett., 1970,* 2719.

187. C. J. W. Brooks and I. MacLean, *J. Chromatogr. Sci., 9,* 18 (1971).

188. P. Capella, C. Galli, and R. Fumagalli, *Lipids, 3,* 431 (1968).

189. C. J. Argoudelis and E. G. Perkins, *Lipids, 3,* 379 (1968).

190. R. D. Batt, R. Hodges, and J. G. Robertson, *Biochim. Biophys. Acta, 239,* 368 (1971).

191. G. Petersson, *Tetrahedron, 26,* 3413 (1970).

192. G. Petersson, *Org. Mass Spectrom., 6,* 565 (1972).

193. G. Petersson, *Org. Mass Spectrom., 6,* 577 (1972).

194. G. Petersson, *Carbohydrate Res., 43,* 1 (1975).

195. K. Biemann, *Mass Spectrometry, Organic Chemical Applications,* McGraw-Hill, New York, 1962.

196. P. Capella and C. M. Zorzut, *Anal. Chem., 40,* 1458 (1968).

197. R. E. Wolff, G. Wolff, and J. A. McCloskey, *Tetrahedron, 22,* 3093 (1966).

198. T. Ariga, E. Araki, and T. Murata, *Chem. Phys. Lipids, 19,* 14 (1977).

199. B. M. Johnson and J. W. Taylor, *Org. Mass Spectrom., 7,* 259 (1973).

200. D. E. Minnikin, P. Abley, F. J. McQuillin, K. Kusamran, K. Maskens, and N. Polgar, *Lipids, 9,* 135 (1974).

201. S. Blum, S. Gertler, S. Sarel, and D. Sinnreich, *J. Org. Chem., 37,* 3114 (1972).

202. B. Å. Andersson and R. T. Holman, *Lipids, 9,* 185 (1974).

203. B. Å. Andersson, W. W. Christie, and R. T. Holman, *Lipids, 10,* 215 (1975).

204. N. K. Kochetkov and O. S. Chizhov, *Adv. Carbohydrate Chem., 21,* 39 (1966).

205. T. Radford and D. C. DeJongh, "Carbohydrates," in *Biochemical Applications of Mass Spectrometry* (G. R. Waller, Ed.), Wiley-Interscience, New York, 1972.

206. S. Hanessian, "Mass Spectrometry of Natural Products Containing Sugars," in *Methods of Biochemical Analysis* (D. Glick, Ed.), Interscience, New York, 1971, vol. 19.

207. J. A. McCloskey, "Mass Spectrometry," in *Basic Principles in Nucleic Acid Chemistry*, vol. 1, Academic Press, New York, 1974.

208. K. G. Das and B. Thayumanavan, *Org. Mass Spectrom.*, *6*, 1063 (1972).

209. J. Moor and E. S. Waight, *Biomed. Mass Spectrom.*, *2*, 36 (1975).

210. D. C. DeJongh, T. Radford, J. D. Hribar, S. Hanessian, M. Bieber, G. Dawson, and C. C. Sweeley, *J. Am. Chem. Soc.*, *91*, 1728 (1969).

211. D. Horton, E. K. Just, and J. D. Wander, *Org. Mass Spectrom.*, *6*, 1121 (1972).

212. S. Morgenlie, *Carbohydrate Res.*, *41*, 285 (1975).

213. R. A. Laine and C. C. Sweeley, *Carbohydrate Res.*, *27*, 199 (1973).

Note Added in Proof

For addition to the above references, the reader should refer to the following general reviews which have been published since the completion of this manuscript:

1. K. Blau and G. S. King, *Handbook of Derivatives for Chromatography*, Heyden and Son, Ltd., London, 1977.

2. D. R. Knapp, *Handbook of Analytical Derivatization Reactions*, J. Wiley and Sons, New York, 1979.

Chapter 3

ANALYTICAL APPLICATIONS OF TWO-DIMENSIONAL MASS SPECTROMETRY

P. F. BENTE, III

Research and Development Department
Hewlett Packard Co.
Avondale, Pennsylvania

F. W. McLAFFERTY

Department of Chemistry
Cornell University
Ithaca, New York

I. INTRODUCTION

The need for analytical methods that are capable of detecting and identifying components of complex matrices is encountered in many biomedical, environmental, clinical, and industrial chemical problems. A successful general approach to these problems has been to use a combination of analytical techniques to increase selectivity [1]. The widespread use of gas chromatography-mass spectrometry (GC-MS) and the increasing interest in liquid chromatography-mass spectrometry (LC-MS) illustrates the utility of combining techniques of separation and identification. Here, we discuss a similar separation-identification system that substitutes ionization and mass separation of the mixture components for chromatographic separation. The mass separated ions are then characterized further by examining their mass spectra. This general approach is designated two-dimensional mass spectrometry (MS-MS) and is illustrated schematically in Fig. 1.

A mixture of sample molecules is ionized and the ions are separated according to their mass-to-charge (m/z) ratio to form the primary mass spectrum. To the extent that the structures of the ions produced are related to the structures of the molecules present in the sample, the ions formed carry information about the identities of the sample molecules. The method of ionization must therefore be chosen to minimize structural rearrangement of the ions formed. For complete characterization of a mixture, all components should be ionized uniformly and, ideally, in such a way that one type of ion is produced for each component of the mixture. For analysis of a single component in a complex matrix, the ionization technique can be tailored to ionize that component selectively.

After this initial mass spectrometric separation, the mass spectrum of a selected precursor ion is obtained by its fragmentation with observation of the products. These reactions may occur unimolecularly for ions of suitable half-life (commonly approximately 10^{-5} sec) or as a result of further excitation of the precursor ion. The former are metastable ion (MI) decompositions; their utility in the elucidation of ion structures, their application to problems of molecular

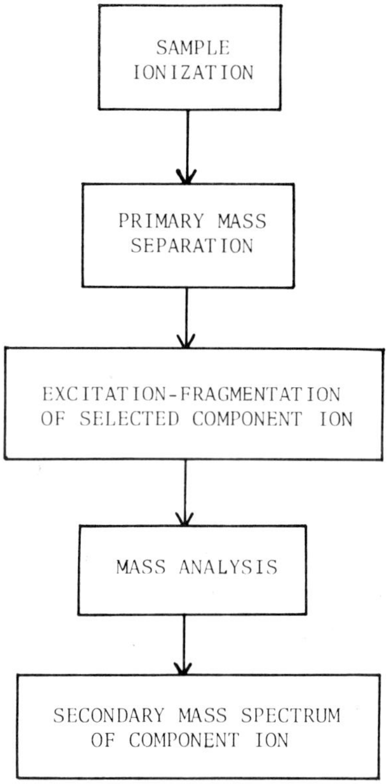

FIG. 1. Schematic illustration of two-dimensional mass spectrometry (MS-MS).

structure determination, and their usefulness in analytical problems has been thoroughly reviewed [2]. Unfortunately, those ions which have sufficient internal energy to fragment unimolecularly often show significant structural isomerization before fragmentation occurs. Information about the structure of the molecule before ionization is therefore lost. Further, only the lowest-energy reactions are observable, yielding few peaks in the MI spectrum, further limiting the applicability of metastable ions to analytical problems. If, however, ions that are stable on a mass spectrometric time scale are excited by collision, which can produce a large increase in the internal energy, the fragmentation reactions are found to be characteristic of the ion structure, and many more are observable. This method is known as collisional activation (CA) mass spectrometry [3].

The CA mass spectrum is obtained by colliding a beam of precursor ions having translational kinetic energies of several thousand electron volts with an inert collision gas such as helium. The collision transforms some of the translational kinetic energy of the ion into internal excitation energy. The resulting internal energy distribution after collision is comparable to that from electron ionization, and the resulting CA mass spectrum (excluding peaks found in the MI spectrum) is independent of the average internal energy of the ion prior to collision [3]. Consistent with the quasi-equilibrium theory, the CA mass spectrum is related to the precursor ion structure in the same way that the ordinary electron-impact mass spectrum is related to the structure of the molecule. This has allowed CA to be used in solving molecular structure problems by identifying substructural units in the molecule through the determination of the structures of fragment ions. A large amount of work has demonstrated the successful application of this technique to the determination of ion structures, and has been recently reviewed by Levsen and Schwarz [4] and by McLafferty [5].

Work from a number of laboratories indicates that CA and MI secondary mass spectra also can be used in MS-MS separation-identification systems for a wide variety of analytical problems. Specific applications have included characterization of mixtures and identification of selected compounds in complex matrices [6-8]. In comparison with GC-MS and LC-MS, this two-dimensional mass spectrometric technique should show complementary advantages in selectivity, sensitivity, and speed of analysis, and therefore may prove to be an important addition to the laboratory that must solve today's complex analytical problems.

As has occurred in other areas of mass spectrometry, when a number of workers have simultaneously developed related techniques, nomenclature has become somewhat confusing [9]. The following is a listing of terms that have been used to describe variations of two-dimensional mass spectrometry.

MIKES. In reversed-geometry double-focusing mass spectrometers (see below), the secondary mass spectrum of a selected precursor ion is generated by kinetic energy analysis of the fragment ions. This technique is called mass-analyzed ion kinetic energy spectrometry (MIKES) to distinguish it from ion kinetic energy spectrometry (IKES) in which no primary mass separation is performed and the fragmentations of all ions are observed simultaneously [2].

DADI. One commercial manufacturer of reversed-geometry double-focusing instrumentation has termed the secondary mass spectrum "direct analysis of daughter ions" (DADI), applied to both unimolecular and collision-induced fragmentations [10].

Tandem MS. The term tandem mass spectrometry has been applied generally to sequential coupling of mass spectrometers for mass separation of selected ions followed by mass analysis of their reaction products.

MI. The ions produced by unimolecular fragmentation reactions without secondary excitation are known as metastable ions. The display of product ions produced by decomposition of a metastable ion precursor is called the MI spectrum.

CA, CID, CIM. The process of fragmentation of precursor ions by collision with neutral gas molecules is referred to as collisional activation (CA), collision-induced decomposition (CID), or, less frequently, collision-induced metastable formation (CIM) [2].

+E, -E, 2E, E/2. In addition to fragmentation, reactions involving a change in charge can take place on collision between an ion and a neutral molecule. If the collision process results in charge stripping (removal of an additional electron from a positive ion), the product ions are focused at one-half the electric sector potential and are termed E/2 spectra [11,12]. Doubly charged ions that gain an electron on collision produce 2E spectra [13,14]. Positive ions that gain two electrons to produce negative ions yield -E spectra [15], while charge stripping of two electrons from a negative ion produce +E spectra [16,17].

II. INSTRUMENTATION

A wide variety of instrument configurations have been used to accomplish MS-MS. The instruments used for pioneering studies in this area utilized conventional MI techniques [18], were constructed from component parts [2], or were substantially modified commercial instruments [19,20]. However, many existing instruments can be easily adapted to obtain MS-MS data, and several commercial instruments are now available with this capability.

A. Instrument Geometry

1. Reversed Geometry

The most commonly used and convenient instrumental configuration for MS-MS is a magnetic sector instrument of "reversed" geometry in which the magnetic field precedes the electrostatic analyzer, as illustrated in Fig. 2. In this configuration, ions of charge z are produced in

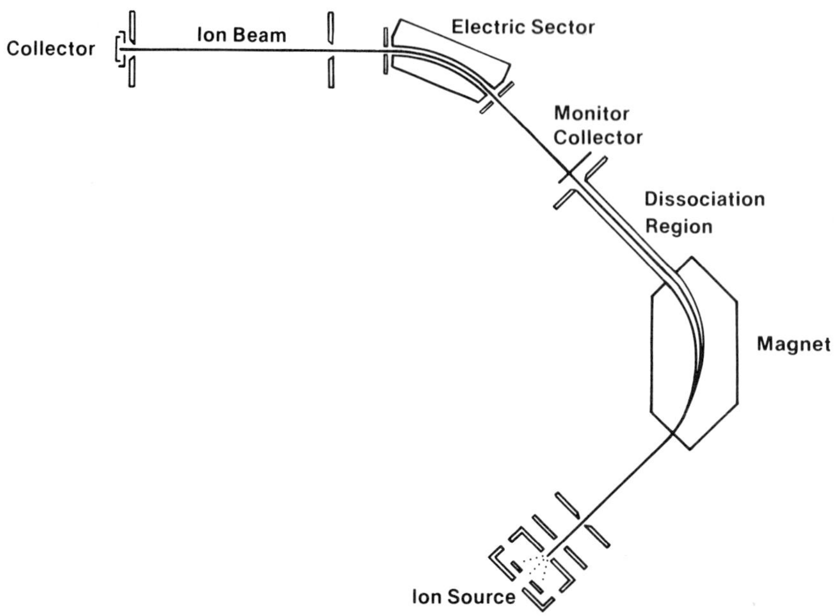

FIG. 2. "Reversed-geometry" double-focusing mass spectrometer suitable for MS-MS.

the ion source and are accelerated to a constant kinetic energy, zV, through an accelerating potential, V. These ions are dispersed according to their momentum-to-charge ratio in the magnetic field, B. If the electrostatic analyzer is adjusted to its focused potential, E_o (which will pass ions of kinetic energy zV), then the ordinary mass spectrum of the ions produced in the ion source is observed at the collector by sweeping the magnetic field. However, by fixing the magnetic field, an ion of selected mass-to-charge ratio m_1^+ can be focused on the dissociation region. Fragment ions m_2^+, formed in this field-free drift region, have kinetic energy $(m_2/m_1)zV$ and can be focused on the collector by adjusting the electrostatic analyzer to a value $(m_2/m_1)E_o$. Sweeping the electrostatic analyzer potential from zero to E_o will produce a mass spectrum of all the ions produced by fragmentation of the selected precursor ion m_1^+. A typical CA spectrum is shown in Fig. 3.

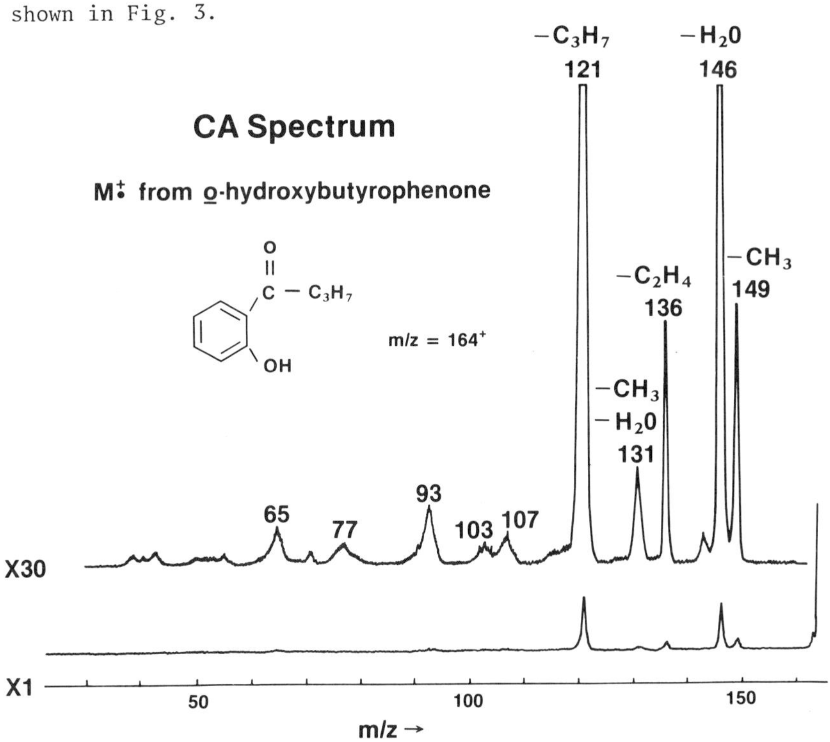

FIG. 3. Collisional activation mass spectrum of the molecular ion of orthohydroxybutyrophenone. Data obtained by Dr. M. S. Kim, Cornell University.

When a collision gas is used to induce fragmentation of m_1^+, this is referred to as the collisional activation (CA) mass spectrum of the ion m_1^+. When no collision gas is used, the unimolecular metastable ion (MI) mass spectrum is observed. Since the electrostatic analyzer separates ions according to their translational kinetic energy, this technique has also been referred to as mass-analyzed ion kinetic energy spectrometry (MIKES). Details of the construction of several instruments using this general configuration have been published [2,10,20-23]. Typical accelerating potentials are in the range of 2-10 kV with unit mass resolution (or even resolution >5,000 [23]) in selecting the precursor ion. In CA studies, the precursor ion beam is attenuated to 10-30% of its initial intensity by the collision gas to achieve the maximum intensity for the secondary mass spectrum, which is typically 0.1-1% of that of the precursor ion.

2. *Conventional Geometry*

For instruments of conventional double-focusing design in which the electrostatic analyzer precedes the magnet, "linked scan" [24] modes of operation have been successfully applied to both Nier-Johnson [25-27] and Mattauch-Herzog [28] geometries to obtain the mass spectrum of a selected ion. In one mode of linked scan operation, the selected precursor ion m_1^+ is focused on the collector with an accelerating potential V_1 and an electrostatic analyzer potential E_1. With the magnetic field held constant, V and E are then scanned simultaneously while holding V^2/E constant. Daughter ions, m_2^+, formed by fragmentation of m_1^+ in the first field-free drift region, are brought successively into focus according to the relationship $m_2^+/m_1^+ = E_1/E_2 = V_1{}^2/V_2{}^2$. In this technique, the range of observable daughter ion masses is limited by the minimum and maximum accelerating potentials that can practically be used. For $V_1 = 1$ kV and $V_{max} = 8$ kV this corresponds to m_2 in the range m_1 to 0.34 m_1 [25]. Daughter ion peaks observed in this way are much narrower than those observed with reversed-geometry instruments because the energy-focusing characteristics of the mass spectrometer compensate in part for the distribution of kinetic energies of the daughter ions.

A second mode of linked scan operation, B/E, is possible and does not have this inherent mass range limitation. With the accelerating voltage kept constant, m_2^+ ions formed in the first field-free drift region are brought successively into focus according to the relationship $m_2^+/m_1^+ = E_2/E_1 = (B_2/B_1)^{1/2}$. The chief disadvantage of this method is that it requires an accurate, continuous measurement of B from which the instantaneous setting of E must be calculated in real time [29].

In the Mattauch-Herzog geometry, which does not contain an energy-resolving slit between the electrostatic and magnetic sectors, much higher intensities of the collisional activation spectrum were obtained because ions scattered off-axis were not removed by the energy-resolving slit. A resolution (m/Δm) of several hundred in the secondary mass spectrum has been reported [25]. Disadvantages of this approach are that the accelerating voltage must be varied and hence the kinetic energy of the precursor ion on collision varies while the secondary spectrum is obtained. However, collisional activation spectra obtained this way correlated well with those obtained with reversed-geometry instruments at constant kinetic energy. Also, information on the energy release in unimolecular metastable decompositions is not available. The simultaneous scan rates can be controlled digitally by relatively inexpensive modern minicomputers. With this simple modification, then, two-dimensional collisional activation mass spectra data can be readily obtained from conventional double-focusing instrument geometries.

Other scanning methods for observing metastable ions, unimolecular and collision-induced, have been discussed by Beynon [30,31].

3. *Tandem Quadrupole*

Preliminary results [32] from an instrument configuration incorporating tandem quadrupole mass filters [33] have been published. The first quadrupole is used to select the precursor ion and focus it into a collision region. The second quadrupole is used for mass analysis of the daughter ions produced. Although the kinetic energy of the precursor ions was quite low (10 eV) compared with that of magnetic sector instruments, an intense fragmentation spectrum was obtained because of the long residence time in the collision chamber; similar ion dissociation in a high-pressure drift tube after chemical ionization has been described [34]. This system could also be used to study MI decompositions. The overall collision-induced dissociation efficiency of this system appears to be considerably higher than that reported for double-focusing instruments [7] because of the wide aperture of the secondary quadrupole compared with the narrow energy analyzer focusing slits used in reversed-geometry magnetic instruments.

4. *Other Geometries*

A number of more specialized combinations of mass analyzers illustrates the fact that instrumentation for MS-MS is limited only by the ingenuity and resources of the chemists involved. A combination of magnetic and time-of-flight mass spectrometers has been used to study dissociations of selected precursor ions [35]. In this configuration, the magnetic mass spectrometer is used to produce a beam of selected precursor ions perpendicular to the axis of a time-of-flight mass spectrometer. The precursor ion beam is focused on a metal target and the product ions produced by collision with the target are mass-analyzed. The secondary mass spectrum obtained correlated with that obtained using collisional activation.

A tandem mass spectrometer incorporating a magnetic field followed by a quadrupole mass filter with an intermediate collision region also has been described, and was used primarily to study kinetic energy thresholds for collision-induced fragmentation processes [36].

Recently, Tuithoff et al. [37] described the preliminary design of an impressive tandem mass spectrometer for studies of collision-induced dissociations that provides higher transmission and sensitivity and increased mass resolution. This is achieved by using a 10-kV accelerating potential and a magnetic analyzer to image the precursor ion beam on the slit of a collision chamber. Application of a strong accelerating potential (up to 30 kV) after the collision chamber and use of a magnetic quadrupole lens between two electrostatic quadrupole lenses preceding a magnetic sector provides the capability of focusing all daughter ions within a mass ratio of four onto a channel electron multiplier array for simultaneous rapid detection of the secondary mass spectrum. A "zoom" capability is incorporated in which the mass range of the secondary ion spectrum focused on the detector array can be varied. Simultaneous recording of the entire secondary mass spectrum will be particularly useful in investigating short-lived phenomena such as flash pyrolysis, field desorption, and laser-induced ionization [38].

The preliminary design characteristics of tandem double-focusing mass spectrometers with computer control for MS-MS have been described [39]. They will incorporate greatly increased sensitivity, a resolution of 1:50,000 in selecting the primary ion, a resolution of 5,000 in the secondary spectrum, and collision energies up to 30 kV.

B. Sample Introduction

Conventional means of sample introduction into mass spectrometer ion sources can be used with this technique. For quantitative analysis and for compound identification, sample introduction should be done so that the concentration of components in the ion source remains constant during the time necessary to obtain the required secondary (CA or MI) mass spectra. For milligram quantities of relatively volatile materials, the conventional heated batch inlet system with a controlled sample leak rate into the ion source is satisfactory. For low-volatility samples, a solids introduction probe can be used with good success. The technique of "direct chemical ionization" (DCI) [40,41] provides volatility enhancement that can be used to advantage along with the generally simpler primary mass spectra produced by chemical ionization. In these ways a substantial range of samples can be run by MS-MS that are not sufficiently volatile or are too reactive for GC or GC-MS analysis.

The technique for direct introduction of liquids into an ion source, developed as a liquid chromatography/mass spectrometry interface, could also be used for sampling liquids, particularly those containing components of low volatility. The use of the solvent as the chemical ionization reagent gas will produce simpler primary spectra. Similarly, moving wire and related systems provide a means for introducing relatively involatile components after evaporation of a volatile solvent.

Pyrolysis has proved a useful technique for complex and very involatile samples such as bacteria and plant materials and is also compatible with MS-MS analysis [42]. Several instruments currently under construction will take advantage of this technique [37,38].

C. Ionization Techniques

Any one of the present range of ionization techniques can be used to ionize the sample prior to the first stage of mass analysis. Recent reviews [43,44] have covered the wide variety of ionization techniques used in analytical mass spectrometry. As mentioned earlier, in applying two-dimensional mass spectrometry to analytical problems, it is important that the precursor ions be structurally related to the molecules so that the secondary mass spectrum will also be characteristic of the structure of the original molecule. Second, it is important to minimize the number of ions produced; ideally, only one ion should be formed from each component sought, and none from others of the mixture. The so-called "soft" ionization techniques deserve special consideration because they can exhibit selective ionization, they often minimize structural rearrangement of the ions produced relative to electron-impact ionization, and they generally induce less fragmentation, producing mass spectra with few peaks, often with only a pseudo-molecular ion. These soft ionization techniques include field desorption, chemical ionization for production of positive or negative ions, and negative ion formation by secondary electron attachment. However, as is the case with conventional mass spectrometry, many useful results can be obtained with an ordinary electron-impact source; low-energy operation of this source also achieves soft ionization and selectivity, but at the price of reduced sensitivity.

D. Dissociation Region

Usable collisional activation spectra can be produced simply by reducing the pumping efficiency in the field-free drift region, by heating the walls of the drift region to cause adsorbed molecules to outgas [18], or by the addition of a collision gas. Under these conditions, a background pressure of 10^{-5} to 10^{-4} torr is sufficient to produce a detectable collisional activation spectrum. However, significantly improved sensitivity and resolution in the secondary spectrum can be obtained with special collision chambers; details of several for use with reversed geometry instruments have been published [24,45,46].

A collision chamber containing high gas density in the field-free drift region serves to localize the collisions of the ion beam so that all collisions occur near the focal point of the electrostatic analyzer, resulting in high-resolution secondary (CA) spectra. If the collision chamber precedes the focal point, the effect of angular scattering of the main beam can be reduced significantly, allowing more precise measurement of secondary mass spectral peaks resulting from loss of one or two mass units from the primary (precursor) ion. Incorporation of a differential pumping region between the β-slit and the collision chamber helps to reduce the collision probability after ions have passed through the slit and thus minimizes angular scattering and loss of resolution in the electrostatic analyzer [46].

Applying an intermediate potential to the collision chamber can discriminate against ionic fragmentations that occur elsewhere in the field-free drift region. This is particularly useful in distinguishing between MI decompositions, which occur uniformly through the drift region, and CA decompositions, which are localized in the collision chamber [47].

The use of grazing collisions of the ion beam with a metal surface has been reported to effect dissociation of selected precursor ions. The reaction $CH_4^+ \rightarrow CH_3^+ + H$ along with a variety of charge-stripping reactions have been attributed to excitation from the collision with the metal surface [48]. However, a related study showed no detectable secondary ions for a variety of organic precursor ions and a variety of metal surface configurations when the metal surface had been thoroughly baked out [49]. Note that CA spectra are produced by a 90° scattering from a metal surface, however [35].

E. Computerized Data Acquisition

Signal-to-noise ratios on observed secondary mass spectra can be improved by decreasing the scan rate or selectively spending more time acquiring the signal for low-intensity product ions. In order to compensate for variations in sample concentration in the ion source as well as variations in source conditions, rapid repetitive scanning of the secondary mass spectrum can be used and the resulting scans averaged. For reversed-geometry magnetic instruments, this is accomplished by repetitively scanning the electrostatic analyzer potential. Several systems for accomplishing this under computer control have been described [46,50]. When the signal is shot-noise limited, the signal-to-noise ratio will increase in proportion to the square root of the number of scans. However, if constant-frequency noise (such as 60 Hz) is present, multiple asynchronous scans at random phase angles will reduce this noise much more rapidly [51]. When multiple electric sector scan averaging is incorporated, the instrument must be sufficiently stable (e.g., electric sector potential) to prevent peak broadening and loss of resolution due to drift.

Such computer data acquisition for CA spectra not only increases sensitivity but has been particularly valuable for improving the quantitative accuracy of relative abundances in the CA spectrum. Presently achievable reproducibilities of approximately ±2% [46] are necessary to distinguish between the CA spectra of particular isomeric ions and to allow more accurate quantitative determination of isomeric ion mixtures [52].

Computer control of the instrument functions, as is common with many GC-MS systems, provides significant improvement in the ease of operation of the instrument. A similar system for two-dimensional mass spectrometry, which would automatically select precursor ions and record the secondary mass spectra, storing them in memory for later recall and comparison, would prove useful [39].

III. THEORY

The similarity between fragmentation reactions resulting from collisional activation of a molecular ion and the fragmentations of the electron-impact mass spectra of the corresponding molecules has led to a mechanism for CA processes similar to that for formation of EI mass spectra [53]. Additional evidence in support of this mechanism and a more detailed discussion of its limitations have been presented [30], and recently a method was reported for partial calculation of CA spectra [49]. For ion kinetic energies greater than approximately 1 keV, the collision-induced dissociations occur largely via a two-step process in which collision of the incident ion with a neutral gas molecule results in electronic excitation of the ion via a Franck-Condon process. After excitation, the fragmentation reactions follow the predictions of the quasi-equilibrium theory [54], according to which the ion reaches the ground electronic state via potential energy surface crossings. Fragmentations occur from the ground electronic state as a result of the excess internal vibrational energy of the ion.

The fragmentation reactions of a collisionally activated ion should thus be the same as those of electron-impact-generated ions of the same internal energy. The internal energy distribution in the ions before collision has the greatest effect on the relative abundance of ions formed via low-energy fragmentation pathways. These pathways are important for MI decompositions as well. The products of high-energy reactions, whose relative abundances show the least effect of internal energy of the ion before collision, typically arise from simple bond cleavages and hence are the most useful for deduction of ion structures. Figure 4 shows typical internal energy distributions before fragmentation resulting from electron-impact

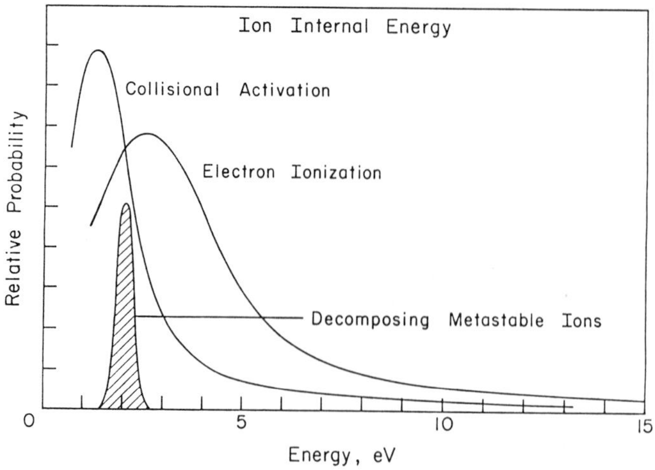

FIG. 4. Representative internal energy distributions before fragmentation for an ensemble of ions after electron ionization or collisional activation. The internal energy distributions for metastable ions that fragment unimolecularly with an activation energy of 2 eV is also shown.

ionization and collisional activation [49]. Also shown is the range of internal energies for metastable ions that decompose unimolecularly for a hypothetical case in which the lowest-energy ion decomposition has an activation energy of 2 eV.

The effect of ion kinetic energy on the amount of energy transferred by collision (ΔE) can be estimated using the Massey "adiabatic criterion" [55], which states that the highest probability for the transfer of ΔE occurs when $\Delta E = hv/a$, where h is Planck's constant, v is the velocity of the ion, and a is an adiabatic interaction parameter of molecular dimensions (approximately 10 Å). Although a wide range of energies can be transferred on collision, the higher the velocity of the ion the higher the average amount of energy transferred. As shown earlier by Jennings [18], >12 eV can be transferred by collision, although this amount of energy transfer is much less probable.

The abundance of the collision-induced fragmentations increases linearly with the collision gas pressure as long as the probability that an ion will experience a second collision is small. At higher collision gas pressures, different scattering cross sections of precursor and daughter ions cause the relative abundance of fragmentation products (the collisional activation secondary mass spectrum) to change. Experimentally, adjusting the collision gas pressure to give a fixed attenuation of the primary ion beam to 10-30% of its initial value provides the highest sensitivity and satisfactory spectra.

The nature of the collision gas also has an effect on the relative abundance of the secondary spectrum. Small molecules such as helium have been observed to be generally more effective than larger molecules [53], and a linear correlation between ionization potential of the target gas and the cross section for collisional activation is observed [49], indicating that helium is the collision gas of choice.

IV. INTERPRETATION OF SECONDARY MASS SPECTRA

Molecular structure information on particular components of the sample is derived from the *m/z* value of the component's ion (or ions) separated in the first MS and from the secondary spectrum of the component ion obtained in the second MS. As in GC-MS, the secondary spectrum is used in two general ways to characterize the unknown component; namely, by comparison with reference spectra of known compounds or by interpretation of the spectrum to deduce the structure of the unknown component. For both ways, CA spectra are usually more valuable than MI spectra. If a reference spectrum is found to match the secondary spectrum sufficiently well, this provides a direct identification of the component's ion; for this the CA spectrum is usually much more definitive than the MI spectrum because of its much larger number of peaks. The reproducibility of CA spectra from instrument to instrument is similar to that for EI spectra so CA spectra should be suited to automated file searching procedures [56] when an appropriate reference file of CA spectra is available. To date, however, this has not been accomplished. Attempts at matching CA spectra of molecular ions with electron-impact mass spectra using a probability-based matching system (PBM) [56] have been relatively unsuccessful [8].

Even if a matching reference spectrum is not found, structural information can be obtained by interpretation of the secondary spectrum. Because the products displayed by the MI spectrum arise from the lowest-energy ion dissociation processes, rearrangements are more probable, and structural interpretation is often ambiguous (for further information on MI spectra, see Ref. 2). However, the CA spectrum displays products of high- as well as low-energy dissociation reactions; formation of these is governed by the quasi-equilibrium theory. Thus the utility of the CA spectrum for characterizing a component molecule in MS-MS is generally similar to that of an electron ionization (EI) spectrum for GC-MS [57]. The extensive correlations that have been developed for interpretation of EI mass spectra [58] can be applied to interpretation of CA spectra, but there are specific limitations. Before discussing these, however, the possible relationships of the secondary mass spectrum to the molecular structure of the mixture component should be examined.

The problem of using MS-MS for identification and quantitation of unknown components in mixtures is quite different from its use for the selective detection and quantitation of specific known compounds in complex mixtures. A great deal has been written concerning the application of GC-MS or LC-MS to the latter problem using techniques such as "multiple-ion monitoring." For the former problem, a general approach to the characterization of unknown mixtures has been described [8]. The primary mass spectrum of a mixture presents potential ambiguities in that (1) a single component can generate multiple ions; (2) an ion at a single *m/z* value can contain contributions from more than one mixture component; and (3) with "soft ionization" some mixture components may not be represented by any ions. The molecular ion of a component is probably the most valuable for producing a characteristic CA spectrum, as this will most closely match the component's EI spectrum. However, primary ionization with electrons does not always produce a molecular ion (for a large data base of EI spectra, 85% exhibit $M^{+\cdot}$), and 70-eV EI usually also produces dozens (or hundreds) of fragment peaks that can interfere with molecular ions of other components (see Sec. III). Lowering the ionizing electron energy in EI spectra can aid in distinguishing between fragment ions and molecular ions, but this discriminates against components of high ionization potentials. Field ionization (FI) is probably the best way for producing molecular ions from a mixture while minimizing fragmentation, at least for major components [59]. To obtain the maximum amount of structural information from a totally unknown mixture, it is best to employ various primary ionizing conditions, including chemical ionization (CI) with a variety of ionizing reagent gases [8]. Reagents with the highest proton affinities (e.g., NH_3) will give $[M+H]^+$ ions only from mixture components of similarly high proton affinities; identification of these from their CA spectra should thus simplify the identification of the more complex array of $[M+H]^+$ ions produced by CI with ionizing reagents of lower proton affinity. Note, however, that in particular cases the $[M+H]^+$ ion from a molecule can have a different structure under different CI conditions due to a change in the site of protonation [8]. It is also possible to obtain useful structural information from one or more fragment ions formed by primary electron ionization of a particular mixture component; the CA spectrum of its molecular ion will give a qualitative indication of its EI fragment ions. The problem of several isobaric ions of different elemental composition can be solved by high-resolution spectra, if available. Since the CA spectrum of a mixture is a linear combination of the individual spectra, the CA spectrum can also aid in identification of ions of different elemental composition at the same nominal mass and also can indicate the presence of two ion structures of the same elemental composition, if CA reference spectra of the individual ions are known.

Experience in interpretation of CA spectra is limited, especially in comparison with that of EI spectra. As discussed recently [8], most of the correlations developed for EI spectra are qualitatively applicable to CA spectra. Major differences include: (1) the CA spectrum gives no indication of molecular (or other precursor) ion stability, as these ions by necessity are in large excess; (2) the fragment ions do not exhibit isotope peaks in natural abundance ratios, as precursor ion selection also involves isotope separation;

(3) the different distribution of ion internal energy values for decomposing ions in CA vs EI spectra not only produces quantitative differences in fragment ion abundances but in isolated cases unusual fragmentation pathways (such as the loss of $:CH_2$) [60]; and (4) the primary fragmentations of EI spectra must involve the odd electron ($OE^{+\cdot}$) molecular ion, while the precursor ion of a CA spectrum can be either an even-electron (EE^+) or an $OE^{+\cdot}$ species. A substantial amount of information on the unimolecular decompositions of EE^+ ions has been derived from EI mass spectral correlations [58], but our knowledge of these is much more limited than that of $OE^{+\cdot}$ decompositions (thus, CA spectra offer an excellent tool to reduce these limitations in EI interpretation). It is well known that the decomposition of an EE^+ ion to give an EE^+ product ion plus a neutral molecule is energetically favored over the complementary reaction to give an $OE^{+\cdot}$ ion and a neutral radical. Since the former reaction is often accompanied by rearrangement, this can introduce ambiguities into the interpretation of CA spectra of EE^+ ions, such as the $[M+H]^+$ and $[M+Na]^+$ ions formed by CI or field desorption. For example, the first through third largest peaks in the CA spectrum (*not* eliminating peaks also formed by MI) of the EE^+ ion $C_2H_5CH{=}OH^+$ formed by protonation of propionaldehyde are from the losses of C_2H_4, H_2O, and H_2 [61]; the loss of C_2H_4 can also give the largest peak in the CA spectra of the isomeric ions $CH_2{=}\overset{+}{O}C_2H_5$, $CH_3CH{=}\overset{+}{O}CH_3$, and $(CH_3)_2C{=}\overset{+}{O}H$ (protonated acetone). However, such isomers can easily be distinguished by comparison with their reference spectra, even for mixtures, and further correlation studies will surely decrease the ambiguities of spectral interpretation.

V. SELECTIVITY

The selectivity of an analytical technique such as MS-MS is a measure of the likelihood of false positive determinations; i.e., the ability of the technique to distinguish the compound or structure of interest from all interfering compounds, refered to as background. The more potentially interfering compounds present in a sample, the higher the selectivity required to discriminate against them. Any quantitative estimation of selectivity for an analytical technique will depend highly on both the compound and the sample matrix in question. However, some semiquantitative measures, based on statistical assumptions, are useful in comparing MS-MS with a related technique such as GC-MS.

In a gas chromatographic separation, assuming the stationary phase has been chosen, the selectivity can be defined as the number of peaks that can be resolved in a chromatogram. Each peak position can be considered as a selectivity bin. According to this definition, typical values for selectivity range from approximately 50 peaks for an isothermal, packed-column chromatogram to 100 for a temperature-programmed packed-column chromatogram to 500 or more for a temperature-programmed, high-performance, open tubular (capillary) column chromatogram [62-64]. In a chromatographic analysis, the compound of interest occurs at a known retention time corresponding to a particular bin. If there are N bins and q possible interfering compounds occurring randomly in the chromatogram with equal probability of falling in any bin, then the probability of an interference with the compound of interest is $P(\rho) = e^{-\rho}$ where $\rho = q/N$ as given by Klein [65]. Thus, a 50% probability of an interference will occur at $\rho = 0.7$, a 10% probability at $\rho = 0.11$, and 5% at $\rho = 0.05$. Such a purely statistical treatment of selectivity does not account for the fact that a skilled chemist will use chemical information about the sample to choose the best stationary phase for the separation. However, it is worth noting in this regard that in a typical GC-MS analytical service laboratory, more than 90% of all GC-MS analyses are performed using a methyl silicone stationary phase [66].

The selectivity of the initial ionization and mass separation step in two-dimensional mass spectrometry can be estimated in a similar manner. At unit mass resolution, the number of possible masses in the primary mass spectrum is of the order of 500, depending on the molecular-weight range of the sample components. For electron-impact mass spectra there are, on the average, 12 ions of greater than 9%

relative abundance per compound [67]. This will have the effect of increasing the probability that a background ion will interfere with a selected precursor ion by a factor of 12, reducing the effective number of bins under these conditions from 500 to approximately 42. Since there are an increasing number of peaks of lower abundance in a typical mass spectrum, the selectivity of the primary mass separation stage will decrease as the relative concentration of potentially interfering substances increases. Figure 5 shows the effective selectivity of electron-impact MS-MS with a selected precursor ion as a function of the minimum significant abundance level of ions in the mass spectrum of the interfering compounds, based on the data in Table 1, Ref. [67]. The data show that the average spectrum in the file has approximately 50 peaks of abundance >1%, 12 peaks >9%, and 3 peaks >38% relative abundance. For an assumed total number of mass values of 500, the effective number of bins per compound and hence the selectivity would vary from 167 to 10 if potentially interfering peaks of >38% to >1% relative abundance were considered. Based on this analysis, the selectivity of the primary mass separation step with electron-impact ionization is comparable to that of a packed-

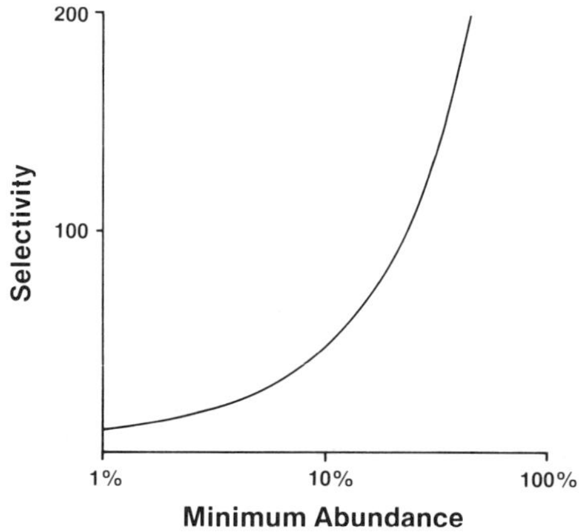

FIG. 5. Effective selectivity for electron-impact mass spectrometry as a function of the minimum significant abundance of ions in the mass spectra of potentially interfering compounds. Calculated from the data of Ref. 67 assuming a maximum selectivity of 500.

column gas chromatographic separation, although the comparison is increasingly less favorable in situations involving trace analysis. In contrast to electron-impact primary mass spectra, the selectivity of gas chromatography is relatively insensitive to the relative abundance of potentially interfering compounds. Assuming gaussian peak shapes at a resolution of 1.0 [68], a 600-fold increase in abundance is required to double the effective width of a peak and hence to halve the effective selectivity; in actual practice the figure is somewhat less because of peak tailing.

In any particular application, the choice of which ion in the primary mass spectrum of the compound of interest to select as the precursor ion can be aided by reference to the uniqueness and abundance criteria established for probability based matching [69] of mass spectra. These criteria provide a quantitative estimation of the probability that a randomly selected compound will cause an interference as a function of *m/z* of the ion and its relative abundance in the spectrum of the compound of interest. This choice of ion can be modified by further knowledge about the chemical nature of potential interfering compounds. Additional specificity not found in GC-MS is available in MS-MS in that the mass of the fragment ions and the kinetic energy released during their formation (as measured by peak broadening) are characteristic of ion structure.

The selectivity of the primary mass separation with soft ionization methods such as chemical ionization is considerably higher than that for electron-impact ionization. Although fewer data are available on the statistical occurrence of peaks in chemical ionization mass spectra, the fact that such spectra typically contain only one to three major peaks and significantly fewer peaks of lower abundance yields estimations of selectivity in the range of 100-200 and a smaller decrease in selectivity from consideration of potentially interfering lower-abundance ions. This predicts that the soft ionization techniques would be the techniques of choice for trace analysis applications, with selectivity comparable to that of open tubular columns in GC-MS.

The inclusion of high-resolution mass spectrometry, capable of separating isobaric ions of differing elemental compositions, will provide a further quantum jump in selectivity of the primary mass separation. In this way each unit mass can be separated into 10-100 bins, depending on the mass and number of possible elemental compositions, so that the selectivity should significantly exceed that available with high-performance open tubular column chromatography [39].

VI. SENSITIVITY

The sensitivity of two-dimensional mass spectrometry is controlled by a combination of the following factors:

1. Ionization efficiency of the sample in the ion source
2. Relative abundance of the selected precursor ion in the mass spectrum of the compound of interest
3. Ion transmission efficiency through the first mass spectral separation
4. Conversion efficiency to form the secondary mass spectrum
5. Ion transmission through the second stage of mass separation
6. Ion detection sensitivity--ultimately limited by statistical fluctuations in the rate of ions reaching the detector and the background ion count

As in GC-MS, sensitivity in MS-MS is maximized where a single precursor ion is selected and a single ion in the secondary mass spectrum is monitored continuously, as would be the case when looking for a single trace component in a mixture. Under these conditions a limit of detectability of p-nitrophenol of 10^{-11} gram has been reported [6]. It was estimated that sensitivity an order of magnitude better could be obtained in cases where the intensity of ions in the secondary mass spectrum is greater. Cooks has reported similar sensitivities, using +E spectra of negative ions [70].

In comparison with the sensitivity of GC-MS, MS-MS shows limits of detection, based on absolute amount of material required for detection, that average some 100-fold higher. This difference can be attributed chiefly to factor 4 above. The efficiency of production and detection of the secondary mass spectrum by collision with neutral gas molecules is of the order of 1%, particularly in present instruments of reversed-geometry design that tend to be optimized for higher resolution rather than higher sensitivity. Preliminary results with a tandem quadrupole instrument indicate that significantly higher conversion efficiencies can be achieved at lower collision energies [32].

In applications involving trace analysis but without restrictions on the total amount of sample, as, for example, in water-pollution studies or analysis of trace impurities in gasoline, minimum detectable levels can be improved by averaging repetitive scans of the desired portion of the secondary mass spectrum or by using the major sample component(s) as the CI ionizing reagent gas [8,71]. In cases where the minimum detectable amount is limited by interference from other components in the mixture, performance can be improved by increasing the selectivity of the primary mass separation stage, either by selecting a different precursor ion or by using a different ionization technique, if possible. On the other hand, the cross section for higher-energy dissociation reactions increases rapidly with ion collision energies [49], so substantial sensitivity improvement should be possible in this way [39]. Also, the background ion current in tandem mass spectrometers is substantially reduced vs a single MS, so that single ion monitoring using ion counting could also yield a substantial sensitivity improvement. The increased selectivity would be particularly useful when chemical background, rather than ultimate instrument sensitivity, determines the detection limits in complex samples.

As is the case with GC-MS, scanning the secondary mass spectrum to confirm the identity of an unknown component rather than monitoring a single ion in the secondary spectrum results in higher minimum detectable quantities because of the requirement of obtaining adequate ion statistics for all the peaks of interest.

VII. SPEED

MS-MS has a significant advantage over GC-MS in speed of analysis because the chromatographic separation is made in the time domain while the primary mass separation is made in the spatial domain. The time required for a chromatographic separation is fixed, regardless of the amount of sample, and is limited by the time required to elute all components from the chromatographic column, before the next analysis can begin. Attempts to increase the speed of chromatographic analysis are limited by the corresponding decrease in selectivity. On the other hand, with MS-MS, sample ions are transmitted virtually instantaneously (approximately 10^{-5} sec) through the first separation stage. Scan rates for the secondary mass spectrum are limited by the amount of sample and by the desired signal-to-noise ratio. In cases where the detection of trace components in large samples is required, repetitive scanning of the secondary mass spectrum or portions of it would have the same effect as concentration of the sample before analysis.

VIII. ANALYTICAL APPLICATIONS

While most of the work with MS-MS has been applied to the elucidation of ion structures, several groups have begun exploiting the analytical capabilities of the technique [2,6-8,37,39,70,72-84]. The general characteristics of mixture analysis by MS-MS have been treated using results on synthetic mixtures to illustrate features of the approach [7,8]. A general approach to identifying components in unknown mixtures has been outlined and illustrated using three different five-component mixtures to show the effect of ionization technique and the utility of known mass spectral correlations in interpreting the primary and secondary mass spectral data [8].

In an earlier example, ions from mixture components of derivatized oligopeptides were separated by mass spectrometry, and amino acid sequence information on the individual components was deduced from their MI spectra [74]. Similar applications have been reported by Russian workers [85]. Both MI and CA spectra are useful for distinguishing leucine and isoleucine residues in oligopeptides, which usually show no characteristic differences in EI spectra [86]. Levsen and Schulten [87] identified six major components of a mixture of pyrolysis products of deoxyribonucleic acid using MS-MS with low-energy electron ionization and CA secondary mass spectra.

Broad applicability of MS-MS to steroid mixtures has been shown by the Stanford group [75]. Using a primary EI spectrum and secondary MI spectra, applications were found for component identification and for the differentiation of structural isomers.

Gallegos [78] has shown that MS-MS can be used for the analysis of specific compound classes that produce a characteristic peak in their EI mass spectra, such as *m/z* 149 of phthalates or *m/z* 191 of terpenes. By scanning the ion-accelerating voltage of a double-focusing MS, all precursors of this characteristic daughter ion can be recorded, thus displaying the distribution of such compounds in the sample by molecular weight (this assumes that the main MI transition yielding the characteristic daughter ion is from the molecular ion).

Results on a mixture of three isomeric pentanones and a five-component mixture of ketones showed quantitative accuracy of 5-10% using an internal standard, EI, to form the primary mass spectrum, and observing either the CA or MI spectrum of the molecular ion [6]. A reaction mixture of the methylation products of 2-butanone and methyl iodide in the presence of strong base was also analyzed using EI ionization and both CA and MI spectra. In this case, the secondary mass spectra of all the expected reaction products were determined beforehand. The reaction products were detected and quantitated in the presence of at least a 100-fold excess of methylating agent and solvents. Semiquantitative results were obtained in a similar study of nine reaction products of the methylation of 2-pentanone in which the CA and MI spectra of the expected products were not determined beforehand [6].

In a different study [8], detection limits for thiophene in gasoline of 25 ppm and tetrahydropyran in gasoline of 500 ppm were determined using primary EI spectra and CA secondary spectra of the molecular ions *m/z* 84 and 86 respectively. In the case of thiophene, the CA spectrum of the *m/z* 84 peak in gasoline showed relatively little interference with the CA spectrum of *m/e* 84 from thiophene. On the other hand, the CA spectrum of *m/z* 86 ions from gasoline showed some interference with the CA spectrum of *m/z* 86 from tetrahydropyran, resulting in a considerably larger detection limit. However, by using chemical ionization with the gasoline itself as the reagent gas, a much larger amount of the sample could be introduced, and the detection limit was improved by more than an order of magnitude. This illustrates the advantage gained by selecting the proper ionization conditions for formation of the primary mass spectrum to optimize the selectivity for a particular component of interest. The results of MS-MS of several synthetic mixtures were studied to determine the effect of changing parameters such as ionization conditions. Identification of a maximum number of components in a complex mixture by MS-MS in most cases will require a combination of EI, CI, and other ionization methods.

The ability of this technique to identify individual selected components in the presence of massive amounts of contamination is further illustrated by the detection of the alkaloid coniine in a sample of freshly cut poison hemlock. Sample preparation consisted of grinding the plant tissue under liquid nitrogen to facilitate cell breakage and introduction of the tissue sample into the ion source via the direct insertion probe. Isobutane chemical ionization was used with nitrogen as the collision target gas. In a similar manner, morphine and papaverine were identified in raw opium from *Papaver somniferum*, with an estimated sensitivity of 1-10 ng. Recently, the same laboratory identified hippuric acid in 2 μl of untreated urine using the +E spectrum of its $[M-H]^-$ ion formed by CI [70].

The technique has been applied to the identification of the alkaloids ubine and mescaline in crude cactus extracts by observing the CA spectrum of the pseudo-molecular ion formed by methane or isobutane chemical ionization [88]. Hordenine and N-methyl tyramine were identified in cacti extracts from the CA spectrum of the molecular ions found by electron impact ionization. The presence and preliminary structure of a new alkaloid, later determined to be uberine, were deduced from similar experiments.

IX. CONCLUSIONS

Two-dimensional mass spectrometry, using collisional activation to produce the secondary mass spectrum, is a promising new analytical technique. It is being used by an increasing number of workers as a complement to GC-MS in applications where speed of analysis is important. Several exciting new developments in instrumentation promise to expand the sensitivity and selectivity of the technique [32,37,39] as well as the range of problems to which it is applied.

Several areas should prove fruitful for future development of the technique. Further exploitation of selectivity in the initial ionization step will yield benefits in sensitivity and selectivity. Additional studies of the relationship between CA spectra and ion structures for both odd- and even-electron ions will add to the data base necessary for spectral interpretation. Applications of the technique to monitoring specific compounds at low levels in complex samples, as is frequently required in environmental and clinical problems, will make good use of the speed and selectivity of the technique.

In all, the technique is a valuable addition to the array of tools available to the analytical chemist.

REFERENCES

1. H. A. Laitinen, *Anal. Chem., 47,* 1225 (1975).
2. R. G. Cooks, J. H. Beynon, R. M. Caprioli, and G. R. Lester, *Metastable Ions,* Elsevier, Amsterdam, 1973.
3. F. W. McLafferty, P. F. Bente III, R. Kornfeld, S.-C. Tsai, and I. Howe, *J. Am. Chem. Soc., 95,* 2120 (1973).
4. K. Levsen and H. Schwarz, *Angew. Chem. Int. Ed. Engl., 15,* 509 (1976).
5. F. W. McLafferty, in *High Performance Mass Spectrometry* (M. L. Gross, Ed.), American Chemical Society, Washington, D.C., 1977, p. 77.
6. T. L. Kruger, J. F. Litton, R. W. Kondrat, and R. G. Cooks, *Anal. Chem., 48,* 2113 (1976).
7. R. W. Kondrat and R. G. Cooks, *Anal. Chem., 50,* 81A (1978).
8. R. W. McLafferty and F. M. Bockhoff, *Anal. Chem., 50,* 69 (1978).

9. J. H. Beynon, *Pure Appl. Chem., 50,* 65 (1978).

10. J. H. Maurer, C. Brunee, G. Kappus, K. Habfast, U. Shröder, and P. Schulze, *Paper K-9,* 19th Annual Conf. on Mass Spectrom. and Allied Topics, Atlanta, Ga., 1971.

11. R. G. Cooks, J. H. Beynon, and T. Ast, *J. Am. Chem. Soc., 94,* 1004 (1972).

12. T. Ast, J. H. Beynon, and R. G. Cooks, *J. Am. Chem. Soc., 94,* 6611 (1972).

13. J. H. Beynon, R. M. Caprioli, W. E. Baitinger, and J. W. Amy, *Org. Mass Spectrom., 3,* 455 (1970).

14. J. H. Beynon, A. Mathias, and A. E. Williams, *Org. Mass Spectrom., 5,* 303 (1971).

15. T. Keough, J. H. Beynon, and R. G. Cooks, *J. Am. Chem. Soc., 95,* 1695 (1973).

16. J. H. Bowie and T. Blumenthal, *J. Am. Chem. Soc., 97,* 2959 (1975).

17. J. H. Bowie and T. Blumenthal, *Aust. J. Chem., 29,* 115 (1976).

18. K. R. Jennings, *Int. J. Mass Spectrom. Ion Phys., 1,* 227 (1968).

19. W. F. Haddon and F. W. McLafferty, *Anal. Chem., 41,* 31 (1969).

20. T. Wachs, P. F. Bente III, and F. W. McLafferty, *Int. J. Mass Spectrom. Ion Phys., 9,* 333 (1972).

21. J. H. Beynon, R. G. Cooks, J. W. Amy, W. E. Baitinger, and Y. T. Ridley, *Anal. Chem., 45,* 1023A (1973).

22. Technical Literature, Nuclide Corp., State College, Pa.

23. Technical Literature, V. G. Micromass, East Grimstead, Sussex, England.

24. J. H. Beynon and R. G. Cooks, *Int. J. Mass Spectrom. Ion Phys., 19,* 107 (1976).

25. A. F. Weston, K. R. Jennings, S. Evans, and R. M. Elliott, *Int. J. Mass Spectrom. Ion Phys., 20,* 317 (1976).

26. M. J. Lacey and C. G. MacDonald, *Org. Mass Spectrom., 12,* 587 (1977).

27. A. F. Weston, K. R. Jennings, R. M. Elliott, and S. Evans, *Org. Mass Spectrom.* (in press).

28. D. L. Kemp, R. G. Cooks, and J. H. Beynon, *Int. J. Mass Spectrom. Ion Phys., 21,* 93 (1976).

29. J. E. Coutant and F. W. McLafferty, *Int. J. Mass Spectrom. Ion Phys., 8,* 323 (1972).

30. J. H. Beynon and R. K. Boyd, in *Advances in Mass Spectrometry* (N. R. Daly, Ed.), Heyden & Son, Philadelphia, Pa., 1978, vol. 7B, p. 1115.

31. J. H. Beynon in *High Performance Mass Spectrometry* (M. L. Gross, Ed.), American Chemical Society, Washington, D.C., 1977, p. 103.

32. R. A. Yost and C. G. Enke, *J. Am. Chem. Soc., 100,* 2274 (1978).

33. J. D. Morrison and D. C. McGilverey, *Int. J. Mass Spectrom. Ion Phys.* (in press).

34. P. C. Price, H. S. Swofford, Jr., and S. E. Buttrill, Jr., *Anal. Chem., 50,* 1127 (1978).

35. R. M. Gandy, R. Ampulski, J. Prusaczyk, and R. H. Johnson, *Int. J. Mass Spectrom. Ion Phys., 24,* 363 (1977).

36. A. Giardini-Guidoni, R. Platania, and F. Zocchi, *Int. J. Mass Spectrom., Ion Phys., 13,* 453 (1974).

37. H. H. Tuithoff, A. J. H. Boerboom, P. G. Kistemaker, and H. L. C. Meuzelaar, in *Advances in Mass Spectrometry* (N. R. Daly, Ed.), Heyden and Sons, Philadelphia, Pa., 1978, vol. 7B, p. 838.

38. M. A. Posthumus, P. G. Kistemaker, H. L. C. Meuzelaar, and M. C. T. N. de Brauw, *Anal. Chem., 50,* 985 (1978).

39. F. W. McLafferty, in *Analytical Pyrolysis* (G. E. R. Jones and C. A. Cramers, Eds.), Elsevier, Amsterdam, 1977, p. 39.

40. M. A. Baldwin and F. W. McLafferty, *Org. Mass Spectrom., 7,* 1353 (1973).

41. G. Hanson and B. Munson, *Anal. Chem., 50,* 1130 (1978).

42. H. L. C. Meuzelaar, P. G. Kistemaker, and M. A. Posthumus, *Biomed. Mass Spectrom., 1,* 312 (1974).

43. H. M. Fales, G. W. A. Milne, H. U. Winkler, H. D. Beckey, J. N. Damico, and R. Barron, *Anal. Chem., 47,* 207 (1975).

44. G. W. A. Milne and M. J. Lacey, *Crit. Rev. Anal. Chem., 4,* 45 (1974).

45. J. H. Beynon, R. G. Cooks, and T. Keough, *Int. J. Mass Spectrom. Ion. Phys., 14,* 437 (1974).

46. T. Wachs, C. C. van de Sande, P. F. Bente III, P. P. Dymerski, and F. W. McLafferty, *Int. J. Mass Spectrom. Ion Phys., 23,* 21 (1977).

47. T. Wachs, C. C. van de Sande, and F. W. McLafferty, *Org. Mass Spectrom., 11,* 1308 (1976).

48. R. G. Cooks, D. T. Teriwilliger, T. Ast, J. H. Beynon, and T. Keough, *J. Am. Chem. Soc., 97,* 1583 (1975).

49. M. S. Kim and F. W. McLafferty, *J. Am. Chem. Soc., 100,* 3279 (1978).

50. D. Teriwilliger, J. H. Beynon, and R. G. Cooks, *Int. J. Mass Spectrom. Ion Phys., 14,* 15 (1974).

51. J. W. Cooper, *Anal. Chem., 50,* 801A (1978).

52. B. van de Graf, F. M. Bockhoff, and F. W. McLafferty (to be published).

53. J. Durup, in *Recent Developments in Mass Spectrometry* (K. Ogata and T. Hayakawa, Eds.), University Park Press, Boston, 1970, p. 921.

54. H. M. Rosenstock, M. B. Wallenstein, A. L. Wahrhaftig, and H. Eyring, *Proc. Nat. Acad. Sci. U.S.*, *38*, 667 (1952).

55. H. S. W. Massey and E. H. S. Burhop, *Electronic and Ionic Impact Phenomena*, Oxford University Press, London, 1952, p. 513.

56. G. M. Pesyna, R. Venkataraghavan, H. E. Dayringer, and F. W. McLafferty, *Anal. Chem.*, *48*, 1362 (1976).

57. F. W. McLafferty, R. Kornfeld, W. F. Haddon, K. Levsen, I. Sakai, P. F. Bente III, S.-C. Tsai, and H. D. R. Schuddemage, *J. Am. Chem. Soc.*, *95*, 3886 (1973).

58. F. W. McLafferty, *Interpretation of Mass Spectra*, 2nd ed., W. A. Benjamin, Inc., Reading, Mass., 1973.

59. K. Levsen and H. D. Beckey, *Org. Mass Spectrom.*, *9*, 570 (1974).

60. B. van de Graf, P. P. Dymerski, and F. W. McLafferty, *Chem. Commun.*, 978 (1975).

61. F. W. McLafferty and I. Sakai, *Org. Mass Spectrom.*, *7*, 971 (1973).

62. J. C. Giddings, *Anal. Chem.*, *39*, 1027 (1967).

63. E. Grushka, *Anal. Chem.*, *42*, 1142 (1970).

64. S. P. Cram and T. H. Risby, *Anal. Chem.*, *50*, 213R (1978).

65. P. D. Klein and S. A. Tyler, *Anal. Chem.*, *37*, 1281 (1965).

66. T. Wachs, Cornell University, personal communication.

67. G. M. Pesyna, F. W. McLafferty, R. Venkataraghavan, and H. E. Dayringer, *Anal. Chem.*, *47*, 1161 (1975).

68. L. R. Snyder, *J. Chromat. Sci.*, *10*, 200 (1972).

69. F. W. McLafferty, R. H. Hertel, and R. D. Villwock, *Org. Mass Spectrom.*, *9*, 690 (1974).

70. R. W. Kondrat, G. A. McClusky, and R. G. Cooks, *Anal. Chem.*, *50*, 1222 (1978).

71. I. C. Wang, H. W. Swofford, P. C. Price, D. P. Martinsen, and S. E. Buttrill, Jr., *Anal. Chem.*, *48*, 491 (1976).

72. G. G. Wanless and G. A. Glock, Jr., *Anal. Chem.*, *39*, 2 (1967).

73. F. W. McLafferty and T. A. Bryce, *J. Chem. Soc.*, *Chem. Commun.*, 1215 (1967).

74. H. K. Wipf, P. Irving, M. McCamish, R. Venkataraghavan, and F. W. McLafferty, *J. Am. Chem. Soc.*, *95*, 3369 (1973).

75. D. H. Smith, C. Djerassi, K. H. Maurer, and U. Rapp, *J. Am. Chem. Soc., 96,* 3482 (1974).

76. J. H. Beynon, D. F. Brothers, and R. G. Cooks, *Anal. Chem., 46,* 1299 (1974).

77. U. P. Schlunegger, *Angew. Chem. Int. Ed. Engl., 14,* 679 (1975).

78. E. J. Gallegos, *Anal. Chem., 48,* 1348 (1976).

79. K. Levsen and H. D. Beckey, *Org. Mass Spectrom., 9,* 570 (1974).

80. K. Levsen and H. R. Schulten, *Biomed. Mass Spectrom., 3,* 137 (1976).

81. T. L. Kruger, J. F. Litton, and R. G. Cooks, *Anal. Lett., 9,* 533 (1976).

82. J. H. McReynolds and M. Anbar, *Int. J. Mass Spectrom. Ion Phys., 24,* 37 (1977).

83. P. P. Dymerski and F. W. McLafferty, *J. Am. Chem. Soc., 98,* 6070 (1976).

84. C. Köppel and F. W. McLafferty, *J. Am. Chem. Soc., 98,* 8293 (1976).

85. R. R. Razakov, K. A. Aslanov, A. S. Sadykov, and B. V. Rosynov, *Biorg. Khim., 3,* 600 (1977).

86. K. Levsen, H.-K. Wipf, and F. W. McLafferty, *Org. Mass Spectrom., 8,* 117 (1974).

87. K. Levsen and H. R. Schulten, *Biomed. Mass Spectrom., 3,* 137 (1976).

88. T. L. Kruger, R. G. Cooks, J. L. McLaughlin, and R. L. Ranieri, *J. Org. Chem., 42,* 4161 (1977).

89. R. W. Kondrat, R. G. Cooks, and J. L. McLaughlin, *Science, 199,* 978 (1978).

Chapter 4

NEGATIVE ION CHEMICAL IONIZATION MASS SPECTROMETRY*

J. RONALD HASS

Laboratory of Environmental Chemistry
National Institute of Environmental Health Sciences
Research Triangle Park, North Carolina

*A collection of papers presented at the Negative Ion Chemical Ionization Mass Spectrometry Workshop held at NIEHS, Research Triangle Park, N.C., May 26-27, 1977.

I. INTRODUCTION

MARLIN D. FRIESEN and J. RONALD HASS

Laboratory of Environmental Chemistry
National Institute of Environmental Health Sciences
Research Triangle Park, North Carolina

A. Formation of Negative Ions

Traditionally, analytical mass spectrometry has involved the observation of positive ions formed by electron bombardment of sample molecules in the gas phase. But useful negative ions, which are often characteristic of molecular structure, are formed at the same time as the positive ions. Comparatively little attention has been paid to negative ions in mass spectrometry, partly due to a lack of the instrumental capabilities necessary for their detection, but mainly due to fundamental differences in the nature of their production. Under electron-impact conditions, negative ion mass spectra are typically much less intense than are positive ion mass spectra. The fundamental processes by which negative ions are formed have been the subject of a number of recent publications [1]. They are summarized below.

At low ion source pressures, i.e., 10^{-7} to 10^{-5} torr, the positive ions that make up a mass spectrum result from a series of competing unimolecular decompositions of a molecular cation that has been created and excited by electron impact. The electrons responsible for such decompositions generally range in energy from about 10 to 100 eV, the lower value representing the positive ionization potential for a "typical" organic ion. Less fragmentation is observed when lower-energy electrons are involved but positive ion mass spectra are relatively uniform across the entire electron energy range mentioned.

Negative ions can also be produced in the mass spectrometer ion source at low pressures by electron impact. However, depending on the energy of the electron and the nature of the molecule, negative ion formation will occur by one of the three processes shown in Fig. 1.

Resonance capture, as the name implies, occurs only for a very narrow band of electron energies near 0 eV. Electrons possessing such low energies are called thermal or thermalized electrons. A stable negative ion can result from capture of a thermal electron if the resulting excited anion can emit its excess energy of formation as radiation. Obviously, few thermal electrons are present in the 70-eV electron beam often used in positive ion electron-impact mass spectrometry and such molecular anions are rarely seen.

Dissociative resonance capture occurs for electron energies between 0 and 15 eV. In this case, the excited anion formed by electron capture fragments to accommodate its excess energy of formation. Such peaks, resulting from the loss of small neutral fragments, make up the bulk of the peaks observed under negative electron-impact conditions. Such peaks are generally 10^1-10^3-fold less intense than peaks observed under positive electron-impact conditions.

$$AB + e^- \longrightarrow AB^{\cdot -}$$ RESONANCE CAPTURE

$$AB + e^- \longrightarrow A^{\cdot} + B^-$$ DISSOC. RESONANCE CAPTURE

$$AB + e^- \longrightarrow A^{+\cdot} + B^{\cdot -} + e^-$$ ION-PAIR PRODUCTION

FIG. 1. Fundamental processes leading to the formation of negative ions by interaction of an electron and gaseous molecule.

Ion-pair production is a nonresonance process and occurs with electrons having energies of more than 10 eV. In this case, the electron is not captured, but merely acts as a source of energy for the event. Electron-impact excites the molecule into such a high-energy vibrational state that fragmentation into $A^{+\bullet}$ and $B^{-\bullet}$ occurs. Ion-pair production can be expected to be the dominant process for negative ion formation under conditions where the traditional 70-eV electron beam is used.

If the ion source is raised above ca. 10^{-5} torr by the addition of some nonreactive reagent gas such as methane or nitrogen, other kinds of reactions can begin to come into play. Two important types of reactions observed at high pressure are shown in Fig. 2. The formation of low-energy secondary electrons takes place as a part of the reagent gas ionization reactions encountered in conventional positive chemical ionization mass spectrometry. The resulting electrons are at thermal or near thermal energies and are available for capture by a sample molecule. The excited anion resulting from this process can relax to a more stable vibrational energy level by collision with reagent gas molecules.

In 1968, Dougherty and Weisenberger [2] reported that the addition of nitrogen to the ion source enhanced peaks resulting from dissociative resonance capture processes in the spectra of organic aromatics. However, at the pressures they used (6×10^{-5} torr) collisional stabilization of any molecular ions formed was not observed.

$$CG + e^- \longrightarrow CG^{+\cdot} + 2e^- \qquad \text{LOW ENERGY ELECTRON FORMATION}$$

$$e^- + AB \longrightarrow AB^{*-\cdot}$$
$$AB^{*-\cdot} + CG \longrightarrow AB^{-\cdot} + CG \qquad \text{COLLISION STABILIZED ELECTRON CAPTURE}$$

FIG. 2. Mechanisms by which a nonreactive collision gas (CG) enhances the formation of negative ions from a sample molecule (AB).

At higher source pressures, various kinds of ion-molecule reactions can also occur to complicate mass spectra. Several of these reaction types are outlined in Fig. 3.

Simple charge-transfer reactions are characterized by transfer of an electron from an anion to a molecule during a collision. For charge transfer to occur, the electron affinity of the neutral must be greater than that of the neutral corresponding to the molecular anion. Indeed, such reactions have been studied to gain information about electron affinities [3].

Dissociation reactions occur when the negatively charged molecular ions collide with other atoms and decompose. Such collision-induced decompositions usually require relatively high-energy ions and are consequently not commonly observed under chemical ionization conditions. Bowie [4], however, has observed fragment anions and cations representative of molecular structure created by accelerating ions out of the ion source at 3.6 kV and colliding them with nonreactive gas molecules in a collision chamber held at 10^{-5} torr.

CHARGE TRANSFER

$$AB^- + CD \longrightarrow AB + CD^-$$

DISSOCIATION

$$AB^- + CD \longrightarrow A + B^- + CD$$

DISPLACEMENT

$$AB^- + CD \longrightarrow AC^- + B + D$$

ABSTRACTION

$$AB^- + CD \longrightarrow ABC^- + D$$

$$AB^- + CD \longrightarrow A + BCD^-$$

TRANSFER

$$AB^- + CD \longrightarrow ABC + D^-$$

ASSOCIATIVE DETACHMENT

$$AB^- + CD \longrightarrow AB + CD + e^-$$

FIG. 3. Types of ion-molecule reactions typically observed under negative chemical ionization conditions.

Displacement reactions occur quite readily under high-pressure negative ion conditions. Jennings [5] reports that many aromatic compounds give a peak at $[M+15]^-$ due to displacement of a hydrogen atom in a reaction involving the radical species, $O^{\overline{\cdot}}$. In addition, there are numerous reports of chlorine atom displacement by O_2 leading to a prominent ion at $(M-19)^-$ [6-8].

Ionic abstraction involves the transfer of an ion from either the reactant ion to the neutral or from the neutral reactant to the ion. Typical of this reaction type is the production of $[M+Cl]^-$ ions as reported by Dougherty et al. [6] for polychlorinated pesticides with methylene chloride as enhancement gas.

The principal transfer reaction observed under high-pressure negative ion conditions involves transfer of H· from a neutral to an anion. Such reactions have been widely used to determine gas phase proton affinities of negative ions [9,10]. At high pressures, the collisional stabilization of the proton-bound dimer intermediate of this reaction can give rise to prominent ions.

Associative detachment reactions occur when an anion and a neutral collide to produce a neutral and an electron. Although such uncharged species cannot be directly observed in the mass spectrometer, such reactions do play an important part in the gas phase chemistry of negative ions [5] and might be an important process limiting analytical sensitivity.

We can summarize by observing that high-pressure negative ion mass spectrometry differs from conventional positive ion chemical ionization mass spectrometry in several important respects.

1. Initial sample molecule ionization results from interaction with electrons rather than cations. The energy of the ionizing electron greatly affects the type of ion produced, which in turn affects the mass spectrum.
2. The sensitivity with which a sample can be detected depends upon the extent to which newly formed and excited anions can be stabilized by the reagent gas molecules present. Since collisional processes can also lead to fragmentation of the moleculer anion and/or electron detachment, there will generally be an optimal source pressure for maximum sensitivity. Further, the relative importance of collisional stabilization vs electron detachment

will depend upon the internal energy of the reagent gas so that sensitivity will also depend strongly on source temperature.

3. A wide range of ion-molecule reactions can occur, thus leading to chemically interesting products. Such reactions have rates that depend on such factors as the choice of reagent gas, the energies of the ions and neutrals, and even on the values of ion source potentials. This wide range of variables requires that great care be exerted if negative ion spectra are to be reproducible on a day-to-day basis.

In view of the fundamental differences between the two techniques, it does not seem reasonable to refer to such high-pressure negative ion mass spectrometry as negative ion chemical ionization. A less misleading name would be gas-enhanced negative ion mass spectrometry as proposed by Dougherty et al. [6]. However, the term "negative chemical ionization" is in widespread use so that we will propagate this until a suitable body determines the proper nomenclature.

B. Negative Chemical Ionization Mass Spectrometry

Historically speaking, analytical negative chemical ionization mass spectrometry is a new field of research and few results have been published. In general, work done to this point has been concerned with one or the other of two useful aspects of the technique: (1) characteristic ion-molecule reactions, or (2) increased sensitivity relative to other methods of sample ionization.

1. *Ion-Molecule Reactions*

Following the report of negative ion enhancement with the addition of a nonreactive buffer gas, Dougherty et al. [6] explored the possibility of using methane, isobutane, and methylene chloride as enhancement gases to characterize a series of polychlorinated pesticides. These three gases served mainly to create low-energy electrons. The major ion-molecule reaction involved the formation of $[M+Cl]^-$ anion in the case of methylene chloride.

The most interesting ion-molecule reactions have been those involving oxygen. When employed as a reagent gas, it can exist in three active forms: O_2, $O_2^{\cdot-}$, or $O^{\cdot-}$. For example, Hunt et al. [8] have shown that O_2 can react with the molecular anion of 2,3,7,8-dibenzo-p-dioxin to produce ionic fragments that are highly indicative of molecular structure. They observed an ionic product that is most likely a dichloro-ortho-diquinone; Eq. (1). This anion is not subject to interference by polychlorinated biphenyls (PCBs) or a variety of chlorinated pesticides.

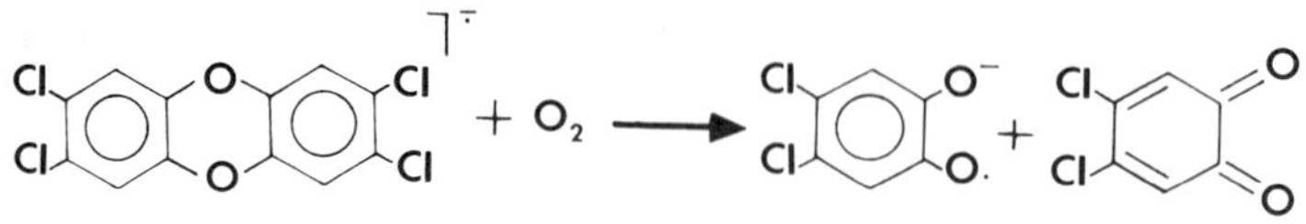

(1)

This is an unusual ion-molecule reaction in that the neutral product formed is capable of capturing an electron to form an ion identical to the ionic reaction product. Obviously, a very complicated situation might be encountered when one is determining instrumental response for analytical applications of this reaction. Atmospheric pressure ionization (API) methods have been used to study the reactions of the superoxide ion, $O_2^{\overline{\cdot}}$, with various chlorobenzenes and the complementary reaction of chlorobenzene molecular anions with O_2 to form the phenoxide anion by replacement of a chlorine by an oxygen atom [11]. Finally, Jennings [5] has reported the use of $O^{\overline{\cdot}}$, formed using a 9:1 N_2-N_2O mixture at 0.5 torr, to study a series of isomeric aromatic compounds. $[M-H]^-$, $[M-2H]^-$, and $[M-H+O]^-$ were the principal ions observed.

2. *Sensitivity*

Both Dougherty [6] and Hunt [8] have reported that, under certain conditions and for specific compounds, negative chemical ionization mass spectrometry is more sensitive than either electron-impact or positive chemical ionization methods. Dougherty et al. [12] used the technique to detect and quantitate low levels of pentachlorophenol residues in urine and samples of environmental interest. Hunt et al. [8] reported that, using oxygen as a reagent gas, 2,3,7,8-TCDD could be detected at the low picogram level. His group has subsequently reported [13] detection of various fluoro-substituted compounds at levels less than 100 femtograms ($<10^{-13}$ gram).

3. Simultaneous Acquisition of Positive and Negative Chemical Ionization Spectra

Positive chemical ionization mass spectrometry has been found useful in a large number of structural problems. Positive CI spectra are generally characterized by abundant ions indicative of molecular weight. This is the major advantage, and also the limitation, of the technique. The importance of molecular-weight measurement for structural analysis needs no elaboration. Similarly, the usefulness of electron-impact-induced fragmentation reactions is largely responsible for the current popularity of analytical mass spectrometry. Unfortunately, the deficiency of fragment ions in the positive CI spectra generally results in both CI and EI analysis being performed in the course of a structural identification. The analysis of complex mixtures requiring elaborate gas chromatographic separation in conjunction with the mass spectrometric analysis becomes tedious, time-consuming, and expensive. Recognition that negative CI spectra will usually reflect different structural features of a molecule not present in the positive CI spectrum has led to development of instrumentation to permit detection of both positive and negative CI spectra in one analysis [13]. In the initial publication, the positive and negative ion spectra of a variety of compounds demonstrated that the quantity and quality of information available for sample identification is substantially increased when both types of spectra are available. For example, consider the example of the positive and negative CI spectra of a permethylated tetrapeptide obtained using CH_4/CH_3ONO as the reagent gas mixture (Fig. 4). The negative ion spectrum of

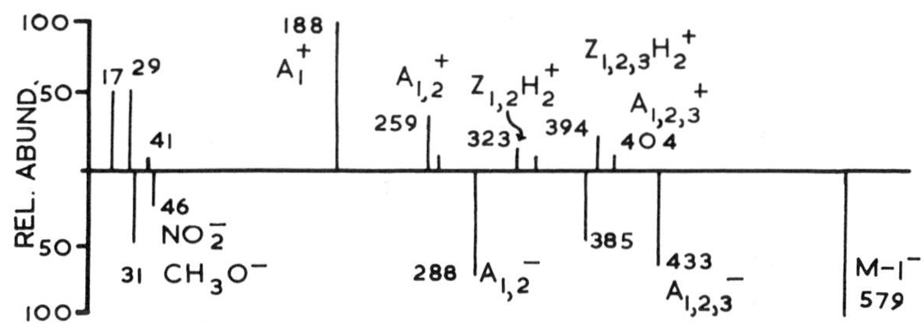

FIG. 4. Pulse positive-negative ion chemical ionization mass spectrum obtained under Brönsted-acid-Brönsted-base electron-capture conditions of the N-acetyl-permethylated-methylester of the polypeptide Met-Gly-Met-Met (adapted from Ref. 13, used by permission of the author).

methyl nitrite consists primarily of CH_3O^-. This particular spectrum was obtained under conditions such that the CH_3ONO had not depleted the available supply of thermal electrons. Thus, the negative ion spectrum consists of products of the reactions of CH_3O^- with the sample molecule and the collision-stabilized molecular anion. Examination of spectra and their interpretation (Fig. 5) clearly demonstrated the utility of simultaneous recording of both sample ion currents.

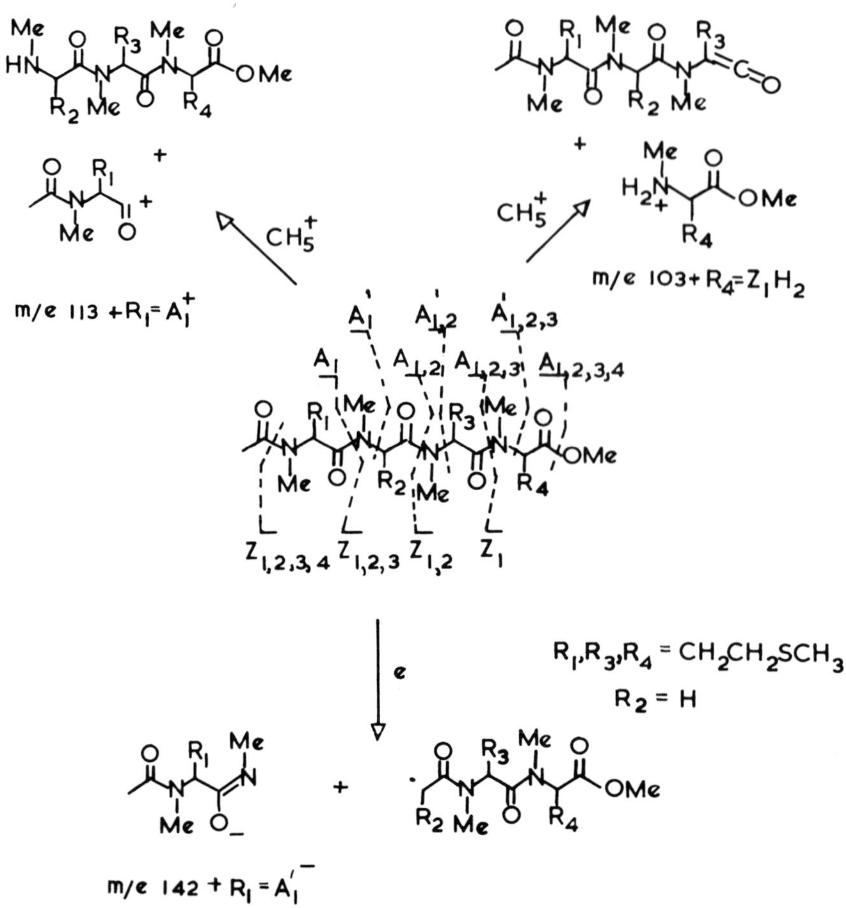

FIG. 5. Fragmentation of the N-acetyl-permethylated-O-methyl ester of the tetrapeptide Met-Gly-Met-Met under Brönsted-acid electron-capture conditions (adapted from Ref. 13, used by permission of the author).

Hunt et al. [14] have reported a novel application in which the enhanced production of negative ions in a fluorinated compound is used to provide an internal calibration for exact mass measurements in the positive ion beam. They find that a mass measurement accuracy of 10 ppm can be attained. This represents an important factor-of-two improvement over the 20 ppm we found using instrumental background peaks and spectrum averaging for positive EI spectra obtained on similar mass spectrometers [15]. Presumably both techniques can be improved by decreasing high-frequency noise on the mass scan functions of the instruments.

II. THE SYMMETRY OF CI/MS TOWARD NEGATIVE AND POSITIVE IONS

M. W. SIEGEL

Extranuclear Laboratories, Inc.
Pittsburgh, Pennsylvania

The intent of this contribution is to view the chemical ionization process in very general terms, and in this context to emphasize that negative ions are as natural and as expected a product of the CI environment as are positive ions. Whether it happens in any specific case that sample molecules are more efficiently or more informatively converted to positive ions or to negative ions of course depends on the nature of the sample species and the choice of reagent gas and operating conditions, but we will simply assume that all the right choices have been made.

That the subject of sample analysis by mass spectrometric detection of negative ions produced by ion-molecule reactions and thermal electron attachment is of sufficient novelty that it merits separate discussion is probably attributable more to historical accident than to any recent conceptual or technological breakthrough. Of all the ionization technologies that might have evolved for mass spectrometry, the mainstream approach happened to be electron bombardment at low pressure. Any one of the many alternative ionization processes that might have become highly developed instead would surely have made the analytical significance of negative ions obvious at an earlier time. This conclusion is easily (and frequently) reached inductively by enumerating negative ion formation mechanisms, but it is more interestingly reached deductively by considering the "environments" in which ionization might be done.

It is by now well known that, with the exception of the rare gases and the alkaline earths, the ground states of all but a few atomic species have positive electron affinities in the range of 0.2 to 3.6 eV [16]. Similarly, most molecular species either have positive electron affinities or are otherwise convertible to structurally significant negative ions by the removal or addition of simple side groups.

It follows that in an environment whose temperature is sufficiently low that negative ions are not rapidly destroyed by collisional detachment (recall that 1 eV = 23.06 kcal/mole = 11,600K), given an external supply of energetic particles as an ionization mechanism, and given sufficient confinement time for thermalization of the secondary electrons, most of the negative charge can be expected to reside on negative ions rather than free electrons. Conditions in the CI ionization volume (and similar high-pressure environments such as atmospheric pressure ionization) are easily seen to satisfy these requirements. Exceptions that prove the rule are those situations where elaborate precautions are taken to provide a reagent gas that does not itself form negative ions, and that is free of even trace level impurities that do. Examples include the extremely high-purity N_2, Ar, or Ar-CH_4 used in the API or electron capture detector.

In contrast, the EI ionization volume clearly fails to meet the environmental requirements for supporting a negative ion population: the diffusion time is far too short (because EI volumes are far too small) to allow the thermalization and attachment of the secondary electrons accompanying positive ion production. An appropriately scaled low-pressure "EI" source several kilometers in diameter could be expected to produce a CI mass spectrum.

Before developing a model of CI negative and positive reagent ion and sample ion production, it is necessary to review some fundamental properties of ionized gas behavior. These properties are derived from the observation that at even fairly low ion densities if a significant departure from net charge neutrality occurs then the electrostatic energy of a charged particle in the resulting electric field exceeds its thermal energy. Under these conditions charged particle motion is dominated by mobility rather than thermal diffusion effects, and the direction of the motion is to restore net charge neutrality even under circumstances where the two signs of charge carrier have very different masses, e.g., positive ions and free electrons.

The situation in a partially ionized gas is similar to that in an ionic solution as described by the Debye-Hückel theory [17]. Local accumulations of charge imbalance are prevented by the neutrality-restoring electric fields set up by such imbalances. Since this neutrality-restoring effect is counteracted by the thermal motion of the charge carriers, the spatial extent over which imbalances are actually possible is temperature-dependent, the characteristic distance for this possibility being the Debye length [18]:

$$\lambda = \left(\frac{4\pi\varepsilon_0 kT}{e^2 n}\right)^{1/2} \qquad (2)$$

where kT is the Boltzmann constant times the absolute temperature, e is the electron charge, n is the charge density of either sign, and $4\pi\varepsilon_0$ is a constant that puts the result into rationalized MKS units.

Similarly, any attempt to impose, by means of electrodes, an electric field within the ionized gas will fail: negative charge carriers will accumulate around the anode and positive charge carriers will accumulate around the cathode in "sheaths" of characteristic thickness λ, the effect of these charge layers being to shield the bulk of the ionized volume from the electrodes.

Thus, in any ionization volume that is large compared with the Debye length, charge neutrality will be the rule. Under these circumstances the ionized gas fits Langmuir's definition of a plasma [19]. If we assume a CI source producing an exiting ion current (including reagent ions) of about 0.1 μA (or 6 x 10^{11} ions/sec) with a gas outflow of 100 cm^3 sec^{-1} (the pumping speed of a 1-mm-diameter aperture), it follows that the ion density in the source is about 6 x 10^9 cm^{-3}. At a temperature of 400 K, Eq. (2) gives a Debye length of about 60 μm, or about 1% of the characteristic source dimensions. It follows that plasma effects must be considered in discussing the CI source of this example, and in discussing any CI source producing an exiting ion current in excess of about 0.1 nA.

Phenomenologically, plasma effects enter the rate equations balancing ion production against ion loss via the concept of "ambipolar diffusion" [20]. The electrostatic forces working to maintain charge neutrality must have the effect of making all charge carriers, whatever their masses, diffuse with the same velocity, provided only that all charged species have the same temperature. Any deviation from equal diffusion velocities would clearly result in local charge imbalances, which would be counteracted by the accompanying electric fields. The result, when treated mathematically by considering the interrelationship between mobility and diffusion, is that in a positive ion-electron plasma the ambipolar diffusion rate is twice the electrostatically free positive ion diffusion rate. This means that the positive ions diffuse twice as fast as when electrostatically free, but the electrons diffuse about 500 times more slowly than when free. In the case of a positive ion/negative ion plasma, the ambipolar diffusion rate is equal to the concentration weighted mean of the diffusion rates of all species present. This behavior, wherein in any plasma all charged species are characterized by a single diffusion constant, greatly simplifies the rate equations governing CI source behavior.

These concepts apply not just to the ion-electron pairs produced by the primary ionization; the charge bookkeeping must also take the high-energy ionizing electrons into account. The details are complicated by the fact that the primary electrons are nonthermal, but it is easy to see that since each high-energy electron will make several ($\lesssim$ 10) positive ion-electron pairs, the high-energy electrons can be thought of as a perturbation rather than a dominating influence. We would thus expect only a slight ($\lesssim$ 10%) excess of positive ions over thermal electrons and negative ions in the CI plasma.

To emphasize that this essential symmetry between positive and negative charge carrier concentrations in CI not only is true, but that it is highly desirable, it is of interest to ask the circumstances under which we could break the symmetry. We could, for example, imagine supplying the CI volume with a copious source of thermal energy electrons. Mutual charge repulsion would then cause an unreplenished charge density n to decay in time according to the relation

$$\frac{dn}{dt} = -\frac{e\mu}{\varepsilon_o} n^2 \qquad (3)$$

where μ is the mobility. If the charge is replenished at a rate S, then balancing the gain and loss rates gives a steady state charge density [21].

$$n = \sqrt{\frac{S\varepsilon_o}{e\mu}} \qquad (4)$$

For a typical electron mobility of 10^6 cm^2 sec^{-1} volt^{-1} at 1 torr, maintaining an electron charge density of 10^{10} cm^{-3} requires S = 2×10^{20} electrons cm^{-3} sec^{-1}, or about 30 amperes into a 1-cm^3 source volume! Clearly, the high charge density required for CI is made possible by the charge symmetry, the charged particle cloud of each sign being the glue that keeps the charged particles of the opposite sign from being rapidly expelled by their mutual repulsion.

With these facts in hand, we are in a position to write down a rate-controlled picture of CI behavior and response to sample. For simplicity, we will assume that the primary ionization takes place uniformly throughout the CI volume at a rate of S ion-electron pairs/cm^3 sec. Also for simplicity, we will consider a CI mode such as is obtained with an argon carrier gas, in which the reagent ions for positive sample ionization (by charge transfer) are the primary Ar^+ ions, and in which the reagent ions for negative sample ionization (by electron capture) are the accompanying thermalized electrons.

Extension to more complicated CI systems, in which multiple steps are involved in forming the effective reagent ion species, is straightforward but the rate equations rapidly become sufficiently cumbersome as to obscure the basic principles. The general conclusions reached below are unchanged by the additional complexity.

The relevant rate equation for the positive reagent ion concentration in this example is

$$\frac{d[Ar^+]}{dt} = S - \frac{F}{V}[Ar^+] - \frac{D_+}{r_o^2}[Ar^+] - R_o[Ar^+][e] - R_1[Ar^+][M^-] - A_o[Ar^+][M] \tag{5}$$

where (taking the terms in order)

S is the primary ionization rate in ion-electron pairs/cm^3-sec

$F[Ar^+]/V$ describes the loss of Ar^+ to ventilation at flow rate F through source volume V

$D_+[Ar^+]/r_o^2$ describes the loss of Ar^+ by diffusion to the walls with diffusion constant D_+ and characteristic source dimension r_o, which depends on the geometry

$R_o[Ar^+][e]$ describes the loss of Ar^+ by recombination with electrons

$R_1[Ar^+][M^-]$ describes the loss of Ar^+ by recombination with sample negative ions M^-

$A_o[Ar^+][M]$ describes the only desired loss of Ar^+, the loss by charge transfer with the sample species M which results in the formation of M^+

Similarly, the rate equation for the negative reagent ion (electron) concentration is

$$\frac{d[e]}{dt} = S - \frac{F}{V}[e] - \frac{D_e}{r_o^2}[e] - R_o([Ar^+] + [M^+])[e] - A_1[e][M] \tag{6}$$

where the identification of terms is similar to that detailed above, and where $A_1[e][M]$ is the desirable term describing the loss of electrons by attachment to the sample species resulting in the formation of M^-. By using the recombination rate R_o to apply to M^+ as well as Ar^+ we are assuming, for simplicity, that all positive ion-electron recombination rates are equal.

The rate equation for sample positive ion concentration is

$$\frac{d[M^+]}{dt} = A_o[Ar^+][M] - \frac{F}{V}[M^+] - \frac{D_+}{r_o^2}[M^+] - R_o[M^+][e] - R_1[M^+][M^-] \tag{7}$$

where the term $F[M^+]/V$ is essentially the positive sample ion signal. By reusing the symbol D_+ we are implicitly assuming that all positive ions are characterized by the same diffusion constant, and by using the recombination rate R_1 to apply to M^+ as well as Ar^+ we are assuming that all positive ion-negative ion recombination rates are equal.

The rate equation for the sample negative ion concentration is

$$\frac{d[M^-]}{dt} = A_1[e][M] - \frac{F}{V}[M^-] - \frac{D_-}{r_o^2}[M^-] - R_1[M^-]([Ar^+] + [M^+]) \tag{8}$$

where, as above, the term $F[M^-]/V$ is essentially the negative ion signal, D_- is the diffusion rate for negative ions, and all positive ion-negative ion recombination rates are assumed equal by the use of a single constant R_1.

Under steady-state conditions (that is, a constant flux of sample molecules) each derivative is zero. Some additional simplifications facilitate solution:

- Recombination rates for positive ion-electron processes are comparable to recombination rates for positive ion-negative ion processes, so we can set $R_1 \simeq R_o$.
- All diffusion constants are equal to the ambipolar diffusion constant D_a.
- In the small sample regime $[Ar^+] >> [M^+]$ and $[e] >> [M^-]$, so that by employing this approximation we can determine the ultimate detection sensitivity.

With these well-justified simplications, Eq. (5)-(8) assume the instructive form

$$S \simeq \frac{F}{V}[Ar^+] + \frac{D_a}{r_o^2}[Ar^+] + R_o[Ar^+][e] \tag{9}$$

$$S \simeq \frac{F}{V}[e] + \frac{D_a}{r_o^2}[e] + R_o[Ar^+][e] \tag{10}$$

$$A_o[Ar^+][M] \simeq \frac{F}{V}[M^+] + \frac{D_a}{r_o^2}[M^+] + R_o[M^+][e] \tag{11}$$

$$A_1[e][M] \simeq \frac{F}{V}[M^-] + \frac{D_a}{r_o^2}[M^-] + R_o[Ar^+][M^-] \tag{12}$$

Equations (9) and (10) describe the "zeroth" order characteristics, in which sample molecules are ignored and only the reagent species Ar^+ and e^- are considered. They obviously require $[Ar^+]$ = [e], via the assumption of ambipolar diffusion. If we then set $[Ar^+]$ = [e] = N, the reagent ion density of either sign, we obtain a quadratic equation in N whose solution is

$$N = \frac{D_a/r_o^2 + F/V}{2R_o}\left(\sqrt{1 + \frac{4R_oS}{(D_a/r_o^2 + F/V)^2}} - 1\right) \tag{13}$$

Reasonable estimates for numerical values are

$D_a = 100\ cm^2\ sec^{-1}$ at 1 torr
$r_o = 0.2$ cm
$F = 100\ cm^3\ sec^{-1}$ (1-mm aperture)
$V = 1\ cm^3$
$R_o = 3 \times 10^{-7}\ cm^3\ sec^{-1}$
$S = 3 \times 10^{14}\ cm^{-3}\ sec^{-1}$ (1 μA delivered emission, 5 ion-electron pairs/primary) from which $N \simeq 3 \times 10^{10}\ cm^{-3}$

Multiplying N by the flow rate F gives a reagent ion current of about $3 \times 10^{12}\ sec^{-1}$, or about 0.5 μA.

Equations (11) and (12) describe the first-order characteristics; i.e., the sensitivity in the small-sample limit. They are conveniently rearranged to obtain

$$\frac{[M^+]}{[M]} = \frac{A_oN}{F/V + D_a/r_o^2 + R_oN} \tag{14}$$

and

$$\frac{[M^-]}{[M]} = \frac{A_1N}{F/V + D_a/r_o^2 + R_oN} \tag{15}$$

these two quantities being the efficiencies of conversion of sample molecules to positive and negative ions respectively. The ratio of Eq. (15) to Eq. (14) is the relative efficiency (or sensitivity) for negative and positive ion detection

$$\frac{[M^-]}{[M^+]} = \frac{A_1}{A_o} \tag{16}$$

Equation (16) tells us that there is no fundamental reason to expect either positive or negative CI to be more sensitive than the other; the ratio of the sensitivities is just the ratio of the rate constants for the attachment and charge transfer processes, and in general these rate constants are all about the same size, i.e., about 10^{-9} cm^3 sec^{-1}. This is an exceedingly general statement, and there are of course wide variations that are of extreme importance in any real analytical application. Nevertheless, it is the most reasonable starting point for further discussion.

While the implications contained in Eq. (16) represent the main conclusions sought in this contribution, it is worth looking at Eqs. (14) and (15) a little further in order to obtain some feeling about absolute as well as relative CI efficiencies or sensitivities. For large values of reagent ion density N [large enough so that $R_oN \gg (F/V + D_a/r_o^2)$], we obtain the interesting but only rarely appreciated results

$$\frac{[M^+]}{[M]} = \frac{A_o}{R_o} \tag{17}$$

and

$$\frac{[M^-]}{[M]} = \frac{A_1}{R_o} \tag{18}$$

What Eqs. (16) and (17) tell us is that once sufficient reagent ion concentration has been established, reagent ion losses are dominated by recombination rather than by the "technical" loss mechanisms of flow and diffusion, and the sensitivity is not further increased by increasing the reagent ion concentration. The ultimate ionization efficiency is simply the ratio of the sample ion formation rate constant ($\simeq 10^{-10} - 10^{-9}$ cm^3 sec^{-1}) to the ion loss by recombination rate constant ($\simeq 10^{-7} - 10^{-6}$ cm^3 sec^{-1}) or typically 0.1%. In the small-sample regime (where sample ionization does not significantly deplete the reagent ion concentration) additional reagent ion concentration increases the noise but leaves the signal unchanged. Substitution of the numbers previously estimated in Eqs. (14) and (15) demonstrates that in the "typical" CI system the reagent ion concentration is of such a magnitude that further increases are of marginal but not entirely negligible value; no more than a factor of two could be gained if the reagent ion concentration were made very much larger. When sample molecule concentration becomes sufficiently high that sample ion formation results in a significant decrease in reagent ion concentration, departures from quasi-linear response to sample concentration will be observed. The dynamic range, i.e., the range of quasi-linear response, is increased by increasing the reagent ion concentration. However, it is unfortunately the case that in the recombination-dominated CI source the reagent ion concentration increases only as the square root of the primary ionizing current [compare Eq. (13)], so that large technological efforts to increase the primary current have relatively small practical payoffs in increased dynamic range.

III. DETECTION AND IDENTIFICATION OF TOXIC SUBSTANCES IN THE ENVIRONMENT

RALPH C. DOUGHERTY

Florida State University
Department of Chemistry
Tallahassee, Florida

The application of negative chemical ionization (NCI) mass spectrometry to toxic substance screening extends the use of CI mass spectrometry in rapid screening of crude extracts of biofluids for basic drugs in overdose cases [22]. Our initial work has concentrated on organo-chloride residues because their NCI mass spectra are uniformly intense and are dominated by a single ion cluster, which is easily recognized because of the chlorine isotope patterns [23-29].

Negative chemical ionization mass spectrometry seems uniquely suited for screening partially cleaned-up samples from environmental substrates for contamination with toxic substances. The evidence for this suitability is based on two observations: (1) The molecules of biology are very highly reduced. With the exception of free carboxylic acids and some of the prosthetic groups from the electron transport chain, the molecules of biology are virtually transparent to negative chemical ionization. The reason for this stems from the fact the molecules of biology very generally have negative electron affinities and attach anions in the gas phase only weakly. (2) In contrast, nonspecific toxic substances virtually always give intense negative ion mass spectra. This is true because nonspecific toxic substances generally can be classified as either oxidizing of alkylating agents. These molecules usually have positive electron affinities and very generally attach anions to form complexes that are stable in the gas phase. Thus it is possible to use NCI mass spectrometry to detect nanogram quantities of toxic substances in a matrix that has substantially larger amounts of biomolecules.

The analytical potential of NCI mass spectrometry is much like that of gas chromatography with electron-capture detection. Compounds that give intense electron-capture responses will virtually always give intense NCI mass spectra. Mass resolution in NCI spectra serves the same function as time resolution in gas chromatography and the reduction in chemical noise by mass analysis can lower detection limits by as much as two orders of magnitude as compared with the GC system [29]. NCI mass spectrometry offers major advantages over GC with electron-capture detection: (1) The analysis is fast; NCI mass spectra can be obtained in one-fifth, or less, the time of a conventional GC run. (2) NCI mass spectra do not require derivatization of compounds that must be derivatized prior to GC analysis. (3) NCI mass spectra can be obtained from many compounds that are transparent to an electron-capture detector, e.g., nitrosamines. (4) NCI mass spectra are molecularly specific.

Negative chemical ionization mass spectrometry was first used to obtain spectra of polycyclic chlorinated pesticides [6]. Polycyclic polychloro insecticides of dieldrin class uniformly give intense ions that correspond to chloride adducts of the molecule. The sample requirements for these spectra were small, of the order of nanograms, and thus we decided to use the technique to look for organic polychlorides in extracts of environmental substrates. When NCI mass spectrometry was applied to a series of extracts from sources ranging from birds to human tissues, we were able to detect and identify polychlorobiphenyls [27], mirex, dieldrin, dichlorodiphenyl acetic acid [28], and other DDT metabolites and aldrin. In our initial studies the samples we examined were often known to contain the substances that we found.

In order to extend and develop the utility of NCI screening of environmental substrates, we decided to apply a rudimentary screening technique to human urines [23]. Extraction of urines with methylene-chloride followed by drying and evaporation gave residues that produced very rich NCI mass spectra using methylene chloride as a reagent gas (Fig. 6). Examination of these residues introduced so many contaminants into the ion source that only four or five samples could be run between source cleanings. The cleanup procedure that was finally adopted involved hydrolysis of the urine with sulfuric acid to cleave urinary conjugates and extraction into hexane, followed by drying and evaporation. Since pentachlorophenol (PCP) appeared in every urine sample that we examined as pentachlorophenoxide, m/z 263, we decided to measure order of magnitude of concentrations of PCP urines. The calibration standard that we used was p-chlorobenzophenone.

In a survey of the organic polychlorides in the urines of college-age males [23] we found pentachlorophenol at concentrations ranging from 1 to 80 ppb, a series of ions that corresponded to the chloride adducts of the 2,4,5-trichlorophenoxyacetic acid (2,4,5-T) family of herbicides, and the chloride adduct of trichlorophenol. A limited longitudinal study with two individuals indicated that the concentration of pentachlorophenol in urine varies with time and that the

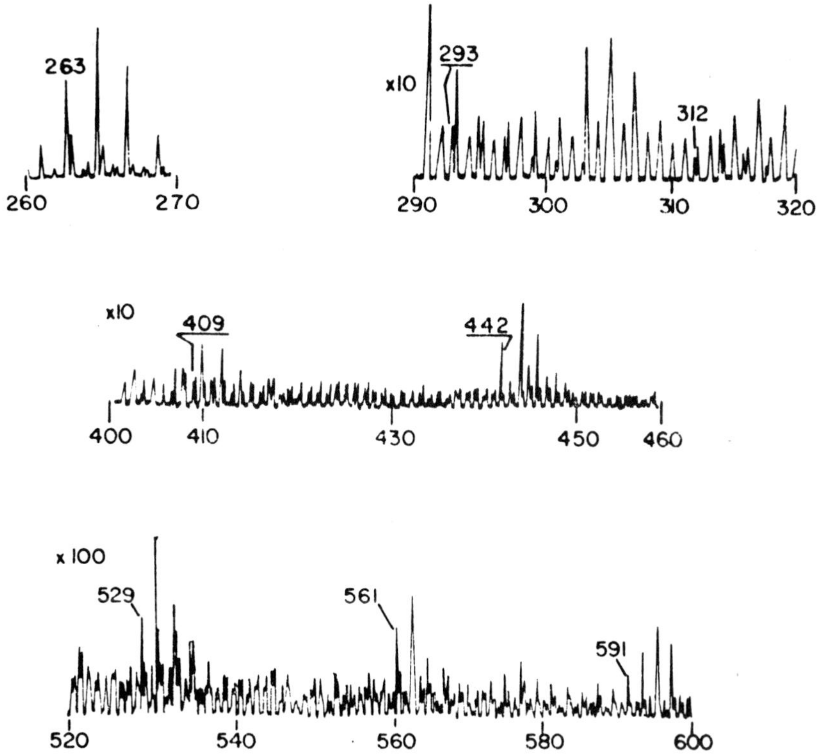

FIG. 6. NCI mass spectrum of a dichloromethane extract of human urine; pH 6; 150°C; CH_2Cl_2, 0.5 torr.

appearance of other polychlorides such as 2,4,5-T varied dramatically with time. This information suggested that diet was one possible source of these molecules. In a spot survey of selected items from the food chain, pentachlorophenol was found in a number of samples with concentrations ranging from 1 to 100 ppb and individual samples often contained one or more of the polyhalo organics that we had seen in urine. The urinary survey indicated the presence of more than 10 organicpolyhalides that have yet to be associated with any structure. Since these ions appeared only infrequently, generally in only one sample, determination of their structures had a low priority.

Since the initial report of our survey of human urines we have been expanding the screening program in two directions: (1) We are attempting to include classes of toxic substances other than organic polychlorides in the screening effort. (2) We have been developing techniques for screening lipid-rich substrates for contamination [24]. Figures 7-12 present the NCI mass spectra of a series of structurally distinct toxic substances obtained with a reagent gas mixture consisting of isobutane, methylenechloride, and oxygen. The reagent gas spectrum has been eliminated in each case in Figs. 7-12. These spectra are representative of the spectra obtained from the more than 100 samples in our toxic substance file and this series of spectra will form the nucleus of our library of NCI mass spectra of toxic substances. All of the compounds in Figs. 7-12 produced recognizable NCI mass spectra with nanogram quantities of material. The fact that compounds as diverse as polychlorinated organics and nitrosamines or epoxides will produce intense NCI mass spectra augurs well for the use of NCI mass spectra in screening environmental substrates for toxic substances.

The choice of reagent gases is critical to the development of a library of standard NCI mass spectra. We propose to use a 5:4:1 mixture of isobutane, methylenechloride, and oxygen. The basis for this choice of gases is threefold. (1) Isobutane is present in the gas mixture primarily to prolong the life of the filament. Chemical ionization mass spectra produced by electron excitation from a hot filament are much more reproducible in our hands than those produced by gas discharge. Both oxygen and methylenechloride substantially reduce filament life if used as reagent gases by themselves; addition of isobutane significantly reduces the adverse effects of oxygen and polyhalogens on filament lifetime. (2) Methylchloride is part of the reagent gas mixture to provide a source of chloride ions. Methylenechloride is easy to obtain in pure form. It produces abundant amounts of Cl^- and the chloride ion is a sufficiently good nucleophile in the gas phase that it will attach to a large number of alkylating and oxidizing agents that would not otherwise produce NCI spectra. (3) It is virtually impossible rigorously to exclude oxygen for an

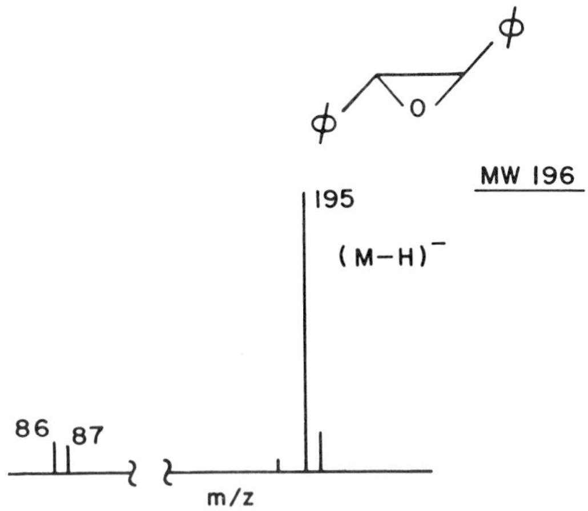

NCI, Isobutane, 175°C

FIG. 7. Trans-stilbene oxide.

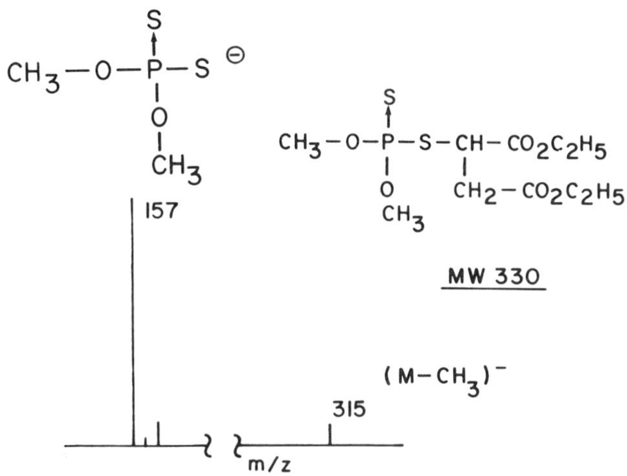

NCI, CH_2Cl_2, 175°C

FIG. 8. Malathion.

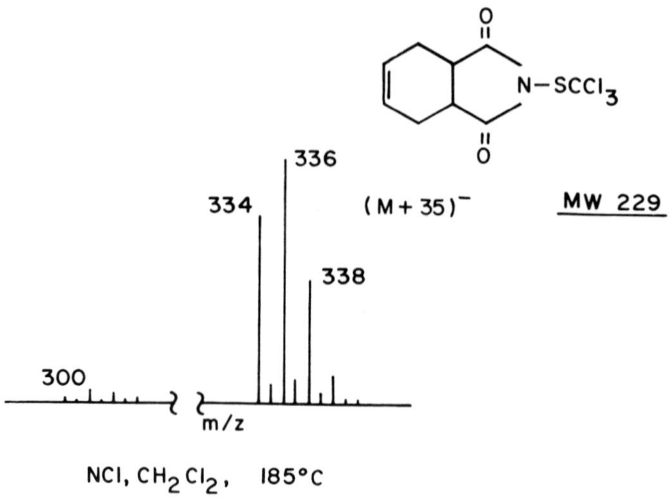

FIG. 9. Captan.

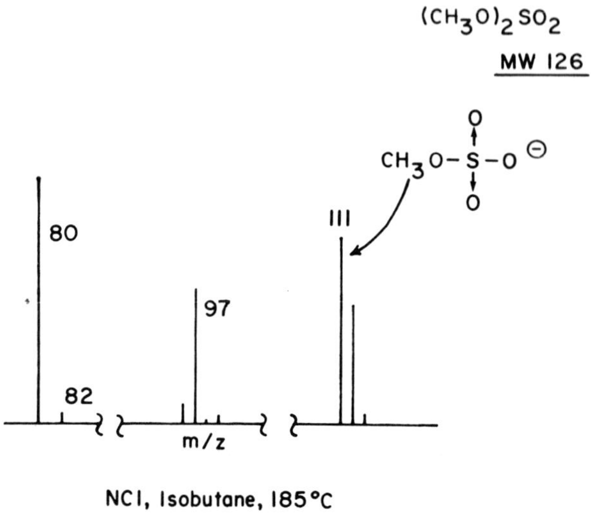

FIG. 10. Dimethylsulfate.

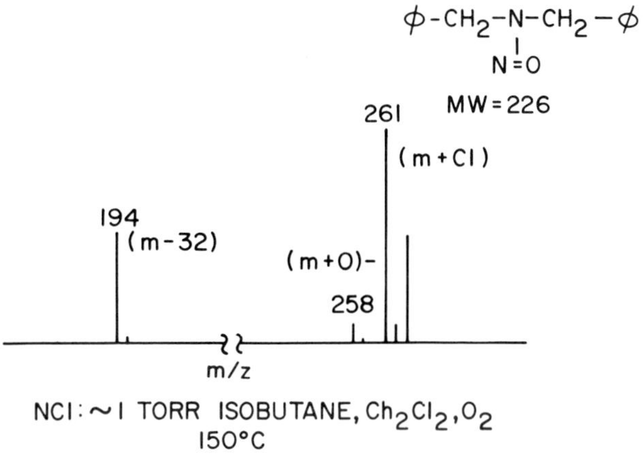

FIG. 11. Dibenzylnitrosamine.

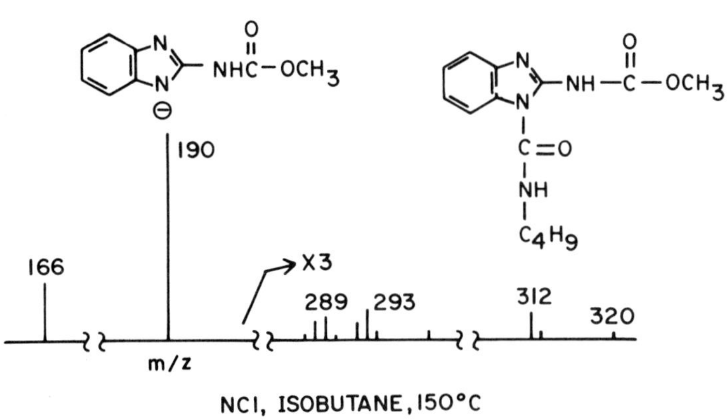

FIG. 12. Benomyl.

NCI experiment. Oxygen is very reactive in anionic mass spectra and produces a number of characteristic ions for exchange with halogens or hydrogens [7,29]. To make the spectra reproducible for samples run at different times and with different instruments, we add controlled quantities of oxygen to the gas mixture and thus the oxygen reaction ions that are produced will be present in a controlled way.

The extension of our screening efforts to lipid-rich substrates has primarily involved the application of gel permeation chromatography to the problem of sample cleanup [24]. The utility of this technique in sample cleanup for toxic substance analysis has been demonstrated by Stalling [30]. The basis for the application of the procedure stems from the fact that neutral and polar lipids generally have molecular weights significantly higher than those of nonspecific toxic substances. In gel permeation chromatography this means that the neutral lipids will elute from a gel permeation column ahead of the toxic compounds of interest.

We have developed procedures for obtaining small samples of breast fluid from nonlactating females. Breast fluid is lipid-rich and has approximately the composition of human milk. It is available in only small quantities, 100-200 μl [31]. It can be obtained by use of a breast pump from approximately 70% of nonlactating Caucasian females. In order directly to address the question of the relationship of toxic substances to human breast cancer, it will be necessary to develop screening procedures for toxic substances in human breast fluid. Our preliminary NCI screening of breast fluid has revealed the presence of of a number of organic polychlorides; an ion cluster at *m/z* 315 has been the most frequently seen in our studies.

The successes of NCI mass spectrometry in detecting a broad spectrum of toxic substances in environmental substrates at ppb and higher levels have demonstrated the power of the technique. One question remains: Does NCI mass spectrometry have sufficient sensitivity to detect toxic molecules at levels that would be of concern in environmental discussions? The answer to this question would seem to be yes, as shown in Tables 1 and 2.

TABLE 1

Numbers of Molecules per Cell

Level	Prefix	Level	Number of Molecules per Cell
10^{-6}	micro-	ppm	3×10^{11} molecules/cell (220 daltons)
10^{-9}	nano-	ppb	3×10^{8}
10^{-12}	pico-	ppt	3×10^{5}
10^{-15}	femto-	ppquad	3×10^{2}
10^{-18}	atto-	ppquint	3×10^{-1}

TABLE 2

Cellular Constituents

DNAs (2)	Code for ~10^6 RNAs proteins	
RNAs ~15,000/cell	~1-10 each, larger for specialized cells	$t_{1/2}$ ~5-10 hr, some longer
Proteins ~15,000/cell	~1-10 each	
Control molecules-RNAs/proteins?	<20/cell	

The data in Table 1 are calculated for the average mass of a human cell. At the ppb level there will be ~10^8 molecules per cell. This must be compared with the numbers of biomolecules per cell. There are roughly 10^5 RNAs per cell at any one time, and they generally turn over fairly rapidly.

It is possible to obtain ppt sensitivity with an NCI mass spectrometry. However, in our experience ppt sensitivity requires an exceptionally clean instrument and a very well cleaned-up sample--two conditions that are not easily met in routine screening. In the case of exceptionally toxic molecules, e.g., 2,3,7,8-tetrachlorodibenz-p-dioxin, that show acute toxicity at ppb levels, the effort to obtain ppt sensitivity is justified. For most toxic substances, however, the numbers presented in Tables 1 and 2 suggest that the probability of critical cellular damage by ppt levels of molecules that are not acutely toxic at ppm levels is small.

ACKNOWLEDGMENT

It is a pleasure to acknowledge the technical assistance of J. D. Roberts and E. J. Hett.

IV. POLYCHLORINATED DIBENZO-p-DIOXINS (PCDD)

J. R. HASS, M. D. FRIESEN, and M. K. HOFFMAN

Environmental Biology and Chemistry Branch
National Institute of Environmental Health Sciences
Research Triangle Park, North Carolina

Potential human health hazards associated with environmental exposure to PCDD [32] have led to interest in their identification and determination. Mass spectrometry provides the presently accepted method for analysis [33]. We wish to describe results from negative chemical ionization studies that demonstrate improved specificity of the mass spectra of these compounds.

A. Experimental

Low-resolution negative chemical ionization mass spectra were obtained using a Finnigan 3300 chemical ionization mass spectrometer modified as described by Hunt et al. [13]. Samples were introduced gas-chromatographically using a 2-mm X 1.5-meter glass column packed with 3% OV-210 on 100/120 mesh Gas Chrom Q operated at 210°C with a source temperature of approximately 100°C. Instrument control and data acquisition were accomplished by means of a System Industries System/150 or Finnigan 6100 data system.

ONCI. When O_2 was employed as the reagent gas, the helium carrier gas flow rate was adjusted to give a source pressure of approximately 0.3 torr (40 Pa). Sufficient O_2 to maintain a discharge (1-2 torr, 100-250 Pa) was then admitted to the source via the probe gas inlet, and an electrical discharge [34] was initiated (1,000-1,500 volts at about 1 μA). During sample injection and solvent elution, carrier gas flow was diverted to an auxiliary pump.

MNCI. Methane negative CI mass spectra were obtained by adjusting the methane carrier gas flow through the column to give an intensity maximum in the negative ion spectrum of perfluoro tri-n-butylamine. Typically indicated source pressures in excess of 2 torr (250 Pa) were employed. A filament current of 0.3 mA and electron energy of 230 eV were used.

MONCI. When methane/oxygen were employed as reagent gases, the methane carrier gas flow rate was adjusted to give a source pressure of from 1 to 2 torr (100-250 Pa). A mixture of methane and approximately 0.1 torr (15 Pa) oxygen was then added to the source via the probe gas inlet in sufficient quantity to maximize the peak at *m/z* 633 in the spectrum of FC43 (filament current = 0.3 mA; electron energy = 230 eV).

B. Results and Discussion

The negative ion spectra resulting from methane chemical ionization (Fig. 13) show predominantely molecular anions that fragment by the loss of H·, Cl· and Cl_2. In addition, several of the spectra contain peaks corresponding to reaction of the dioxin anions with residual oxygen within the ion source. In contrast, the negative ion spectra of the PCDD obtained with an oxygen discharge source (Fig. 14) exhibit few peaks corresponding to $M^{\bar{\cdot}}$ or $[M-Cl_n]^-$. Rather, most of the ion current is carried by products of ion-molecule reactions. In particular, we observe the displacement of Cl by O $[M+O-Cl]^-$ similar to the previously reported reactions of halogenated aromatic compounds with oxygen [11,29] and cleavage of the dioxin ring to give a chloro-orthoquinone (ClOQ) anion in analogy to the reaction reported for 2,3,7,8-TCDD. Note that the ClOQ anion results in the base peak in the spectra of all compounds lacking a peri chlorine while the $[M+O-Cl]^-$ ion gives the base peak for all compounds possessing a peri chlorine. Assuming this observation to be general, the 2,3,7,8-TCDD will have a unique spectrum when oxygen is used as a reagent since it is the only tetra

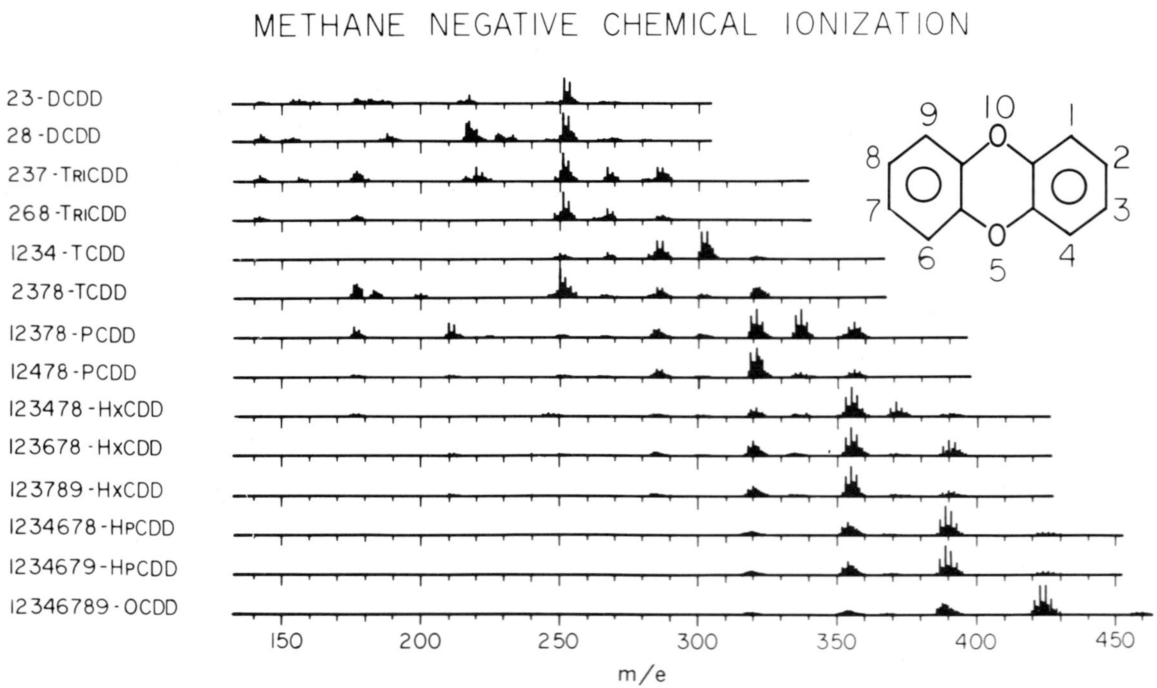

FIG. 13. Methane negative chemical ionization mass spectra of the various PCDD.

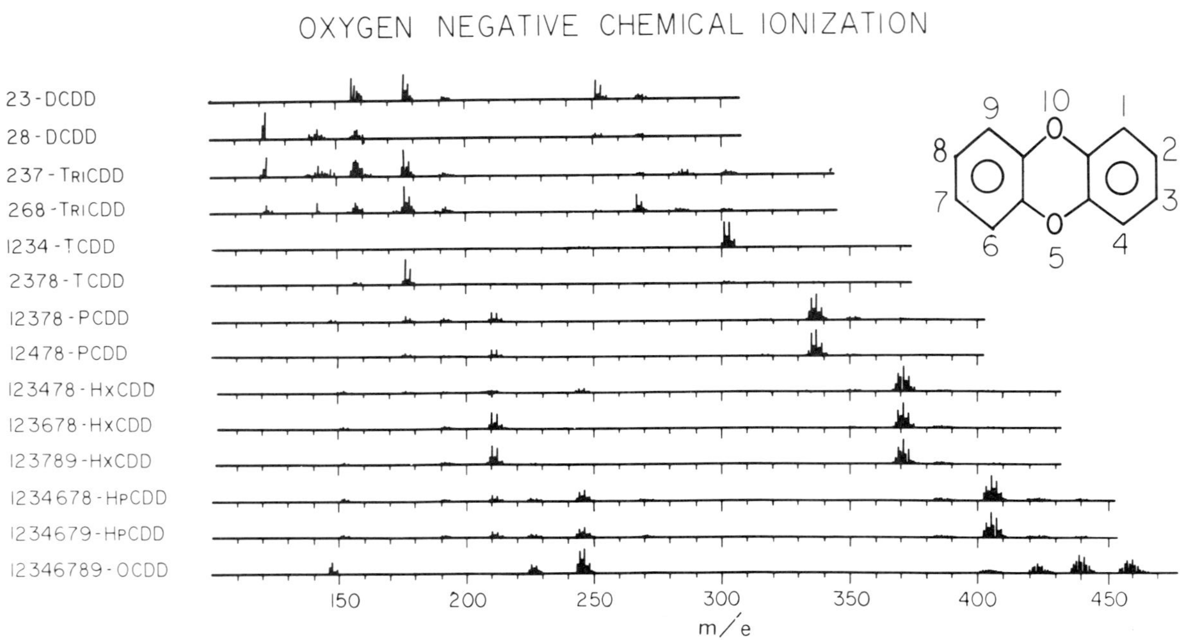

FIG. 14. Negative ion spectra of the various PCDD using an oxygen discharge to achieve ionization.

isomer with no peri chlorines. Thus, the negative ion chemistry of this, the isomer with the highest reported toxicity, can be used to confirm its presence, assuming adequate chromatographic resolution is obtained prior to mass spectral analysis. Unfortunately, in our hands, the oxygen discharge has proved to be less than satisfactory as an ion source for analytical mass spectrometry for three main reasons: (1) fluctuations in the ion current (i.e., instabilities in the discharge); (2) rapid contamination of the quadrupole analyzer; and (3) a curious broadening of the peaks (possibly a space-charge effect from the electron current simultaneously in the quadrupole analyzer).

In an attempt to take advantage of the unique chemistry of the PCDD with oxygen and to improve the analytical performance of negative chemical ionization, the mass spectra of these compounds were obtained using a mixture of 10:1 to 20:1 $CH_4:O_2$ as the reactant gas. These spectra (Fig. 15) are seen to be essentially those that result from the addition of the two techniques taken separately. These spectra contain the greatest information content in that $[m-Cl]^-$ and/or $[M+O-Cl]^-$ reveals the molecular weight, the ClOQ anion gives the number of chlorines on each ring, and the ratio $[M+O-Cl]^-$/ClOQ anion allows one to deduce the presence of a peri chlorine. Although this is not sufficient information to determine the structure of an unknown dioxin solely on the basis of its negative ion spectrum, it is a considerable improvement over the situation for electron-impact mass spectrometry even when high-resolution techniques are employed.

One important mechanistic question is whether the $[M-Cl]^-$ is formed by a process such as dissociative resonance capture or fragmentation of an excited molecular anion. This point was addressed by obtaining a precursor ion scan for the *m/z* 421 ion in the MNCI spectrum of octachlorodibenzo-p-dioxin. In this technique, the magnetic (B) and electric (E) sectors of a double-focusing mass spectrometer are scanned simultaneously so that the ratio B^2/E remains constant. By initially tuning the instrument to transmit main beam daughter ions, a signal is obtained only when a parent ion fragments to give

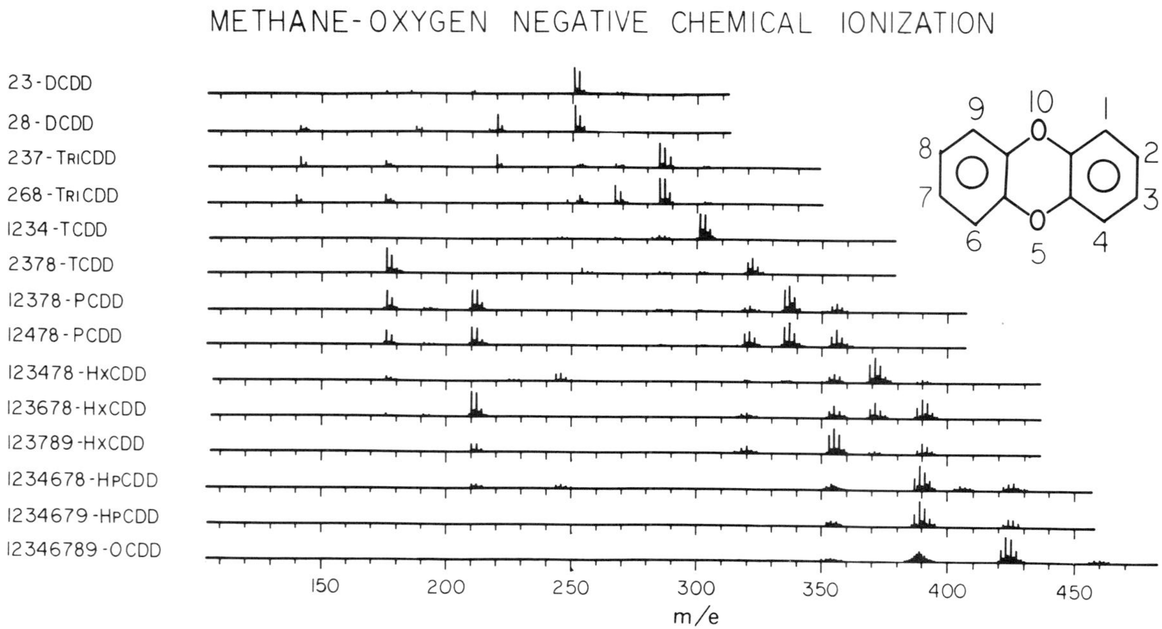

FIG. 15. Negative ion spectra of the PCDD obtained using 20:1 mixture of CH_4/O_2 as the reactant gas with primary ionization by electron bombardment.

that daughter ion in the field-free region between the source and the first sector. Because ions must typically survive for ca. 10^{-6} sec to reach this region of the spectrometer and because dissociative resonance capture generally results in fragment ion formation in $<10^{-9}$ sec, the observation of a signal for the reaction

$$M^{\overline{\cdot}} \rightarrow [M-Cl]^{-} + Cl\cdot$$

would require fragmentation from an excited molecular anion. This signal is in fact observed (Fig. 16) confirming at least a contribution of a simple cleavage reaction to the formation of $[M-Cl]^{-}$.

A more straightforward method of studying the chemistry of excited ions is collisionally to excite a mass-selected ion beam, then measure the kinetic energy of the fragment ions produced [36]. Figures 17 and 18 show the results of applying this technique to $[M-Cl]^{-}$ and $[M-Cl_2]^{\overline{\cdot}}$ in the MNCI spectrum of octachlorodibenzo-p-dioxin. The major process in both cases is the loss of Cl· with less prominent peaks corresponding to the loss of CO, COCl, and Cl_2 from the respective parent ions. These reactions are analogous to the fragmentations observed for the positive ions resulting from electron-

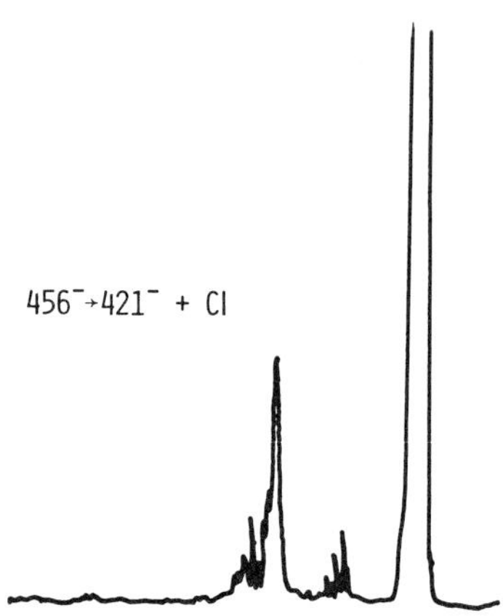

B²/E Scan on m/e 421 in NCI Spectrum of OCDD

FIG. 16. Linked B^2/E scan showing that $[M-Cl^-]$ is formed in part by cleavage of Cl· from the molecular anion of octachlorodibenzo-p-dioxin.

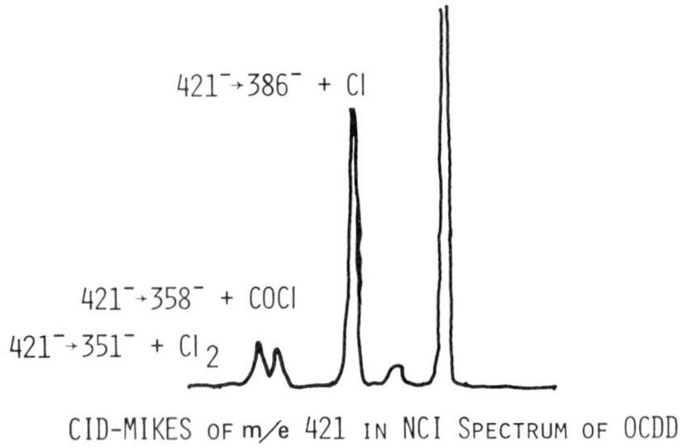

FIG. 17. Negative fragment ion current resulting from the collisional excitation of the [M-Cl$^-$] ion in the MNCI spectrum of OCDD. Note that the fragmentation reactions are similar to those observed for positive ions following electron-impact ionization.

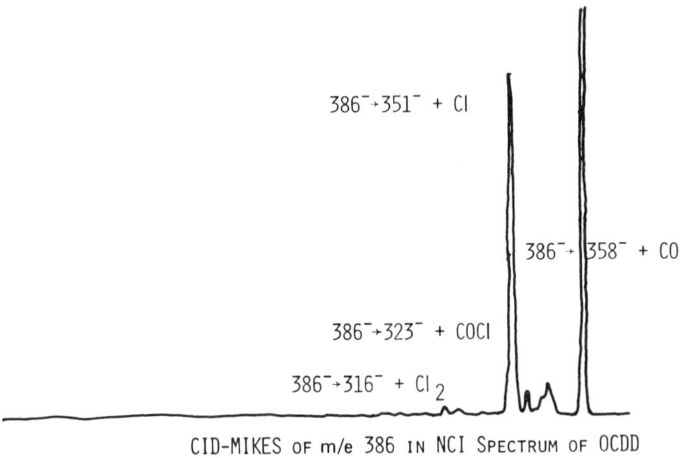

FIG. 18. Negative fragment ion current resulting from the collisional excitation of the [M-Cl_2^-] ion in the MNCI spectrum of OCDD. Again, the fragmentation reactions are similar to those observed for positive ions under electron-impact conditions.

impact ionization of this molecule. They suggest that fragmentation reactions for the negative ions formed under MNCI conditions differ from positive ion reactions mainly as a result of the amount of internal energy available rather than because of, for example, gross differences in ion structure or bond stabilities.

C. Conclusions

The negative ion spectra of 14 PCDD are reported using methane, oxygen, and a methane/oxygen mixture as the reactant gases. The methane/oxygen mixture is found to give the most informative spectra. Studies of field-free region reactions in a double-focusing mass spectrometer indicate that the same concepts that explain the gaseous ionic chemistry of positive ions are applicable to negative ions.

V. DETERMINATION OF CHLORINATED DIBENZO-p-DIOXINS

M. D. FRIESEN, J. R. HASS, D. J. HARVAN, and C. E. PARKER

Laboratory of Environmental Chemistry
National Institute of Environmental Health Science
Research Triangle Park, North Carolina

Polychlorodibenzo-p-dioxins (PCDD) have been introduced into the environment as a result of the widespread use of chemicals derived from chlorinated phenols. These PCDDs have been implicated in a number of human health and agricultural problems [32]. Some of these compounds have demonstrated extreme toxicity while other members of the PCDD family have exhibited toxicity that is much lower [37]. Thus, a highly specific and very sensitive technique is required for the determination of PCDD in environmental samples. In Section IV, we discussed the specificity of negative chemical ionization techniques for PCDD. In this section we shall discuss the application of these techniques to the analysis of biological samples.

A. Experimental

Negative chemical ionization experiments were performed as described in Sec. IV. Electron-impact data were gathered on a similar GC-MS system with the principal difference being the method of sample ionization. Bovine fat (5 grams) and liver (25 grams) samples were subjucted to the extraction and cleanup procedure of Albro and Corbett [38].

B. Results and Discussion

The analytical results for a set of bovine tissue samples are reported in Table 3. The numbers reported represent values that have been corrected for losses during extraction and cleanup as determined by analysis of spiked samples. To facilitate comparisons, results obtained by electron-impact and negative chemical ionization mass

TABLE 3

Dioxin Levels in Bovine Tissue Samples as Determined by MNCI and EI Mass Spectrometry. Concentrations Are in ng PCDD/g Wet Tissue Weight.

Sample	OCDD		HpCDD		HxCDD	
Liver	EI	MNCI	EI	MNCI	EI	MNCI
1	ND	0.23	ND	0.07	ND	0.06
2	ND	0.27	ND	0.03	ND	0.01
3	2.3	3.1	0.4	0.63	ND	0.03
4	3.9	3.9	D	0.93	ND	0.03
5	1.1	1.3	D	0.31	ND	0.03
6	3.0	2.1	0.3	0.61	ND	D
7	5.0	7.7	1.2	2.4	1.4	1.3
8	1.9	1.7	D	0.37	ND	0.01
9	5.7	5.7	0.8	0.81	ND	ND
10	4.6	4.9	1.5	1.3	ND	ND
11	3.6	5.0	D	0.33	ND	ND
12	6.0	3.9	ND	0.44	ND	0.11
13	23	36	8.0	12	ND	ND
14	4.2	4.3	ND	3.5	ND	ND
15	ND	0.15	ND	ND	ND	ND
Fat						
A	D	5.1	ND	0.39	ND	ND
B	26	37	ND	ND	ND	ND
C	18	10	ND	ND	ND	ND
D	50	47	12	0.33	ND	ND
E	37	41	ND	0.44	ND	ND
F	ND	4.4	ND	2.1	ND	0.39

spectrometric analysis of the same samples are included. HpCDD refers to the 1,2,3,4,6,7,8 isomer and HxCDD refers to the 1,2,3,6, 7,8 isomer (other isomers were indicated but could not be confirmed due to the lack of appropriate standards).

The detection limits for the two techniques were determined by injecting known amounts of the various dioxins onto the GC column under the conditions used for the analysis. For electron impact the detection limits were found to range from ca. 0.1 to 1 ng (10^{-10}-10^{-9} grams) for 2,3,7,8-TCDD through octachlorodibenzo-ortho-dioxin (OCDD). For MNCI, the detection limits were found to be less than 0.01 ng (10^{-11} grams) for the more highly chlorinated dioxins to ca. 0.1 ng (10^{-10} grams) for 2,3,7,8-TCDD. Because the GC conditions were the same for all dioxin samples, modest sensitivity increases could be achieved for detection of any particular dioxin by appropriate adjustment of the column temperature.

Although it is interesting to determine the relative sensitivities of EI and MNCI with pure standards, it is of greater analytical significance to study relative abilities of the two techniques to determine minute quantities of dioxins in a biological matrix. In the former case, one is limited by the electronic and ion noise of the mass spectrometer; the limitation in the latter case is frequently "chemical noise" from the matrix material. Of the 63 separate determinations presented in Table 3, there are 25 cases (ca. 40%) in which MNCI made it possible to assign a concentration where it was impossible to do so by EI using a similar mass spectrometer. Furthermore, MNCI offered an improved signal-to-noise ratio on all peaks so that peak areas could be measured with greater precision.

C. Conclusion

The use of MNCI as the ionization technique results in a sensitivity gain of one to two orders of magnitude for the determination of the more highly chlorinated PCDD in bovine tissue samples.

VI. ANALYSIS OF POLYCYCLIC AROMATIC HYDROCARBONS BY PULSED POSITIVE-NEGATIVE ION CIMS WITH OXYGEN AS THE REAGENT GAS

D. F. HUNT, W. C. BRUMELY, G. C. STAFFORD, and F. K. BOTZ

University of Virginia
Department of Chemistry
Charlottesville, Virginia

Interest in polycyclic aromatic hydrocarbons (PAH), which include heterocyclic compounds containing nitrogen, oxygen, and sulfur, centers on their widespread occurrence as environmental contaminants and the carcinogenicity of some PAH. The spectra of these compounds consist largely of the molecular ions and multiply charged species but do not distinguish isomeric PAH or reveal detailed structural information.

Chemical ionization mass spectrometry (CIMS) offers the possibility of increased structural knowledge through the design of ion-molecule reactions. Complementary information from negative ions is available from pulsed positive-negative CIMS (PPNCIMS).

The analytical application of PPNCIMS to PAH using oxygen as the reagent gas produces spectra highly dependent upon molecular structure. The reagent gas consists of an oxygen/hydrogen mixture (9:1) at 1 torr in the CI source. Reagent ions are generated by means of a Townsend discharge [13,34]. The primary reagent ions are $O^{+\cdot}$ and $O_2^{+\cdot}$ in the positive mode and $O_2^{-\cdot}$ in the negative mode. Any $O^{-\cdot}$ ions produced are consumed by hydrogen to produce water and thermal electrons. Thermal electrons are available for electron-capture processes under NCI conditions. The PCI spectrum usually consists exclusively of the $M^{+\cdot}$ ion produced by a charge-exchange reaction of sample with the positive reagent ions. The NCI spectrum consists of one or more of the following negative ion types: M, [M-1], [M+14], [M+15], [M+16], [M+30], [M+31], [M+32], and other ions containing structural information. Several examples illustrate possible applications.

Isomeric PAH of pure hydrocarbon nature may be distinguished using PPNCI(O_2) under favorable conditions. For example, fluorenes may be distinguished from fluoranthenes by means of the NCI(O_2) spectrum. Benzofluorene (MW 216) gives an intense $[M-1]^-$ ion that is characteristic of compounds containing the methylene group as part of a five-membered dienyl ring. It also gives an $[M+14]^{\cdot}$ ion, which presumably results from addition of $O_2^{\cdot}$ followed by loss of H_2O to produce the ketone anion. Fluoranthenes show an $[M+15]^-$ ion ($M+O_2^{\cdot}$ -OH·) as the dominant negative ion. In using PPNCI the negative-positive ratio of specific ions becomes an important parameter of structure. In the fluorenes the $[M-1]^-/M^{+\cdot}$ ratio is typically greater than 10. Another example of isomer distinction is illustrated by benzopyrenes (a) and (e) (MW 252) in which the relative intensities of the negative ions at $[M-1]^-$, $[M+15]^-$, and $[M+30]^{\cdot}$ $[M+O_2^{\cdot}-H_2]$ enable one to distinguish these two isomers from each other as well as from the perylene structure (MW 252), which gives negative ions at $M^{\cdot}$, $[M+15]^-$, and $[M+30]^{\cdot}$.

Five- and six-membered rings containing an acidic N-H proton characteristically give an intense $[M-1]^-$ion. For example, dibenzocarbazole (MW 267) gives an intense $[M-1]^-$ ion, ($[M-1]^-/M^{+\cdot}$) = 6, as well as a weak ion at $[M+15]^-$. Phenazine (MW 180), on the other hand, yields ions at $M^{\cdot}$ ($M^{\cdot}/M^{+\cdot}$=15), $[M+15]^-$, $[M+16]^-$; $[M+31]^-$ $[M+O_2^{\cdot}$ -H·], and $[M+32]^{\cdot}$.

Oxygenated compounds may afford unique spectra as illustrated in the following two compounds. Coumarin (MW 146) has its base peak at $[M+32]^{\cdot}$ in the negative mode. In contrast, 2,3-diphenylisobenzofuran yields a base peak at $[M+16]^{\cdot}$ in the negative mode and $[M+16]^{+\cdot}$ in the positive mode.

A particularly important application of NCI(O_2) involves aromatic sulfides. In the case of dibenzothiophene (MW 184), the negative ion spectrum shows ions at *m/z* 183 and 216 ($[M+32]^{\cdot}$) as well as ions at *m/z* 64 ($SO_2^{\cdot}$) and 98 ($SO_4^{\cdot}$), which are characteristic of sulfur-containing aromatics. These compounds are thus easily distinguished from nonsulfur-containing compounds at the same nominal mass. PPNCI(O_2) also appears useful for the analysis of aliphatic sulfides, sulfoxides, mercaptans, and disulfides.

VII. CONVENIENT STEROL DERIVATIVES

HIDEJI FUJIWARA and AJAY K. BOSE

Stevens Institute of Technology
Department of Chemistry and Chemical Engineering
Hoboken, New Jersey

Negative ionization mass spectrometry has been studied by physicists for the last 20 years or more but organic chemists have started to use this type of analysis only very recently [6]. Encouraged by a report on the dramatic increase in sensitivity in CI spectral measurement on changing from positive to negative ion mode [13], we have modified our Biospect CI mass spectrometer [39]. Accessories have been added that permit reversal of polarity and increase in the voltage amplitude for the negative ion mode.

With our instrument, CI mass spectra in the positive and negative modes cannot be recorded simultaneously; two samples are required and two separate scans are made on the same recorder. We have attached a gas manifold to this spectrometer for the convenience of switching from one reagent gas to another.

Dougherty and co-workers [26] have made the valuable observation that addition of methylene chloride to a test sample results in increased sensitivity for NCI mass spectrometry: the pseudo-molecular ion appears as a lopsided doublet corresponding to $[M+Cl^-]$. For the reagent gas manifold of our CI mass spectrometer, the use of liquid methylene chloride is not convenient. We have found that Freon 12 (CF_2Cl_2) is an excellent reagent gas [40] for producing chloride ion adduct NCI spectra.

We have recorded very satisfactory $[M+Cl^-]$ ions for a wide variety of biologically interesting compounds that are prone to fragmentation when other mass spectral techniques are used. Polyfunctional steroids have been studied. For example, 11β,17α,20α,21-tetrahydroxy-4-pregnen-3-one-21-acetate (<u>1</u>) gives an abundant ion pair for $[M+Cl^-]$ and a weak ion pair for $[M+Cl^- - CH_3CO]$. When scrapings from a commercial prednisone (<u>2</u>) tablet were introduced into the direct inlet system of our spectrometer, we observed an ion pair for $[M+Cl^-]$ for <u>2</u> but no other peaks in the *m/z* 200-400 region. Such uncluttered spectra should be

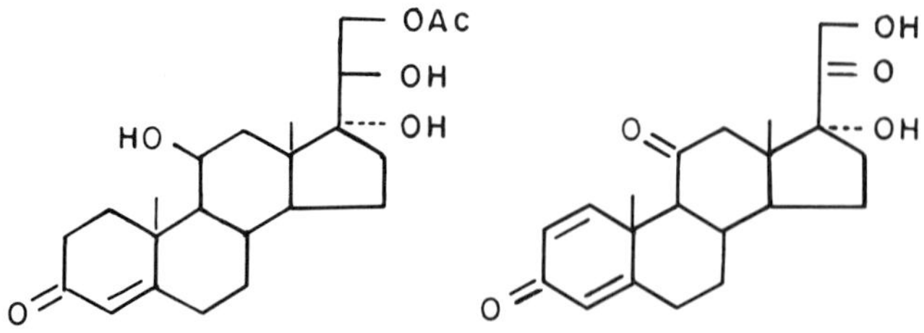

1 2

very useful in the detection or quality control of drugs containing one or more complex steroids.

Chlorine-containing compounds generally produce intense pseudo-molecular ions in the NCI (CH_4) spectra. In the course of our current studies on bile acids from various biological sources, we became interested in the possibility of increasing the sensitivity of NCI spectra by converting the acids to acid chlorides. Accordingly, 3,7,12-triketocholanic acid, 3, and thionyl chloride were dissolved in chloroform. After about 15 min, part of the reaction mixture was evaporated under a stream of warm air (from a hair drier) and the gummy residue containing 4 was used for NCI (CH_4) mass spectral measurement. Chlorine-containing doublet peaks at *m/z* 437 and 439 were observed but these ions correspond to ($RCOCl + OH^-$).

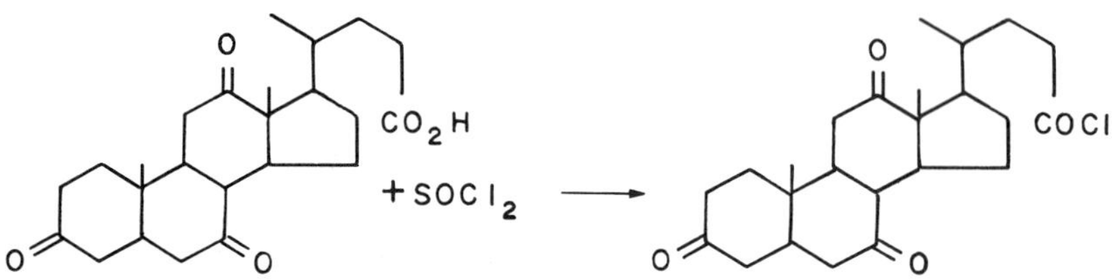

3 4

The ($RCOCl + OH^-$) peaks are coincident with ($RCO_2H + Cl^-$) peaks. Therefore, the question arises as to whether thionyl chloride serves only as a source of chloride ion to produce a chloride ion adduct of the carboxylic acid. To obtain information on this point, phenyl-acetyl chloride, 5, and 4-chlorobutyroyl chloride, 6, were examined in the absence of any thionyl chloride. Each of these acid chlorides produced strong $[M+OH^-]$ peaks.

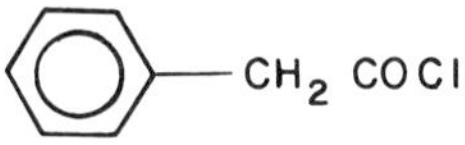

$ClCH_2\,CH_2\,CH_2\,COCl$

5 6

To obtain further evidence for involvement of H_2O, a small amount of D_2O was introduced into the spectrometer before analyzing the crude reaction product from thionyl chloride and 7. The most intense peak now corresponded to $(RCOCl + OD^-)$. A doublet of lower intensity corresponded to loss of water from the pseudo-molecular ion. Interestingly, a doublet of low intensity was observed corresponding to $(RCOCl + HCO_2^-)$. In the presence of D_2O in the spectrometer, this ion corresponded to $(RCOCl + DCO_2^-)$.

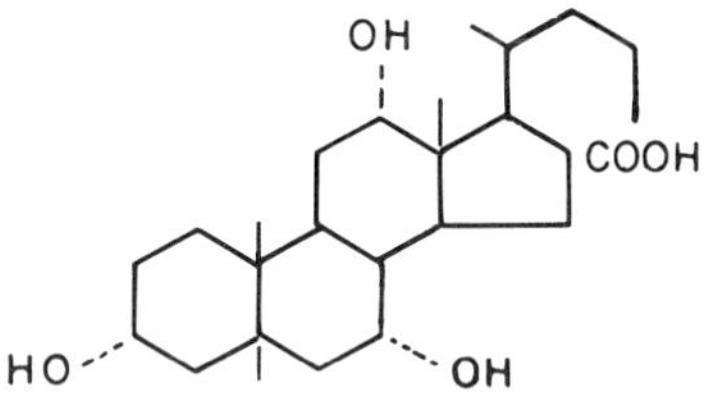

7

The presence of OH^- ions in the CI mass spectrometer is not surprising in view of the reaction:

$$H^- + H_2O \rightarrow OH^- + H_2$$

The mode of formation of the HCO_2^- is not obvious unless one assumes that significant CO_2 is also present in the mass spectrometer. Further work is needed in this area.

Interestingly, the thionyl chloride reaction was successful with hydroxy cholanic acids without protection of the hydroxy groups. Presumably only the acid chloride product led to the strong pseudo-molecular ions that were recorded. We have observed that a steroid chlorinated in the nucleus loses HCl readily and therefore does not give an intense molecular ion in the NCl spectrum.

In connection with studies on atherosclerosis we have been interested in detecting sterols in very small biological samples. A sterol derivative that is formed readily and in quantitative yield is the urethane from a sterol and trichloroacetylisocyanate (TAI). Amines give urea derivatives with TAI.

$$ROH + Cl_3CCON{=}C{=}O \rightarrow Cl_3\underset{O}{\underset{\|}{C}}\text{-}\underset{H}{\underset{|}{N}}\text{-}\underset{O}{\underset{\|}{C}}\text{-}OR$$

$$RNH_2 + Cl_3\underset{O}{\underset{\|}{C}}\text{-}N{=}C{=}O \rightarrow Cl_3\underset{O}{\underset{\|}{C}}\text{-}\underset{H}{\underset{|}{N}}\text{-}\underset{O}{\underset{\|}{C}}\text{-}NHR$$

In the past we have shown that TAI derivatives are valuable for NMR spectral studies [41]. We now find that TAI derivatives show high sensitivity in NCI spectra. It is possible to distinguish between TAI derivatives of alcohols and amines. In the NCl(CH_4) spectra of TAI derivatives of amines intense peaks are found for $[M-1]^-$ and the corresponding chloride ion adduct $[M+Cl^-]$. The TAI adducts from alcohols produce an intense cluster of peaks for $[M-1]^-$ but none for $[M+Cl]^-$.

When the TAI derivative was prepared from estradiol, 8, only the mono-TAI derivative gave a strong peak corresponding to $[M-1]^-$. Using ergosterol, 9, as the internal standard we made quantitative determination of cholesterol as its TAI derivative. Thus, in a sample of human plasma a known amount of ergosterol and then TAI were added; NCI(CH_4) spectral measurement indicated 203 mg of cholesterol per 100 ml of the plasma. This value was in good agreement with the

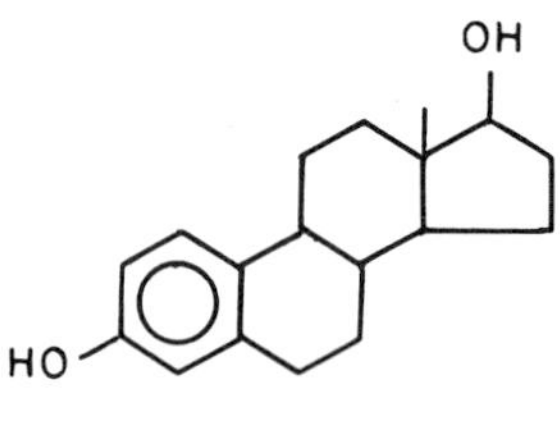

8

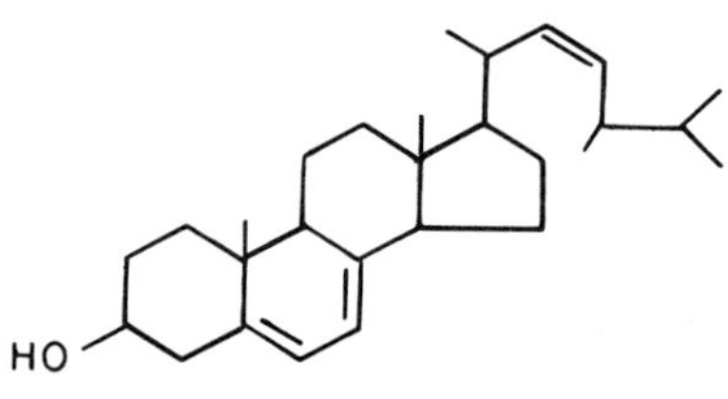

9

value of 190 mg/100 ml reported by a clinical laboratory for the same plasma sample.

Work is in progress in our laboratory for utilizing NCl spectra of TAI derivatives, acid chlorides, and other halogen-containing derivatives of steroids for biomedical studies related to atherosclerosis, acute pancreatitis, gallbladder malfunction, and cancer of the liver.

ACKNOWLEDGMENT

Generous grants from the Schering Plough Foundation and Stevens Institute of Technology for the Cl-MS facilities are gratefully acknowledged. Partial support for this research was provided by the New York Cardiac Center, the U.S. Public Health Services (5ROIAM17921-02), and the Lemcke Bequèst to Stevens Institute of Technology. We wish to thank Dr. B. N. Pramanik, Dr. B. Pujar, and Mary Anne Russo [42] for samples and assistance. We are specially thankful to Dr. A. Yergey, Eric Sharkazy, and Oscar Garcia of Scientific Research Instruments for valuable technical assistance and advice.

VIII. HYDROGEN-DEUTERIUM EXCHANGE REACTIONS OF CARBANIONS WITH D_2O IN THE GAS PHASE

J. H. STEWARD, R. H. SHARPIRO, C. H. DePAY, and V. M. BIERBAUM

University of Colorado
Department of Chemistry
Boulder, Colorado

Using the flowing afterglow technique, we have observed that negative ions participate in sequential hydrogen-deuterium exchange reactions with deuterium oxide. In this technique reactant ions are produced by electron impact and chemical reaction, for example,

$$NH_3 \xrightarrow{e^-} NH_2^- + H$$

$$NH_2^- + CH_3\text{-}CH{=}CH_2 \longrightarrow \overset{-}{C}H_2\text{-}CH{=}CH_2 + NH_3$$

in a flowing helium stream ($P \simeq 0.5$ torr, $\overline{v} \simeq 80$ m/s). The ions are thermalized by numerous collisions with helium and well-defined diffusion and flow properties are developed in the first 30 cm of the 100-cm reaction tube. Deuterium oxide is added through a downstream movable inlet and the anion-D_2O reactions that occur are sampled, analyzed with a quadrupole mass filter, and detected with an electron multiplier.

For the allyl anion, only four of the five hydrogen atoms participate in exchange reactions:

$$C_3H_5^- + D_2O \longrightarrow C_3H_4D^- + HOD$$

$$C_3H_4D^- + D_2O \longrightarrow C_3H_3D_2^- + HOD$$

$$C_3H_3D_2^- + D_2O \longrightarrow C_3H_2D_3^- + HOD$$

$$C_3H_2D_3^- + D_2O \longrightarrow C_3HD_4^- + HOD$$

$$C_3HD_4^- + D_2O \longrightarrow \text{No reaction}$$

However, all seven of the hydrogens of the various isomeric butene anions ($C_4H_7^-$) are exchangeable, as are all hydrogens in the anions from 1-pentene ($C_5H_9^-$) and from 1-hexene ($C_6H_{11}^-$):

$$CH_3CH_2CH_2CH{=}CH{-}CH_2^- \xrightarrow{D_2O} C_6D_{11}^- \text{ etc.}$$

An exchange mechanism consistent with these data is illustrated for the butenyl anion in Scheme 1. We propose that an anion-D_2O reaction complex is formed in which deuteration of the carbanion occurs. However, before the OD^--alkene species dissociates, OD^- abstracts a proton from an allylic position. The carbanion-HOD complex then decomposes and the monodeuterated anion can encounter another D_2O molecule and repeat the process. Some of the anion-neutral complexes may have sufficiently long lifetimes to allow exchange of both deuterium atoms. Therefore, by appropriate protonations at either end of the delocalized anionic system all protons in the butenyl, pentenyl, and hexenyl anions can become allylic and be exchanged. This is not the case for the allyl anion in which the unique central hydrogen always remains vinylic.

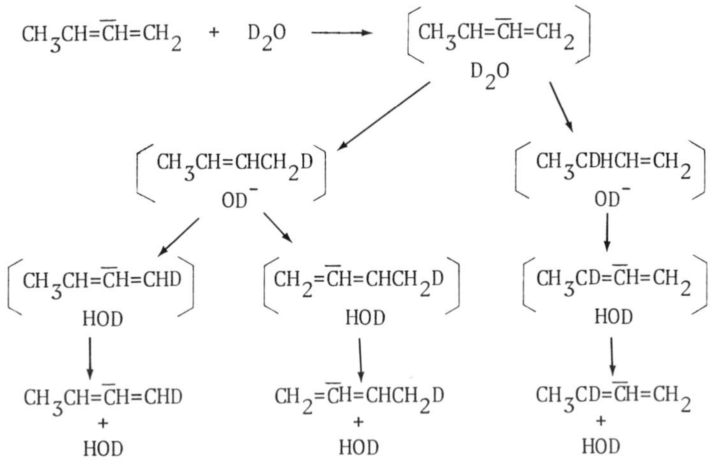

Scheme 1

When the acidity of the organic species and water are very similar, production of OD^- and a deuterated organic neutral competes with production of the deuterated anion and HOD; i.e., dissociation of the OD^--alkene complex of Scheme 1 frequently occurs. For this reason, the undeuterated $C_3H_5^-$ species is always the major organic ion observed in the allyl anion-D_2O system. In contrast, highly deuterated anions dominate the spectrum of the less basic hexenyl anion at large D_2O flows. It should be stressed, however, that any deuterated organic neutrals formed in ion-molecule reactions essentially do not react further since the ratio of undeuterated to deuterated organic neutrals is greater than 10^4. Therefore, the essential mechanistic requirement for the formation of deuterated anions is that the exchange occur with a complex from which the organic species emerges with retention of charge.

Hydrogen-deuterium exchange can be used to distinguish quickly between ions of the same mass or composition. Proton abstraction from methyl acetate and t-butyl alcohol produces anions of the same nominal mass. Upon addition of deuterium oxide, however, the two ions are easily differentiated since the former exhibits two exchanges and the latter clusters rather than exchanges:

$$\bar{C}H_2COOCH_3 \xrightarrow{D_2O} m/z\ 74,75$$
$$m/z\ 73$$
$$(CH_3)_3CO^- \xrightarrow{D_2O} m/z\ 93$$
$$m/z\ 73$$

A second example involves two anions of *m/z* 59 generated by proton abstraction from acetic acid and from glycol aldehyde. In contrast to the last example, these ions are not only isobaric but are truly isomeric and cannot be distinguished even by high-resolution mass spectrometry. However, upon addition of D_2O, acetate undergoes no reaction, whereas glycol aldehyde exhibits a single exchange:

$$CH_3COO^- \xrightarrow{D_2O} \text{No reaction}$$
$$m/z\ 59$$
$$HC(=O)\text{-}CH_2O^- \xrightarrow{D_2O} m/z\ 60$$
$$m/z\ 59$$

The phenomenon of hydrogen-deuterium exchange can be used to determine product ion structures and probe reaction mechanisms. For example, it has been known that amides react slowly with cyclopropane in the gas phase to produce an anion of mass 41. However, whether this species exists in the cyclic or ring-opened form is not known. Upon addition of D_2O, the anion exhibits four exchanges:

$$NH_2^- + \nabla \rightarrow \nabla^- \text{ or } CH_2{=}\bar{C}H{=}CH_2\ (m/z\ 41) \xrightarrow{D_2O} m/z\ 42,43,44,45$$

It can therefore be concluded that the ionic product is the isomeric allyl anion and not the cyclopropyl anion, which, if it exchanges at all, should exchange all five of its hydrogens.

Another example of inferring ion structure and reaction mechanism by use of hydrogen-deuterium exchange involves the reaction of amide with dihydropyran to form an ion one mass unit lower than the parent.

Two possible mechanisms include proton abstraction from the allylic position of β-elimination. The first reaction is characteristic of an alkene, the second of an ether. Upon addition of D_2O, only the cluster at mass 103 is observed:

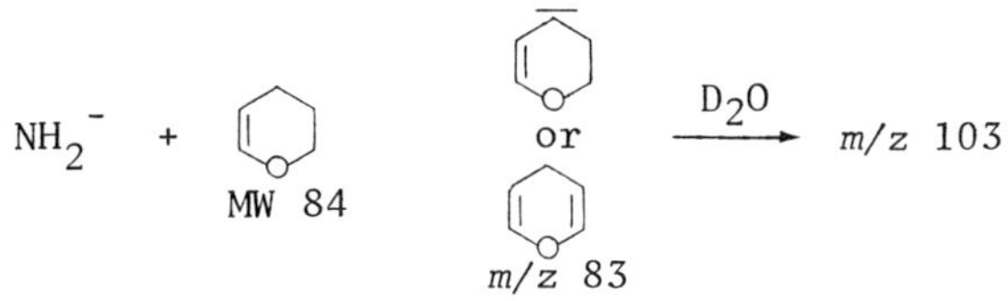

This strongly suggests that β-elimination is occurring to form the enolate anion, which readily clusters. Hydrogen-deuterium exchange would have been expected for an alkenyl anion.

An attempt to distinguish the isomers of $C_3H_3^-$ by hydrogen-deuterium exchange proved interesting. The *m/e* 39 anions formed by proton abstraction from propyne and from allene both exhibited three exchanges:

$$CH_3C{\equiv}CH \xrightarrow{NH_2^-} m/z\ 39 \xrightarrow{D_2O} m/z\ 40,41,42$$

$$CH_2{=}C{=}CH_2 \xrightarrow{NH_2^-} m/z\ 39 \xrightarrow{D_2O} m/z\ 40,41,42$$

These data suggest that in contrast with solution results, at least some proton abstraction from the methyl group must occur. To test this assumption, we studied the reaction of OH^- with monodeuterated propyne:

$$OH^- + CH_3C{\equiv}CD \xrightarrow{40\%} CH_3C{\equiv}C^- + HOD$$

$$\xrightarrow{60\%} \bar{C}H_2\text{-}C{\equiv}CD + H_2O$$

$$\updownarrow$$

$$CH_2{=}C{=}\bar{C}D$$

Preliminary results indicate abstraction occurred from both the acetylene and methyl moieties in a ratio of 2:3. These data stress the difference between gas phase and solution acidities. In solution charge is rapidly dispersed by solvating ligands. In the gas phase where this is not possible, the ability of an anion to delocalize charge through resonance forms becomes an important factor in determining stability.

It is interesting to compare the D_2O reactions for the anions $^-CH_2NO_2$, $^-CH_2COOCH_3$, and $^-CH_2CN$, which have been generated by proton abstraction from a methyl group. The anion from nitromethane exhibits no exchange, only clustering with D_2O. For the anion of methyl acetate, exchange and clustering appear as competitive reactions, whereas for the anion of acetonitrile, exchange reactions dominate. This trend from cluster to exchange reactions reflects the increasing basicity of the anions, or correspondingly, the decreasing acidity of their parent neutrals.

In summary, hydrogen-deuterium exchange of carbanions with deuterium oxide in the gas phase is a simple and effective tool for probing acidities and structures of anions. The phenomenon can frequently distinguish between isobaric or isomeric ions and assist in the elucidation of reaction mechanisms.

IX. DETECTION OF RADICALS AND RADICAL PRODUCTS

HANS KNOF and DIETER KRAFFT

Institut für Experimentalphysik
Universitat Hamburg, Germany

Most ion sources used for electron-impact ionization are operated at 70 eV. These ion sources can be modified to obtain spectra of negative ions (Fig. 19) [43]. In such spectra molecular peaks, M^- and $[M-1]^-$, are frequently observed. This production of molecular ions without loss of hydrogen can be performed by electron attachment. The thermal secondary electrons that are necessary can be produced either by wall reactions of the primary electron beam or by electron-impact ionization:

$$AB + e_p \rightarrow A^+ + B + e_p' + e_s$$

The thermal secondary electrons e_s may react by resonance capture

$$AB + e_s \rightarrow AB^-$$

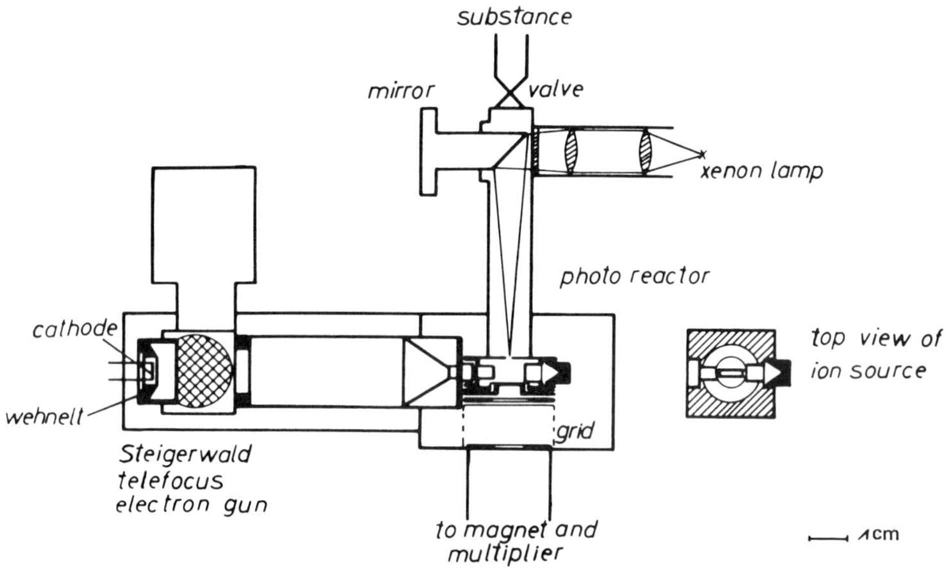

FIG. 19. Schematic view of the electron attachment ion source with photoreactor.

or by dissociative resonance capture

$$AB + e_s \rightarrow A + B^-$$

The latter process is most frequently observed when the neutral fragment is a hydrogen atom. Then the negative ion is one of a paramagnetic molecular fragment. Therefore, electron attachment might be a useful method for ionization of paramagnetic free radicals.

A. Experimental Technique

An ion source was built to detect radicals and radical products via their negative ions (Fig. 20) [44]. The radicals were produced by photochemical reactions without using a matrix gas such as argon or a sensibilization gas such as mercury. In order to achieve this, the ion source had to have a very high ion yield since the radical concentrations that can be produced by photochemistry are minute [45,46].

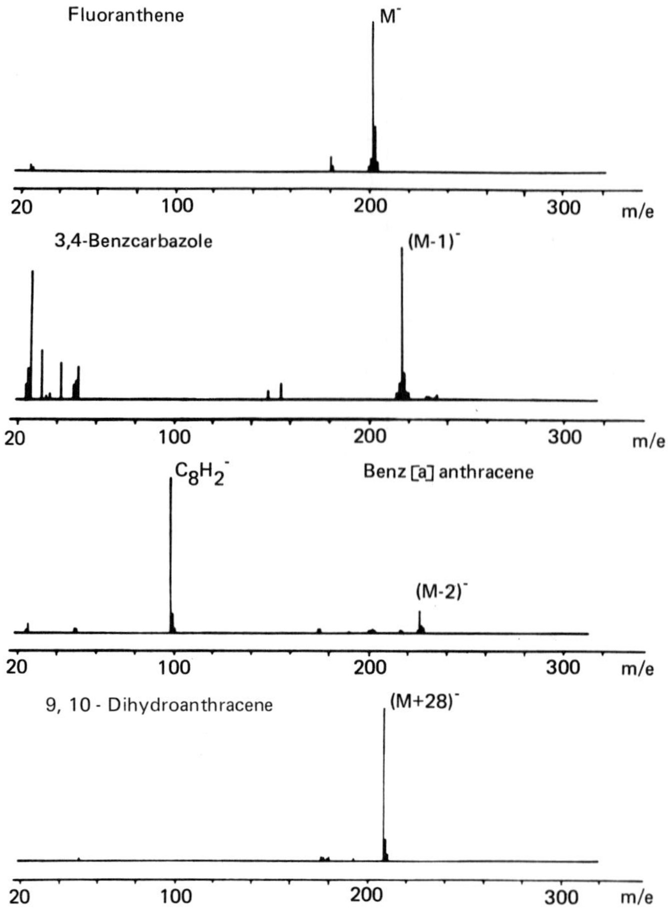

FIG. 20. Negative ions of different types of compounds.

The gas pressure in the ion source is greater than 10^{-2} torr and the beam of primary electrons is about 200 μA. This produces sufficient thermal secondary electrons for electron capture. Pyrolysis can be avoided by use of a telefocus cathode. This cathode, which was introduced originally by Steigerwald for electron microscopy, allows the separation of the hot cathode from the ionization chamber. This arrangement has a fairly high electron current and an electron energy of 200 eV within the ionization chamber. The anode tube between the hot cathode and the ionization chamber is pumped separately and cooled with liquid nitrogen.

The photoreaction chamber is mounted just above the ionization chamber and can be pressurized up to several torr. The ultraviolet light from a 400-watt xenon high-pressure lamp is focused by an optic of special quartz onto the entrance hole of the ionization chamber [47].

B. Results and Discussion

The photochemical reactions of acetone are well known. Primary reactions are

$$CH_3COCH_3 \longrightarrow CH_3COCH_3^* \tag{19}$$

$$\longrightarrow CH_3\dot{C}O + \cdot CH_3 \tag{20}$$

acetyl methyl

$$\longrightarrow 2\ \cdot CH_3 + CO \tag{21}$$

$$\longrightarrow C_2H_6 + CO \tag{22}$$

Although the normal negative ion spectrum of acetone exhibits several peaks in the region of the masses of suspected radicals, only some peaks increase selectively by ultraviolet radiation (Fig. 21). The difference spectrum shows peaks arising from the radicals $\cdot CH_3$, $\cdot CHCO$, and $\cdot CH_3CO$. These identifications were verified by use of deuterated acetone. The detection of $\cdot CH_3$ and $\cdot CH_3CO$ favors Eq. (20). The other two reactions cannot be identified because CO and C_2H_6 do not form negative ions. In addition to the ions of the primary radicals, the ions of the radical products CH_3COCH_3CO and $CH_3COCH_3CH_3CO$ were found, among others (Fig. 22).

Two reactions must be considered in arriving at a mechanism that accounts for the radical product CH_3COCH_3CO:

$$CH_3COCH_3 + CO \rightarrow CH_3COCH_3CO \quad \text{(attachment of CO)} \tag{23}$$

$$CH_3\dot{C}O + CH_3\dot{C}O \rightarrow CH_3COCH_3CO \quad \text{(dimerization of acetyl)} \tag{24}$$

Equation (23) can be excluded by investigating a mixture of acetone and carbon monoxide without ultraviolet irradiation. In this case

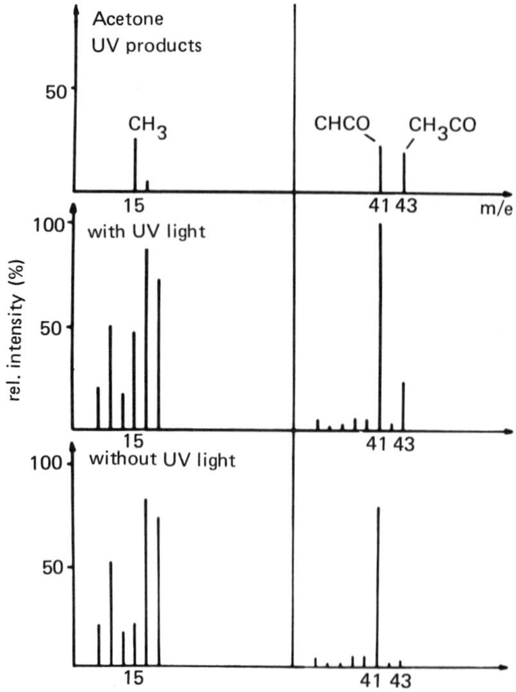

FIG. 21. Mass spectra of acetone in the region of the primary radicals: (a) Spectrum of acetone irradiated with ultraviolet. (b) Spectrum without irradiation. (c) Spectrum b substracted from spectrum a.

CH_3COCH_3CO does not appear in the mass spectrum. The confirmation of Eq. (24) can be obtained by the reaction process in a mixture of acetone and deuterated acetone (Fig. 23). The intensity (I_M) of CH_3COCD_3CO depends on the intensities I_1 and I_2 of the biacetyl and of the deuterated biacetyl, respectively, according to the formula

$$I_M = 2\sqrt{I_1 \cdot I_2}$$

In Fig. 23 the measured peak intensity and that calculated according to this formula differ by only 9%.

The formation of $CH_3COCH_3CH_3CO$ was investigated as well. No peak that could be identified in the same way as the product of a secondary mixture reaction such as

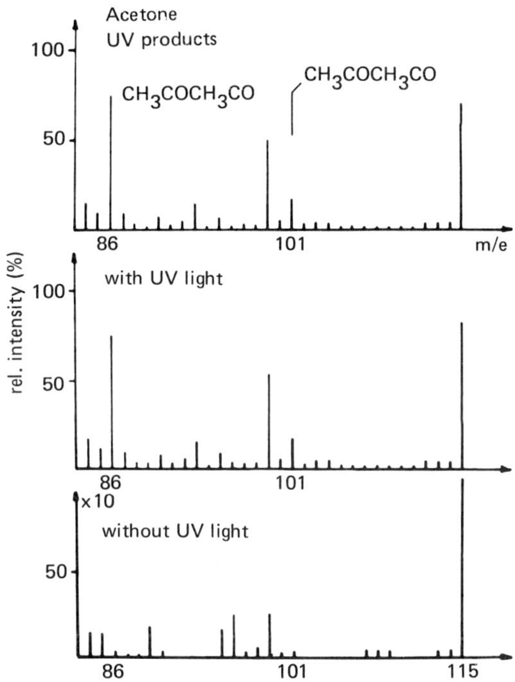

FIG. 22. Mass spectra with acetone in the region of the radical products.

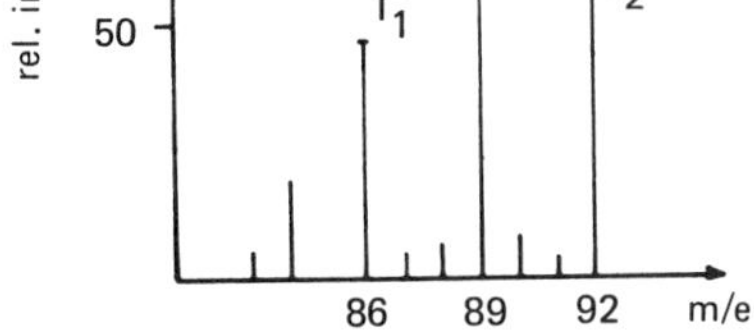

FIG. 23. Spectrum of ultraviolet products from a mixture of acetone (1) and deuterated acetone (2) with the mixture peak (M) compared with the formula $2\sqrt{I_1}\sqrt{I_2}$.

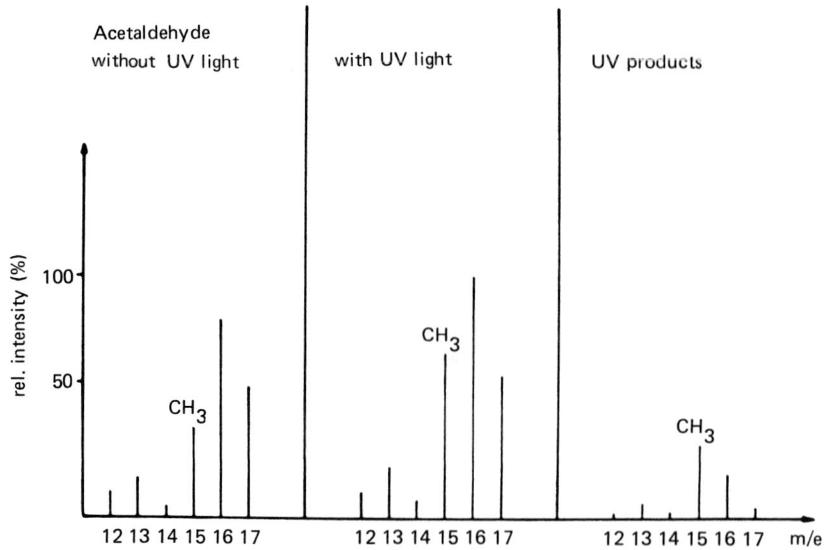

FIG. 24. Mass spectra of acetaldehyde in the region of primary radicals.

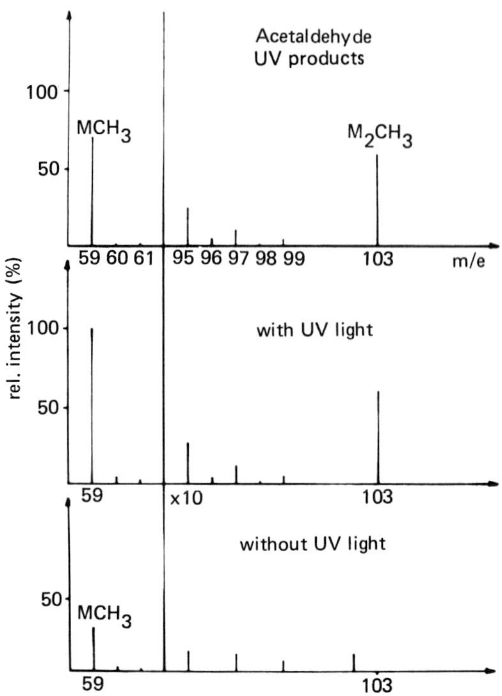

FIG. 25. Mass spectra of acetaldehyde in the region of the radical products.

$$CH_3COCH_3 + CD_3\dot{C}O \rightarrow CH_3COCH_3CD_3\dot{C}O \quad (25)$$

$$CD_3COCD_3 + CH_3\dot{C}O \rightarrow CD_3COCD_3CH_3\dot{C}O$$

Therefore, a photoreaction according to the following primary process involving molecular dimers is more likely.

$$(CH_3COCH_3)_2 + h\nu \rightarrow CH_3COCH_3CH_3\dot{C}O + \dot{C}H_3 \quad (26)$$

Another example is the photolysis of acetaldehyde, which gives rise to the primary radical $\cdot CH_3$ (Fig. 24) and the radical products $CH_3CHO\dot{C}H_3$ and $(CH_3CHO_2)_2\dot{C}H_3$ (Fig. 25). These radical products are formed by the attachment of one or two molecules to the free radical, e.g.,

$$CH_3CHO + \cdot CH_3 \rightarrow CH_3CHO\dot{C}H_3 \quad (27)$$

This reaction can be confirmed with a mixture of acetaldehyde and propionaldehyde. In the mass spectrum $CH_3CHO\dot{C}H_3^-$ and $C_2H_5CHO\dot{C}H_3^-$ peaks can be observed. With pure propionaldehyde no radical product $C_2H_5CHO\dot{C}H_3$ appears. Therefore, $CH_3CHO\dot{C}H_3$ originates from Eq. (27).

C. Conclusions

The detection of free radicals and radical products was made possible by improving the sensitivity of the negative ion source and by using a telefocus cathode. This allowed the investigation of very sensitive compounds, such as free radicals. The same ion source can be used to investigate biological samples, such as steroids.

X. NEW SCANNING TECHNIQUES FOR DOUBLE-FOCUSING MASS SPECTROMETERS

D. S. MILLINGTON

VG-Organic Ltd.
Altrincham, Cheshire, United Kingdom

Recently, new scanning methods have been described that allow analysis of metastable ion reactions occurring in the first field-free region at constant accelerating voltage and without interference from the peaks due to stable ions [35]. From an analytical viewpoint, the

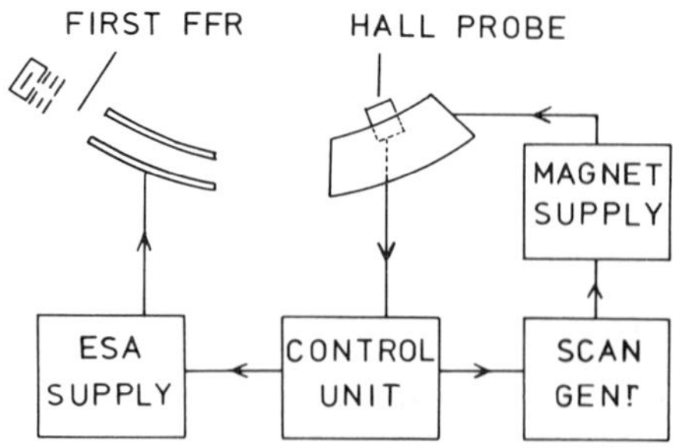

FIG. 26. Schematic of system for generating linked scans.

most useful of these are linked scans of the magnetic (B) and electrostatic (E) fields such that either the ratio B/E or B^2/E is constant. Reference to Fig. 26 will indicate that the theoretically expected information from these scans are as follows:

B/E = fragmentation pattern of a chosen precursor ion
B^2/E = energy spectrum of a chosen daughter ion with kinetic energy release information and precursor ion masses

These new scanning techniques were known to be technically feasible [48] on a VG-Micromass 7070 mass spectrometer using the signal from a Hall-effect probe as a reference voltage to control the fields of the two sectors (Fig. 27).

The powerful B/E scan can be employed to obtain valuable ion structure information, particularly when applied to stable molecular ions that do not undergo extensive fragmentation in the ion source. Under CI conditions, the predominant process for negative ion formation is resonance electron capture, a selective low-energy process that often yields molecular ions without significant fragmentation. An example of the B/E linked scan applied to a stable negative molecular ion, that of benza-α-anthracene-7,12-dione (*m/z* 258), is shown in Fig. 28. The main fragment ions detected can be rationalized in terms of the losses of $\dot{H}$, H_2, $\dot{O}H$, H_2O, and CO from the molecular ion. In the linked scan spectrum, the signal-to-noise ratio is good even for peaks having only 10 ppm of the intensity of the molecular ion, in

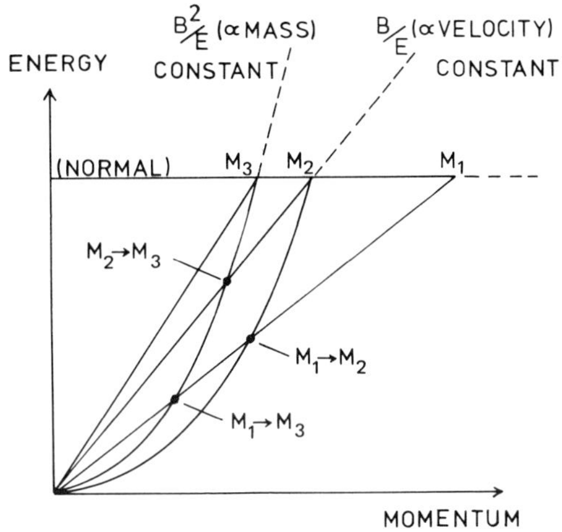

FIG. 27. Energy and momentum diagram showing the effect of B/E and B^2/E linked scans.

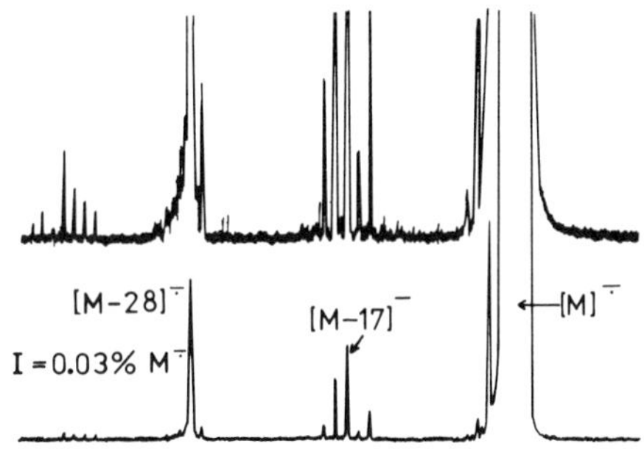

FIG. 28. Fragmentation pattern of the M^{-} ion of benza-α-anthracene-7,12-dione obtained by B/E linked scan.

contrast with the normal mass spectrum in which most of the fragment peaks are lost in the background of other stable ions.

This example shows the potential of B/E linked field scanning for obtaining structural information on negative ions and, clearly, still more information will be provided by applying collision-induced decomposition [49] in the first field-free region.

XI. INSTRUMENTATION FOR NEGATIVE ION DETECTION

M. McKEOWN and R. T. BRACKMANN

Extranuclear Laboratories, Inc.
Pittsburgh, Pennsylvania

The instrumentation required to detect negative ions is partly determined by the source of those ions, but largely by the signal-handling method to be used. That is: Is the signal output in pulse counting or analog current form? Negative ion production is usually via an electron attachment process where a large electron population is thermalized by admixing a high pressure of some diluent gas with the sample molecules. This creates two problems for mass spectrometric analysis and detection. First, the excess electrons have the same charge sign as the ions of interest and a suitable separation must be effected between them before the ions can be detected. Second, the high gas pressure of the ion source must be interfaced to the low-pressure requirements of the mass spectrometer.

Reducing the ionizing chamber's gas pressure to something acceptable to the mass spectrometer means placing a barrier between the two sections pierced with a small-diameter orifice. The characteristic dimensions of this orifice, and the gas flow rate through it, frequently produce conditions where the transit time of the electrons through the orifice is less than their diffusion time to the walls under the prevailing diameter, pressure, and space-charge conditions. The orifice then acts as a partial separator of electrons and negative ions. For electrons that are not so separated, the common practice is to induce perturbing trajectories with small magnets ahead of ion focusing elements, causing the electrons to spiral to the vacuum chamber walls. This is usually sufficient to eliminate electron-induced noise from the system, particularly if the electron multiplier is physically isolated from the chamber in which the excess electrons are being dumped.

A. Detection of Negative Ions

For the detection of negative ions, the importance of establishing complete circuitry, that is, ensuring that dc voltage and current returns and ac bypasses exist, cannot be overemphasized. This precaution is necessary, of course, for positive ion detection but errors seem more common in negative ion circuitry design, presumably because of the unusual bias voltages required.

Detectors fall into two classes: the passive ion collection cup directly connected to an electrometer capable of measuring extremely small currents, and the active electron multiplier in which an incident ion is converted into a large number of electrons. The latter device allows the use of either pulse techniques or fast preamplifier/electrometer combinations, thus increasing the system sensitivity and speed of response.

1. Faraday Cups

The mechanical and electrical characteristics of effective Faraday cups for positive ion detection are established in the literature. The details largely concern the suppression of secondary electrons created by the ions impacting into the surfaces of the cup. When detecting positive ions, loss of these electrons from the measuring circuit produces a false increase in the true ion current. When detecting negative ions the loss causes a false reduction in the true ion current. For a poorly designed Faraday cup, electrons may be ejected and lost at the same rate that negative ions impact. The measured ion current under these conditions would be zero.

Secondary electron suppression may be accomplished by the use of ad hoc magnets or by designing the Faraday cup as a tube with a narrow entry for ions and a closed far end. Apart from ensuring adequate screening of the signal cable, there is little electronic information that is needed to give ion detection.

2. Electron Multiplier

As the name states, this device multiplies electrons and is, therefore, not symmetrical in its biasing requirements for positive or negative ion detection. The energy of the impinging ion must be high to induce secondary electron emission from the cathode or first dynode. But no matter what is done to the multiplier potential at the front surface to induce this ion energy, the electrons must see an increasing positive potential from first dynode to anode or final collector. It should be noted that the potential not only increases from the first to the last dynode but also between the last dynode and collector. This point is a frequent source of error in connecting multipliers for negative ion detection. Since the anode is a Faraday cup detector of electrons, all the aforementioned considerations regarding the Faraday cup are applicable in determining its design.

Figure 29 shows the typical potentials and interconnections applied to an electron multiplier attached to a mass spectrometer having a low ion energy, e.g., a quadrupole mass filter. This elaboration is unnecessary for magnetic mass spectrometers with ion energies of 5 to 8 kV at the mass of interest as the effective ion energy

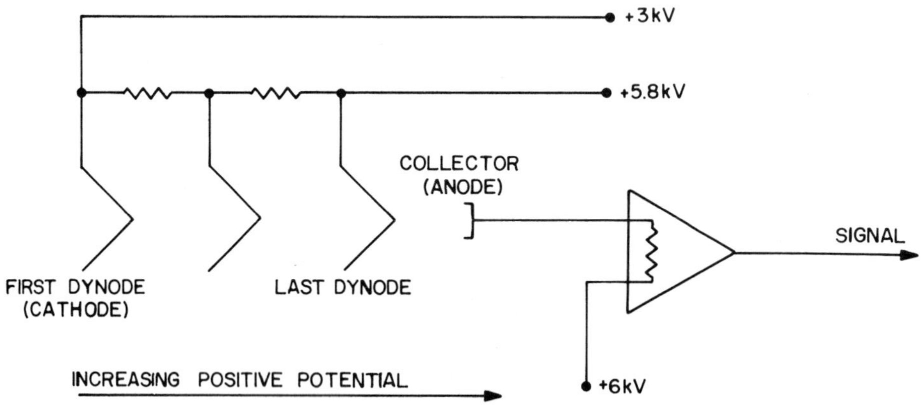

FIG. 29. The typical potentials and interconnections applied to an electron multiplier attached to a mass spectrometer having a low ion energy.

to the multiplier is the algebraic difference between the mass spectrometric ion accelerating voltage and the first dynode potential. As long as this exceeds, say, 1 to 2 kV, with a sign identical to the ion under analysis, ion impact will cause secondary electron emission.

The +3 kV bias shown on the first dynode is larger than required for adequate secondary electron production from negative ions. It has been demonstrated that 600 to 800 volts of impact energy are sufficient. The +3 kV is recommended, however, when the possibility of reduced ion collection exists due to the mechanical arrangement of the mass spectrometer. For example, the off-axis multipliers used by quadrupole manufacturers can lead to lowered collection efficiency at low first-dynode potentials.

The resistor chain establishes a bias of approximately 150 volts per dynode, with the last dynode pegged at +5.8 kV. The collector is then placed at a positive potential above the last dynode by referencing the preamplifier to +6 kV.

B. Decoupling the Signal from the High Potential

Faraday cup detectors are unpolarized in general and the signal due to ion current appears on a lead that can be at virtual ground for either ion sign. That is, the impedance of the signal source is very high but it need not be more than tens of millivolts from ground potential at any time since signal currents are small and the input impedance of the preamplifier is relatively low. But as there are no bias potentials required, the output lead from the Faraday cup can be directly connected to any ground-referenced picoammeter.

Figure 29 shows that for an electron multiplier looking at negative ions, the collector is not only a high-impedance source but is also at a high voltage with respect to ground. Most pulse and current preamplifiers are not manufactured to accept such large common-mode potentials at their inputs and decoupling the signal from the dc voltage powering the electron multiplication process is a major concern in negative ion detection. This remains a problem even if the first dynode is reduced to a few hundred volts, for the collector will still be at a potential in excess of 2,500 volts. The methods used in decoupling are dependent on the signal-handling techniques and before making circuit designs the user must choose either pulse counting or analog current as the preferred detection method. With that primary decision made, the simple schematics shown in Figs. 30 and 31 will provide the necessary decoupling. The choice of the component values depends on the details of the circuit operation and the following explanation, while not electronically complete, allows the calculation to be made.

1. Pulse Counting

Figure 30 illustrates the basic schematic for pulse counting with an electron multiplier. The +6 kV power supply is connected to the multiplier resistor chain via R1, the numerical value of which is determined by the current drawn through the resistor string and the potential difference desired between the last dynode and collector.

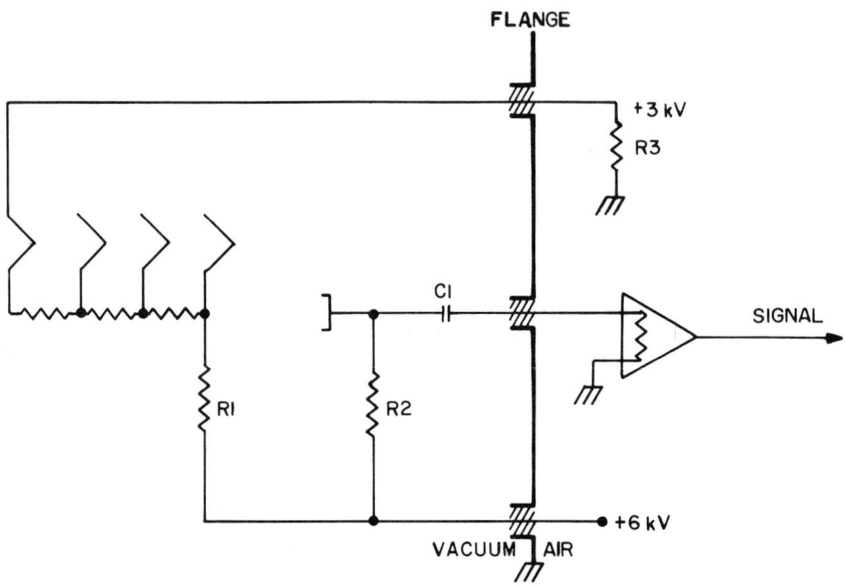

Fig. 30. Basic pulse counting circuit for an electron multiplier.

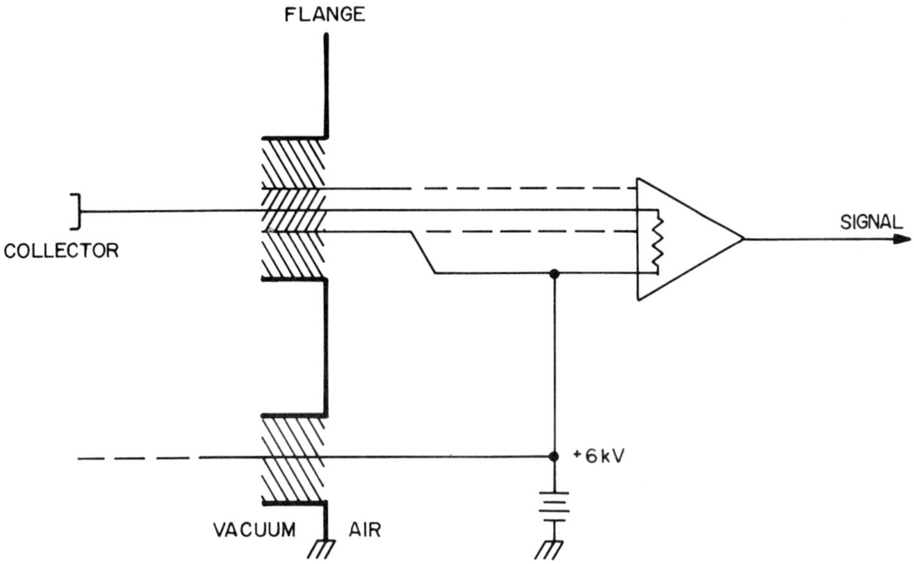

FIG. 31. Correct circuitry applied to analog negative ion detection.

Typically, R1 is 1 MΩ, causing a potential drop of approximately 170 volts. The +3 kV voltage noted for the first dynode can be established either by a power supply or by using a high-ohm resistor (R3) to ground. Usually a resistor is used and its value is made equal to the total resistance of the string across the multiplier including R1.

The determination of the values of R2 and C depends on a knowledge of some fundamental properties of transmission lines. First, let us consider the value of C. Any pulse signal cable between the multiplier collector and the pulse preamplifier needs protection from hum and "glitch" pickup that might be translated into pulses by the preamplifier. The common method is to use coaxial shielded cable to transfer the pulse train from the vacuum system to the preamplifier. Coaxial cables have a specification called the characteristic impedance, which is typically 50 ohms for cable used in scientific instruments. That is, any length of coaxial cable viewed from one end and terminated at the other between central conductor and shield with a 50-ohm resistor looks like a 50-ohm resistor. This is approximately true no matter what frequency the cable is required to carry.

It is important, therefore, to select a value of C that has a low impedance compared with 50 ohms at the frequency of operation because that impedance will act as a series resistor to the pulse train and reduce its voltage amplitude as seen by the preamplifier. But what is meant by the frequency of operation for a pulse-counting circuit? It is not the number of pulses per second but rather that frequency capability necessary to preserve the shape/voltage characteristic of a typical pulse to the preamplifier without excessive distortion. The theory is inappropriate for this text but one can arrive at a rule of thumb, that the required frequency capability is 10 times higher than the maximum pulse counting rate to which the multiplier and preamplifier can respond.

Most multipliers used in mass spectrometers have a maximum counting rate of 1 MHz and the calculation of the value of C using MHz as the required frequency capability can be made using the equation

$$X_c = \frac{1}{2\pi fC}$$

If one assumes that X_c = 2 ohms is sufficiently low to give tolerable losses, this equation gives C = 0.003 μF.

But why not make C very large and so reduce X_c to a still smaller value? This is not done because any relatively harmless discharge that may occur between the multiplier collector and ground will cause the energy stored in C to be rapidly dumped into the input of the preamplifier. This event can easily destroy the input circuit components of the pulse preamplifier. The larger the capacitance, the larger is the stored energy. So C should have the smallest acceptable value and that quoted above is adequate for most circumstances.

One final consideration about this capacitance is its position. Does it have to be inside the vacuum system as shown? In general, the answer is yes. Leakage "pulses" in the insulator, particularly on days of high humidity, are difficult to identify as spurious. It is safer to build a high-voltage, low outgassing capacitor inside the vacuum envelope.

The value selected for R2 depends partly on C but, as a primary concern, it must be regarded as a parallel resistor to the signal load. If R2 is chosen as 50 ohms, clearly it would approximately halve the current spike of the pulse supplied to the input of C. Thus, R2 should be large. However, if it is very large then the dc time constant determined by R2 times C will be long. This time constant does not affect the pulse transfer characteristics significantly, but its effect on the electron flow through R2 would be to reduce the positive potential of the collector as the signal electrons continued to arrive. At some high electron arrival rate at the collector, its potential would be insufficient to attract further electrons and the pulse amplitude would decline until it fell below the discriminator level in the preamplifier. The signal would then disappear.

It is a common fault to make R2 infinite, that is, to forget about it altogether. Clearly, in this case the charge accumulating on the collector does not dissipate at all except for accidental leakage resistances. Even if pulses are seen initially, they quickly disappear as the instrument operates and C charges. At best the performance will be very erratic. The latitude these considerations allow for R2's value is large, however, and it is typically chosen from the range 10^3 to 10^6 ohms. An important proviso to this choice of resistance arises if the output end of the coaxial cable is not terminated in 50 ohms in the pulse preamplifier. If neither end of the cable is terminated in 50 ohms, a considerable number of spurious pulses can arise from wave reflections (ringing) along the cable.

2. *Current Measurement*

Figure 32 shows the circuitry that is commonly but incorrectly applied to negative ion detection using analog or current measurements. Ignoring, for the present, problems associated with preamplifiers referenced to 6 kV, the error in this circuit is illustrated by an arrow showing the leakage current "I" between ground and the signal lead.

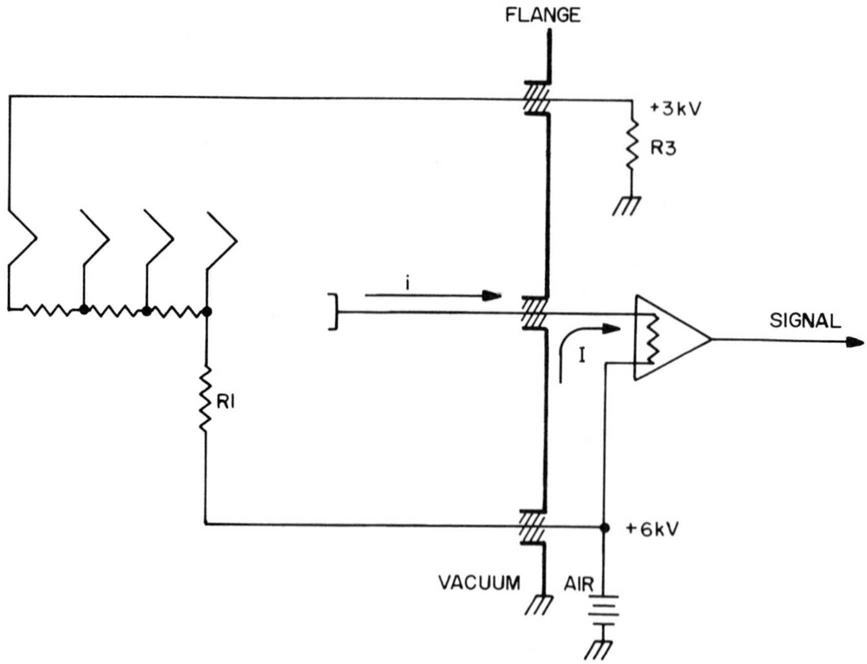

FIG. 32. The circuitry commonly but incorrectly applied to analog negative ion detection.

If the signal lead is fed through the vacuum wall using the normal single insulated feedthrough, 6 kV is imposed across the resistance of the insulator. Making the assumption that the effective resistance of this insulator is 10^{10} ohms, the leakage current is 6 x 10^{-7} amperes.

Upon tracing the circuit, it can be seen that the source of this current is the power supply and that, despite being called leakage, to the preamplifier it is indistinguishable from the signal current "i." If the best possible feedthrough insulators are used, the leakage current may still exceed the signal current by several orders of magnitude, leaving the old problem of measuring a small difference between large numbers to extract the true signal current. The final consequence of choosing this circuit is illustrated by assuming that the leakage current is bucked-out in the preamplifier. The inherent noise of that current flow cannot be bucked out, however, and the signal current as measured contains this extreme noise component together with any drift components that may occur in both leakage or buck-out current sources.

The correct circuitry is shown in Fig. 32 where the triaxial connector has a guard potential on the shield, both through the vacuum wall and on the coaxial cable to the preamplifier. Elimination of the potential difference between signal lead and shield ensures transfer of the signal to the preamplifier without appreciable leakage. Of course, leakage currents driven by the power supply still occur between the shield and ground of the triaxial feedthrough. But now they do not enter the signal current circuit to cause spurious signals.

In tackling the problem of an analog preamplifier referenced to a high potential, one must note that there is nothing intrinsically wrong with this arrangement. FET-input operational amplifiers used in signal preamplifiers can be so connected. The difficulties arise when the output signal needs to be further handled by ground-referenced equipment, as in the cases of a display meter, chart recorder, or computer. The output signal from the operational amplifier is still referenced at high potential with respect to ground. A number of solutions are available to decouple the signal from this potential. For example, one commercial instrument used the high-voltage referenced signal to modulate an RF carrier wave also generated in the high-voltage environment. This modulated RF signal was inductively coupled through a ferrite rod from a transmitting coil at the high-voltage end to the receiving coil at the ground-referenced end. The dc insulation between the two coils provided the necessary isolation and the signal was demodulated and smoothed to remove the carrier. The output was a signal referenced to ground potential that could be interfaced immediately with other instruments. More recent versions of this arrangement use optical isolators as the decoupling devices. Their frequency response and dynamic range is somewhat better than the original "radio" method.

C. Other Methods of Detecting Negative Ions

The techniques and circuitry described above are well established, proven, reliable in operation, and commercially available. But they depend on an electron multiplier having its sensitive dynode surfaces exposed to the dirty vacuum system of the mass spectrometer and the important first dynode suffering ion impact damage. Because the nature of these surfaces changes with time in most vacuum environments, some experimentors seek to avoid, partially or totally, surface phenomena. Partial avoidance is achieved with the Daly doorknob detector. In this method, ions are accelerated toward a metallic but passive surface where secondary electrons are produced. These electrons are then accelerated to very high energies and made to strike a scintillator producing light, which is detected by a photomultiplier housed in its own vacuum envelope. The effects of optical radiation from the initial ion impact are minimized by coating the scintillator with a thin metallic film through which the energetic electrons pass but which is opaque to light. The advantage of this approach is that the multiplier dynodes are not subject to ion impact or deterioration as they are maintained in the exceedingly clean vacuum of the photomultiplier enclosure.

Complete avoidance of the interaction between negative ions and sensitive surfaces may be accomplished by using photodetachment. In this case, an intense light beam strikes a negative ion and the electron is liberated by an interacting photon. The electron is then detected by conventional means. This approach has the advantage that it is a purely gas phase phenomenon insofar as the negative ion is involved. The advent of lasers, whose light intensity is sufficiently high to photo-detach the electron from virtually all negative ions that cross its beam, makes this approach very attractive where surface effects must be avoided.

Other gas phase methods that may have specialized applicability consist of crossing a negative ion beam with a molecular beam or passing the negative ions through a cloud of gas. Collisional detachment and charge transfer can produce either free electrons or negative ions of one single species independent of the nature of the original ion. In general, however, gas phase detectors impose additional vacuum system requirements that are unattractive for simple negative ion detection methods.

One final example is, in itself, facetious but perhaps illustrates that the detection of negative ions may not be limited to physical phenomena. It was reported in 1972 that mice colonies breathing air with negative ion concentration of 3×10^5 ions/cm^3 and zero positive ions were significantly more resistant to death from influenza virus than colonies breathing positive ion-laden or ion-depleted air.

XII. A NOVEL NEGATIVE ION DETECTION SYSTEM FOR A QUADRUPOLE MASS SPECTROMETER

G. C. STAFFORD, J. R. REEHER, and R. S. STORY

Finigan Instruments Division
Sunnyvale, California

Negative ions produced in the ion source of a magnetic sector mass spectrometer are easily detected by a conventional positive ion electron multiplier providing the accelerating voltage is sufficient to convert negative ions to electrons at the cathode (first stage) of a standard electron multiplier. For example, if a negative ion accelerating voltage of -8 kV and an electron multiplier cathode voltage of -3 kV is utilized, a singly charged negative ion will bombard the cathode with 5 keV of kinetic energy.

Ions exiting a quadrupole mass filter will generally have less than 20 eV of kinetic energy. It is obvious that the low-energy negative ions produced on a quadrupole mass spectrometer will not impact the cathode of a conventionally operated positive ion electron multiplier and, therefore, will not be detected.

The conventional approach to the detection of the low-energy negative ions produced on a quadrupole instrument is to isolate electrically an entire electron multiplier at a positive potential such that the cathode is also at a sufficiently positive potential for efficient collection of negative ions (Fig. 33). This methodology

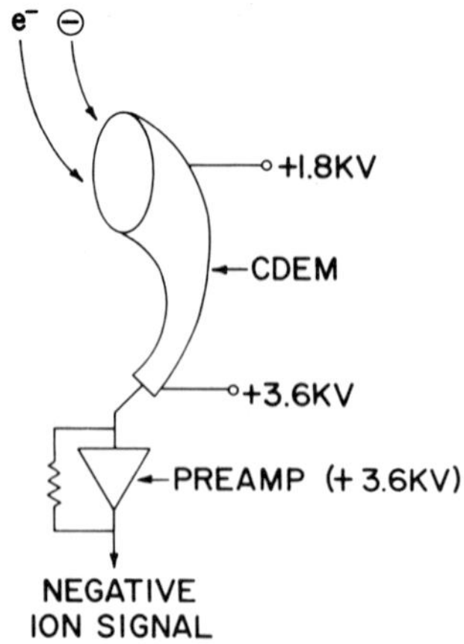

FIG. 33. Off ground negative ion continuous dynode electron multiplier.

has the disadvantage that the anode (signal) of a discrete or continuous dynode electron multiplier (CDEM) and the preamplifier must also be isolated at a high positive potential. Therefore, complex preamplifier circuitry is required to reference the signal-to-ground potential. Also, a special high-voltage floating coaxial signal feedthrough is required on the vacuum envelope to prevent microphonic noise and charge leakage. The floated electron multiplier approach also has the problem that stray electrons in the vacuum system are efficiently detected. Accordingly, an electron multiplier electrically biased for negative ions is approximately 30 times noisier due to stray electrons than a conventional positive ion electron multiplier.

The approach we took to detect negative ions was to convert the negative ions to secondary positive ions, which are subsequently detected by a standard electron multiplier. Negative-to-positive ion conversion can be accomplished by impacting the primary negative ions on a conversion dynode held at a high positive potential. The negative ion detector is shown schematically in Fig. 34A and was constructed by mounting a conversion dynode, in the shape of a cube with two adjacent sides open, opposite the cathode of a Galileo

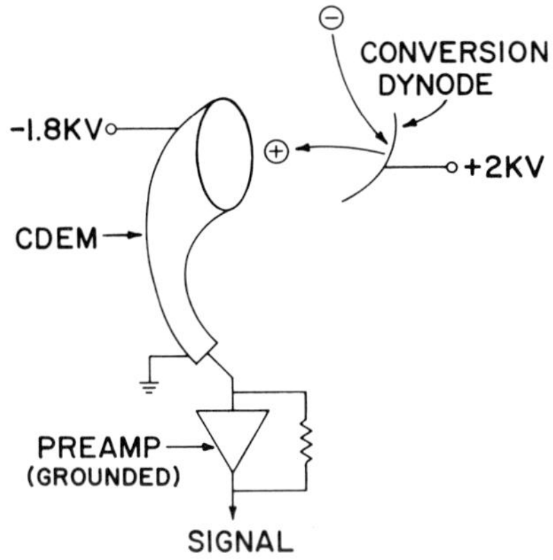

FIG. 34A. Conversion dynode negative ion detector.

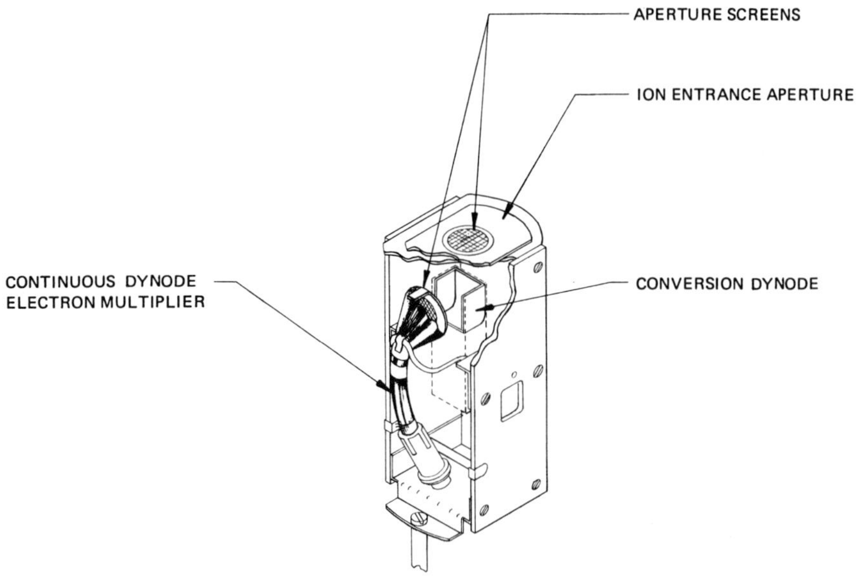

FIG. 34B. Conversion dynode negative ion detector.

model 4770 continuous dynode electron multiplier (Fig. 34B). This methodology eliminates the high-voltage problems associated with floating the anode of an electron multiplier at a high positive potential. A further advantage of this megative ion detector is that stray electrons in the vacuum system do not efficiently produce positive ions when they impact the conversion dynode. Accordingly, the noise background is significantly reduced.

We believe the major mechanisms responsible for this negative-to-positive ion conversion are

1. *Sputtering*. Some of the energy of the bombarding negative ions is transferred to the metal surface, causing metal atoms to be vaporized from the surface. A fraction of these sputtered metal atoms lose electrons to become positive ions.
2. *Fragmentation*. When polyatomic negative ions impact the dynode surface, they undergo extensive fragmentation. These fragments will consist of neutral species, negative ions, and positive ions. We feel that fragmentation becomes the major conversion process for large polyatomic ions (>200 amu).
3. *Charge stripping*. As negative ions are accelerated toward the conversion dynode, an electron may be stripped away to form a neutral species. The neutrals continue traveling toward the dynode, impact the dynode surface, and lose a second electron to form positive ions.

Negative-to-positive ion conversion yield is dependent on the dynode surface and the elemental composition of the ion. Relative yields for the dynodes studied are shown in Table 4. The table shows

TABLE 4

Conversion Yields for Several Dynode Surfaces

Dynode surface	Normalized conversion yield[a]
Cu	1.0
Ag	0.9
Cr	0.7
Al	0.6
304 stainless steel	0.4

[a]Conversion yields were calculated at *m/e* 17 (OH^-) and normalized to Cu. Galileo continuous dynode electron multiplier was operated at -1.4 kV and the conversion dynode was operated at +2.5 kV.

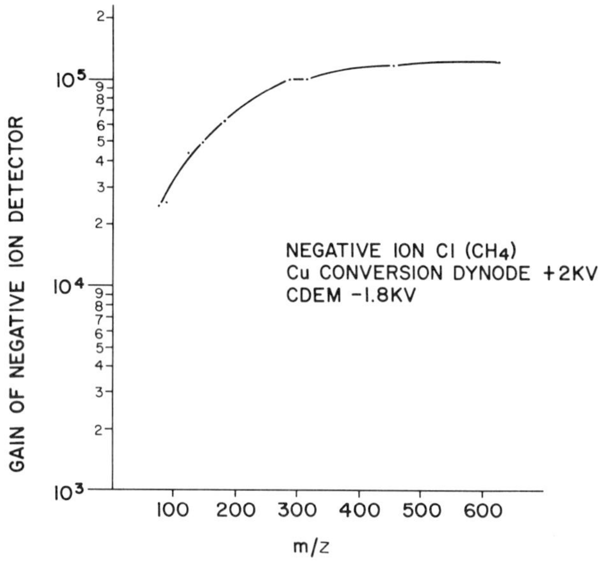

FIG. 35. Negative ion gain vs *m/z* for the conversion dynode negative ion detector.

that secondary positive ion yield is not extremely sensitive to the composition of the dynode surface. The gain versus *m/e* for the negative ion detector using Cu as a dynode surface is shown in Fig. 35. From the figure it is obvious that the negative ion gain and, accordingly, the negative-to-positive ion conversion yield dramatically increases for high mass negative ions.

In conclusion, we see that this methodology for the detection of negative ions enjoys the advantage of increased gain for high mass ions. Since it is not necessary electrically to float the electron multiplier at a high positive potential, feedthrough is eliminated. Generally, this approach to the detection of low-energy negative ions shows significantly less background noise than a floated electron multiplier system.

XIII. THE CONVERSION AND OPERATION OF THE MS30 IN THE NEGATIVE ION MODE: APPLICATIONS IN ORGANIC ANALYSIS

J. R. CHAPMAN, D. DENNE, and G. A. ERROCK

Kratos-AEI Scientific Instruments Ltd.
Manchester, England

A simple negative ion conversion for the MS30 double-beam mass spectrometer (Fig. 36) has been introduced. A switched change of accelerating voltage, electrostatic analyzer voltage, and magnet current polarities allows the instrument to be used routinely in both the positive and negative ion modes.

The program of work to be carried out with the converted MS30 had three objectives: (1) an investigation of the sensitivity enhancements reported for negative ion chemical ionization (NCI) operation; (2) a general appraisal of the analytical applicability of NCI; and (3) the establishment of a chemical mass marking system for the second source of the MS30, suitable for both negative and positive ion operation.

Ionization of methane at 0.5 to 1.0 torr in a tightly sealed chemical ionization source with 70-eV electrons produces a plasma containing positive ions and a large population of secondary electrons having near-thermal energy. The introduction into this plasma

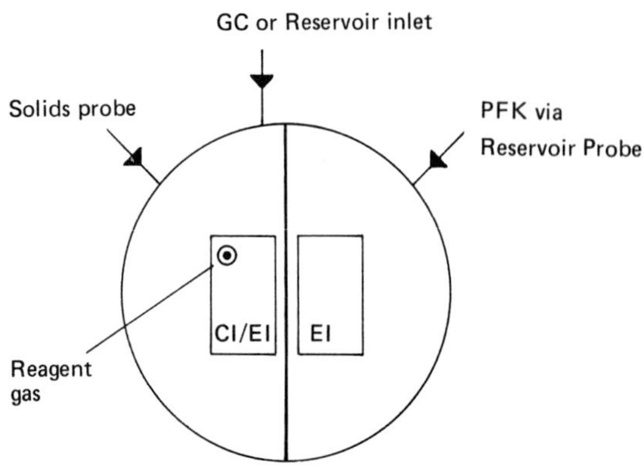

FIG. 36. Combination EI and CI/EI source for the MS30.

of a sample with an appreciable electron capture cross section leads to the formation of a negative molecular ion from the sample. Subsequent fragmentation of this molecular ion is minimal and a very simple spectrum results.

The sensitivity of NCI mass spectrometry can, in favorable cases, be up to two orders of magnitude greater than positive ion CI(CI). This difference appears to be largely due to the fact that if Eqs. (28) and (29) are rate-limiting in producing CI and NCI spectra, respectively, then thermal electrons (e) are much more mobile than thermal ions (XH^+) because of the much smaller mass of the former.

$$XH^+ + A \rightarrow X + AH^+ \tag{28}$$

$$e^- + A \rightarrow A^{\overline{\cdot}} \tag{29}$$

A typical electron-capture spectrum is that of 17-hydroxy-1,4,6-androstatriene-3-one acetate (Fig. 37), which shows no ions other than the molecular ion. Particular note should be taken of the negative ion perfluorokerosene (PFK) spectrum used for mass marking in the second beam. This is produced in the conventional electron-impact second source of the MS30 by a pressure-enhancement technique. This technique, first reported by Dougherty [2] involves the introduction of an intermediate pressure of gas into the source to enhance the secondary electron yield and also to allow collisional stabilization of the negative ions formed. In practical terms, 1 μl of PFK was introduced into the source using a reservoir probe and 4 ml of air was added to this PFK using a gastight syringe. This technique produced a reference spectrum of good intensity for both negative and positive ion operation. In the negative ion mode, PFK alone gives very poor sensitivity. A pressure-enhanced negative ion spectrum of PFK is shown in Fig. 38. As can be seen, many of the major peaks in this spectrum are also major peaks in the positive ion spectrum of PFK. The pressure-enhanced PFK spectrum is also of sufficient intensity to permit extension of the double-beam mass measurement technique [50] to negative ion operation.

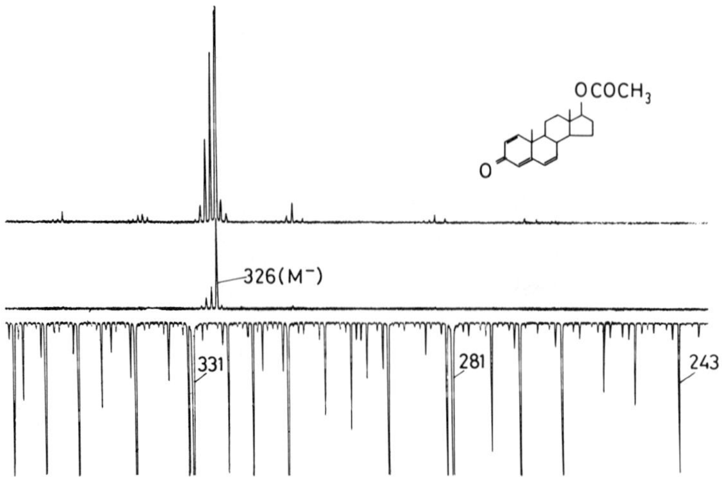

FIG. 37. Electron-capture spectrum of 17-hydroxy-1,4,6-androstatriene-3-one.

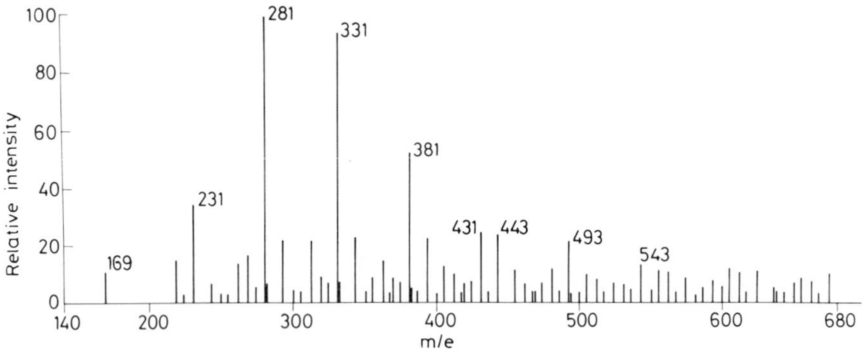

FIG. 38. Pressure-enhanced negative ion spectrum of PFK.

An investigation of the sensitivity afforded by the NCI technique in favorable cases was carried out by measuring the integrated ion current from known amounts of compounds at the base peak in the spectrum. These data were compared with positive ion chemical ionization and electron-impact (EI) sensitivity data from the same compounds and are summarized in Table 5. In all cases, these determinations were carried out with 1 μg of sample. The NCI spectrum of another of the compounds used in this investigation, benzil, is shown in Fig. 39. The total ionization figures were obtained by multiplying the base peak sensitivities by the appropriate factor based on the fraction of the total ion current in the spectrum contained in the base peak.

It will be seen that the NCI total ionization figures for the three compounds investigated are quite similar and that they are of the same order of magnitude as the EI figures. On the other hand, it will be seen that the CI total ionization figures are one or two orders of magnitude smaller, indicating the lower total sensitivity obtainable with this technique. Negative ion chemical ionization has the advantages of high absolute sensitivity in favorable cases and the fact that this sensitivity is concentrated in one major peak. In addition, background levels in NCI operation are usually very low because of the insensitivity of common background materials, e.g., hydrocarbons, under NCI conditions. Therefore, NCI operation has great potential for single- and multiple-ion monitoring experiments where very low levels of compounds are to be determined.

In some cases, the compound to be determined will already have a good electron-capture cross section; in other cases, it will be necessary to make electron-capturing derivatives for the experiment [13]. An example of the former case is malathion, a widely used insecticide. A comparison of the CI and NCI spectra of malathion (Fig. 40) shows that the positive ion spectrum contains a relatively intense quasi-molecular ion (*m/e* 331) as well as fragment ions. However, the negative ion spectrum consists of a single ion at *m/e* 157 corresponding to the formation of a stable phosphorous- and sulfur-containing anion. Preliminary data suggest a sensitivity greater than 5×10^{-9} coulombs per microgram for this ion; thus, it would be ideally suited for

TABLE 5

MS30 Relative Sensitivities in Different Ionization Modes

		Base peak sensitivity C μg^{-1}			Total ionization C μg^{-1}		
Compound	M. Wt	NICI	PICI	EI	NICI	PICI	EI
17-hydroxy-1,4,6-androstatriene-3-one-acetate	326	$2 \cdot 9 \times 10^{-9}$	$5 \cdot 8 \times 10^{-11}$	—	$4 \cdot 2 \times 10^{-9}$	$1 \cdot 7 \times 10^{-19}$	—
Anthracene	154	$3 \cdot 1 \times 10^{-9}$	$5 \cdot 1 \times 10^{-11}$	$1 \cdot 3 \times 10^{-9}$	$3 \cdot 8 \times 10^{-9}$	$1 \cdot 1 \times 10^{-10}$	$2 \cdot 6 \times 10^{-9}$
Benzil	210	$1 \cdot 3 \times 10^{-9}$	$1 \cdot 2 \times 10^{-11}$	—	$1 \cdot 5 \times 10^{-9}$	$3 \cdot 4 \times 10^{-11}$	—
Methyl stearate	298	—	$1 \cdot 0 \times 10^{-10}$	$5 \cdot 0 \times 10^{-10}$	—	$2 \cdot 7 \times 10^{-10}$	$2 \cdot 5 \times 10^{-9}$

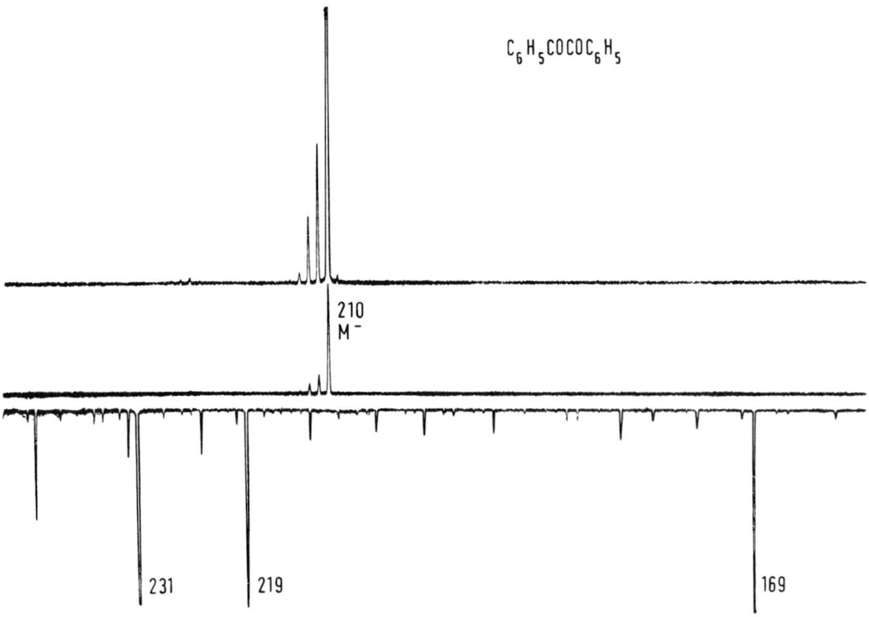

FIG. 39. NCI spectrum of benzil.

single-ion monitoring experiments. In addition, the mass-deficient nature of this ion (accurate mass 156.9547) means that it should be well separated from most materials interfering at this nominal mass.

Examples of electron-capturing derivatives are dansyl derivatives and dinitrophenyl derivatives. The dinitrophenyl derivatives of small peptides (Fig. 41--sample provided by Dr. D. H. Williams, Cambridge University) show molecular ions with excellent sensitivity under NCI conditions and, therefore, by confirming the molecular weight, provide information complementary to the sequence information obtained by electron-impact analysis. In the spectrum of dansyl ephedrine (Fig. 42) the facile addition of $O_2^{\bar{\cdot}}$ (formed from traces of air present in the source) to the molecular ion may be noted.

NCI spectra can be obtained from compounds that do not have high electron-capture cross sections. Thus, the diketo-steroid spectrum illustrated (Fig. 43) shows mainly an $[M-H]^-$ ion but with relatively

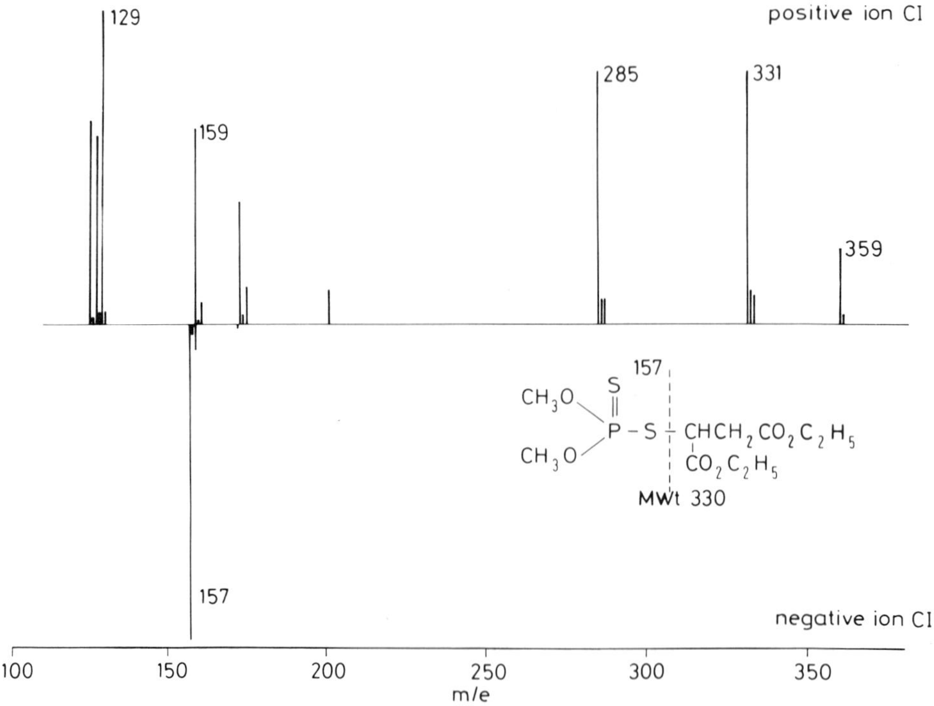

FIG. 40. A comparison of the CI and NCI spectra of malathion.

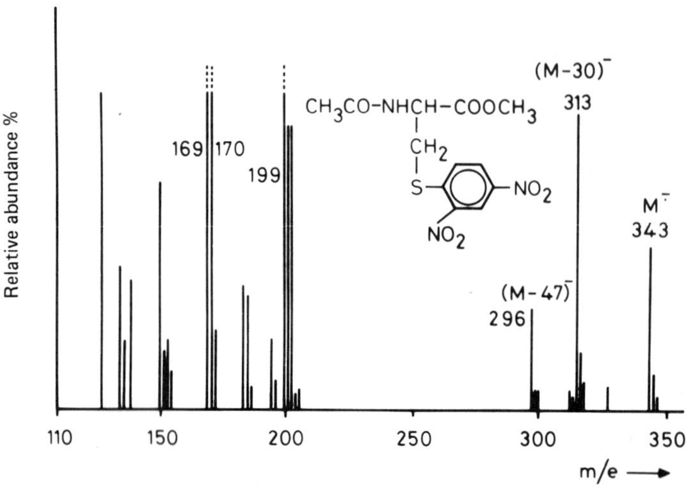

FIG. 41. Electron-capture spectrum of the dinitrophenyl derivative of a small peptide.

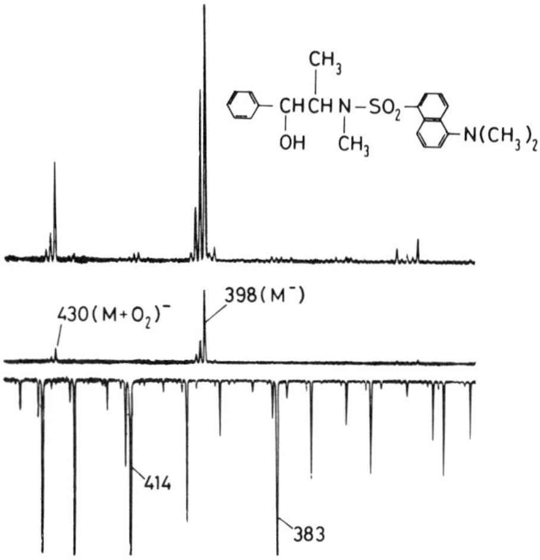

FIG. 42. Electron-capture spectrum of dansyl ephedrine.

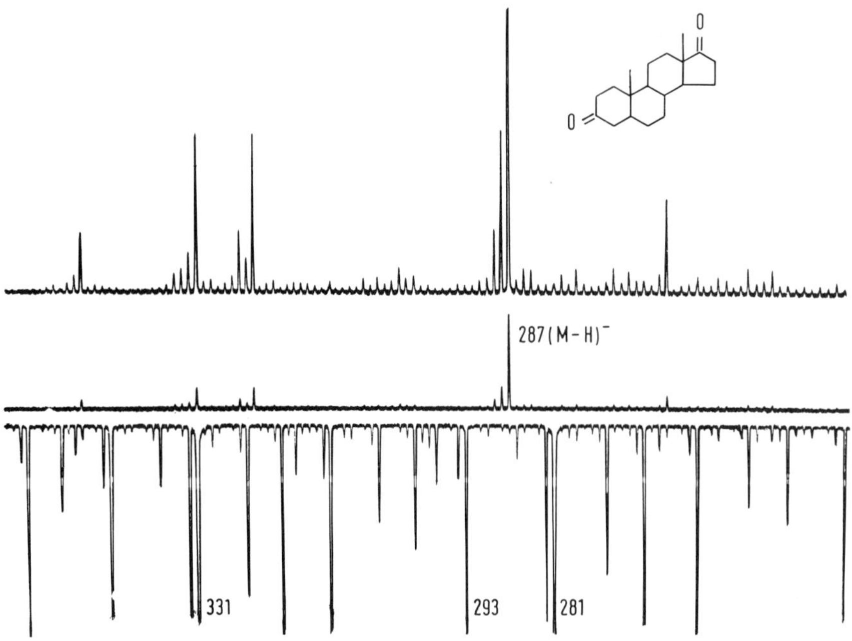

FIG. 43. NCI spectrum of a diheto-steroid.

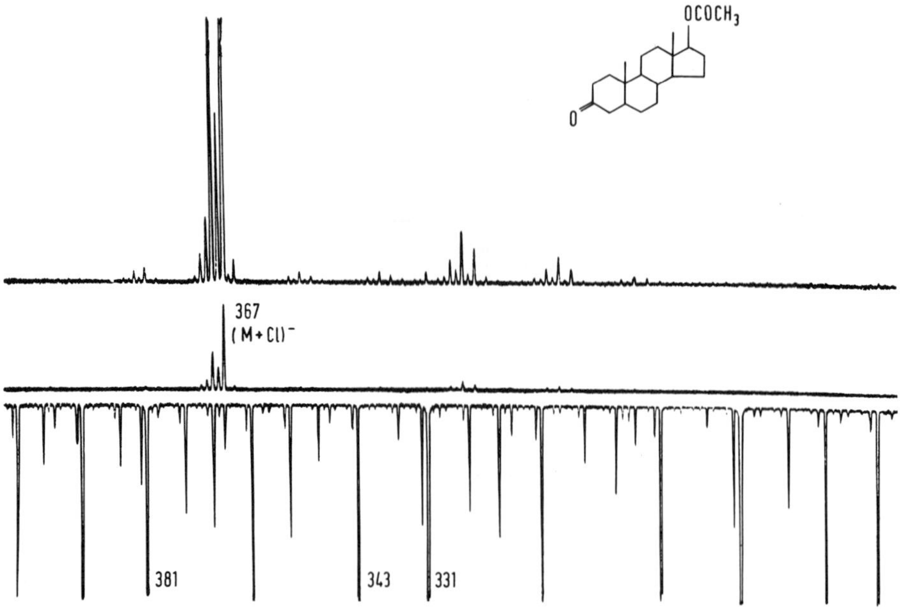

FIG. 44. NCI (chloroform) spectrum of a simple steroid.

low sensitivity as will be evident from the background levels in this spectrum. It is important to appreciate that compounds with a lower electron-capture cross section cannot use up all the secondary electrons available and instead this is done by traces of other strongly electron-capturing materials present in the source, e.g., chlorine-containing compounds give Cl^-, CH_2Cl^-, etc.; oxygen gives O_2^-. Thus, a great many weakly electron-capturing materials show ions in their spectra formed by interaction with trace impurities in the source, e.g., the $[M+Cl]^-$ ion in Fig. 43.

It is also possible to manufacture these nucleophiles deliberately. For example, a steroid was moistened with chloroform before introduction on the solids probe to give the spectrum shown in Fig. 44. The large excess of Cl^- ions produced from the chloroform changes the spectrum so that all that is seen is an $[M+Cl]^-$ ion. Deliberately moistening the sample with a suitable solvent provides an easy way of obtaining a simple NCI spectrum from many molecules that are not themselves strongly electron-capturing. The molecular weight can easily be inferred from the spectrum. Not only is the spectrum simple, but the technique is believed in this case to offer an enhanced sensitivity compared with electron attachment (Fig. 43), although this sensitivity has not been directly measured. This work parallels the earlier work of Dougherty [26] using methylene chloride as a reagent gas, although our method is considerably simpler in practice.

REFERENCES

1. (a) J. H. Bowie and B. D. Williams, *MTP International Review of Science, Physical Chemistry, Series Two*, vol. 5 (A. MacCole, Ed.), Butterworth, London, 1975, p. 89; (b) J. G. Dillard, *Chem. Rev., 73,* 589 (1973); (c) M. van Ardenne, K. Steinfelder, and R. Tummler, *Electronenanalagerungs--Massenspeklrographie Organischer Substanzen,* Springer-Verlag, New York, 1971; (d) A. L. Burlingame, B. J. Kimble, and P. J. Derrick, *Anal. Chem., 48,* 368R (1976); (e) L. G. Christophorou, *Atomic and Molecular Radiation Physics,* Wiley Interscience, New York, 1971, chap. 6.
2. R. C. Dougherty and C. R. Weisenberger, *J. Am. Chem. Soc., 90,* 6570 (1968).
3. T. O. Tiernan, in *Ion-Molecule Interactions* (P. Ausloss, Ed.), Plenum, New York, 1975, p. 353.
4. J. H. Bowie, *J. Am. Chem. Soc., 95,* 5795 (1973).
5. K. R. Jennings, in *Mass Spectrometry--V4, Specialist Periodical Reports* (R. A. W. Johnstone, Ed.), The Chemical Society, London, 1977, p. 203.
6. R. C. Dougherty, J. Dalton, and F. J. Biros, *Org. Mass Spectrom., 6,* 1171 (1972).
7. P. F. Levonowich, H. P. Tannenbaum, and R. C. Dougherty, *JCS Chem. Comm.,* 597 (1975).
8. D. F. Hunt, T. M. Harvey, and J. W. Russell, *JCS Chem. Comm.,* 152 (1975).
9. R. Yamdagni and P. Keborle, *J. Am. Chem. Soc., 95,* 4050 (1973).
10. F. C. Fehesenfeld, in *Ion Molecule Interactions* (P. Ausloss, Ed.), Plenum, New York, 1975, p. 387.
11. I. Dzidic, D. I. Carroll, R. N. Stillwell, and E. C. Horning, *Anal. Chem., 47,* 1308 (1975).
12. R. C. Dougherty and K. Piotrowsha, *Paper L-7,* Proc., 23rd Ann. Conf. on Mass Spectrom. and Allied Topics, Houston, Texas, May 1975.

13. D. F. Hunt, G. C. Stafford, F. W. Crow, and J. W. Russell, *Anal. Chem., 48,* 2098 (1976).

14. D. F. Hunt, G. C. Stafford, J. Shabanouritz, and F. W. Crow, *Anal. Chem., 49,* 1884 (1977).

15. J. R. Hass, unpublished data.

16. H. Hotop and W. C. Lineberger, "Binding Energies in Atomic Negative Ions," in *J. Phys. and Chem. Ref. Data,* vol. 4, 1975, pp. 539-576.

17. D. A. MacInnes, *Principles of Electrochemistry,* Reinhold, New York, 1939.

18. E. W. McDaniel, *Collision Phenomena in Ionized Gases,* Wiley, New York, 1964, app. I.

19. I. Langmuir, *Phys. Rev., 33,* 954 (1929).

20. E. W. McDaniel and E. A. Mason, *The Mobility and Diffusion of Ions in Gases,* Wiley, New York, 1973, p. 25.

21. Ref. 18, p. 519.

22. G. W. A. Milne, H. M. Fales, and T. Axenrod, *Anal. Chem., 43,* 1815 (1971).

23. R. C. Dougherty and K. Piotrowska, *Proc. Natl. Acad. Sci. USA, 73,* 1777 (1976).

24. R. C. Dougherty and K. Piotrowska, *J. Assoc. Off. Anal. Chem., 59,* 1023 (1976).

25. R. C. Dougherty, J. D. Roberts, and F. J. Biros, *Anal. Chem., 47,* 54 (1975).

26. H. P. Tannenbaum, J. D. Roberts, and R. C. Dougherty, *Anal. Chem., 47,* 49 (1975).

27. R. C. Dougherty, J. D. Roberts, H. P. Tannenbaum, and F. J. Biros, in *Mass Spectrometry and NMR Spectroscopy in Pesticide Chemistry* (R. Haque and F. J. Biros, Eds.), Plenum, New York, 1973, p. 33.

28. R. C. Dougherty, A. Bergner, P. Levonowich, and J. D. Roberts, *Advances in Mass Spectrometry in Biochemistry and Medicine,* vol. 1, 1976, p. 181.

29. D. F. Hunt, T. M. Harvey, and J. W. Russell, *Chem. Comm.,* 157 (1975).

30. D. L. Stalling, R. C. Tindle, and J. L. Johnson, *J. Assoc. Off. Anal. Chem., 55,* 34 (1972).

31. L. Mason, R. Lee, B. Sugimoto, S. Pawson, and F. Catehpod, *J. Nat. Cancer Inst., 54,* 829 (1975).

32. (a) J. Kimmig and K. H. Shultz, *Naturwissenschaften, 44,* 337 (1957); (b) D. Firestone, W. Ibrahium, and W. Horowitz, *J.*

Assoc. Off. Anal. Chem., 46, 384 (1963); (c) D. F. Flick, D. Firestone, and J. P. Marliac, *Poultry Sci., 44,* 1460 (1965); (d) J. S. Cantrell, N. C. Webb, and A. J. Mabis, *Acta Cryst., B25,* 150 (1969); (e) A. W. Galston, *Science, 168,* 1067 (1970).

33. (a) D. Firestone, J. Ress, N. L. Brown, R. P. Barron, and J. N. Daminco, *J. Assoc. Off. Anal. Chem., 55,* 85 (1972); (b) R. Baughman and M. Meselson, *Environ. Health Persp., 5,* 27 (1973); (c) R. Baughman and M. Meselson, in *Chlorodibenzodioxins--Origin and Fate, Advances in Chemistry Series,* no. 120, American Chemical Society, Washington, D.C., 1973, p. 92; (d) F. J. Biros, *Anal. Chem., 42,* 537 (1970); (e) H.-R. Buser and H.-P. Bosshardt, *J. Chromatogr., 90,* 71 (1974); (f) H.-R. Buser, *Anal. Chem., 49,* 918 (1977).

34. D. F. Hunt, C. N, McEwen and M. T. Harvey, *Anal. Chem., 47,* 1730 (1975).

35. R. K. Boyd and J. H. Beynon, *Org. Mass Spectrom., 12,* 163 (1977).

36. R. G. Cooks, J. H. Beynon, R. M. Caprioli, and G. R. Lester, *Metastable Ions,* Elsevier, New York, 1973, p. 153 ff.

37. E. E. McConnell, J. A. Moore, J. K. Haseman, and M. W. Harris, *Tox. Appl. Pharmacol.* (in press).

38. P. W. Albro and B. J. Corbett, *Chemosphere, 1977,* 381 (1977).

39. The modifications were made by Scientific Research Instruments Corporation, Baltimore, Md., a division of Chemetron Corporation.

40. Available from Matheson Gas Products, East Rutherford, N.J.

41. A. K. Bose and P. R. Srinivasan, *Tetrahedron, 31,* 3025 (1975).

42. A participant in UPTAM 1977 (Undergraduate Projects in Technology and Medicine) of Stevens Institute of Technology.

43. R. Large and H. Knof, *Org. Mass Spectrom., 11,* 582 (1976).

44. H. Knof, V. Hausen, and D. Krafft, *Z. Naturforsch., 27a,* 162 (1972).

45. H. Knof and D. Krafft, *Adv. Mass Spectrom., 6,* 303 (1974).

46. H. Knof, *Massenspektrometrie von Kondensationskeimen in der Gasphase,* Physik-Verlag, Weinheim, 1974.

47. H. Knof and D. Krafft, *Adv. Mass Spectrom., 7* (in press).

48. D. S. Millington and J. A. Smith, *Org. Mass Spectrom., 12,* 264-265 (1977).

49. F. W. McLafferty et al., *J. Am. Chem. Soc., 97,* 2298, 4613, 4617 (1975).

50. M. L. Aspiral, K. R. Compson, A. A. Dowman, R. M. Elliott, and D. Hazelry, *Paper C-2,* Proc. of the 23rd Ann. Conf. on Mass Spectrom. and Allied Topics, Houston, Tex., May 1975.

AUTHOR INDEX

Numbers in parentheses are reference numbers and indicate that an author's work is referred to although no name is cited in the text. Underlined numbers give the page on which the complete reference is listed.

SUBJECT INDEX

RETURN
LIBRARY
THE STANDARD OIL CO. (OHIO)
WARRENSVILLE LAB